RNA Interference Pathways

RNA Interference Pathways

Guest Editor

Tamas I. Orban

MDPI

Basel • Beijing • Wuhan • Barcelona • Belgrade • Novi Sad • Cluj • Manchester

Guest Editor
Tamas I. Orban
Institute of Molecular Life Sciences
HUN-REN Research Centre
for Natural Sciences
Budapest
Hungary

Editorial Office
MDPI AG
Grosspeteranlage 5
4052 Basel, Switzerland

This is a reprint of the Special Issue, published open access by the journal *Genes* (ISSN 2073-4425), freely accessible at: https://www.mdpi.com/journal/genes/special_issues/4PX02T5HQ8.

For citation purposes, cite each article independently as indicated on the article page online and as indicated below:

Lastname, A.A.; Lastname, B.B. Article Title. *Journal Name* **Year**, *Volume Number*, Page Range.

ISBN 978-3-7258-5151-5 (Hbk)
ISBN 978-3-7258-5152-2 (PDF)
https://doi.org/10.3390/books978-3-7258-5152-2

Contents

About the Editor . . . vii

Preface . . . ix

Radek Malik and Petr Svoboda
CRISPR-Induced Expression of N-Terminally Truncated Dicer in Mouse Cells
Reprinted from: *Genes* **2021**, *12*, 540, https://doi.org/10.3390/genes12040540 . . . 1

Dóra Reé, Ábel Fóthi, Nóra Varga, Orsolya Kolacsek, Tamás I. Orbán and Ágota Apáti
Partial Disturbance of Microprocessor Function in Human Stem Cells Carrying a Heterozygous Mutation in the DGCR8 Gene
Reprinted from: *Genes* **2022**, *13*, 1925, https://doi.org/10.3390/genes13111925 . . . 12

Bence Lázár, Nikolett Tokodyné Szabadi, Mahek Anand, Roland Tóth, András Ecker, Martin Urbán, et al.
Effect of miR-302b MicroRNA Inhibition on Chicken Primordial Germ Cell Proliferation and Apoptosis Rate
Reprinted from: *Genes* **2021**, *13*, 82, https://doi.org/10.3390/genes13010082 . . . 26

Anita Schamberger, György Várady, Ábel Fóthi and Tamás I. Orbán
Posttranscriptional Regulation of the Human ABCG2 Multidrug Transporter Protein by Artificial Mirtrons
Reprinted from: *Genes* **2021**, *12*, 1068, https://doi.org/10.3390/genes12071068 . . . 42

Eva Klimentová, Václav Hejret, Ján Krčmář, Katarína Grešová, Ilektra-Chara Giassa and Panagiotis Alexiou
miRBind: A Deep Learning Method for miRNA Binding Classification
Reprinted from: *Genes* **2022**, *13*, 2323, https://doi.org/10.3390/genes13122323 . . . 56

Pouneh Maraghechi, Maria Teresa Salinas Aponte, András Ecker, Bence Lázár, Roland Tóth, Nikolett Tokodyné Szabadi and Elen Gócza
Pluripotency-Associated microRNAs in Early Vertebrate Embryos and Stem Cells
Reprinted from: *Genes* **2023**, *14*, 1434, https://doi.org/10.3390/genes14071434 . . . 69

Lilla Krokker, Attila Patócs and Henriett Butz
Essential Role of the 14q32 Encoded miRNAs in Endocrine Tumors
Reprinted from: *Genes* **2021**, *12*, 698, https://doi.org/10.3390/genes12050698 . . . 89

Aneta Wiśnik, Dariusz Jarych, Kinga Krawiec, Piotr Strzałka, Natalia Potocka, Magdalena Czemerska, et al.
Role of MicroRNAs in Acute Myeloid Leukemia
Reprinted from: *Genes* **2025**, *16*, 446, https://doi.org/10.3390/genes16040446 . . . 114

About the Editor

Tamas I. Orban

Tamas I. Orban is the Head of the Gene Regulation Research Group in the Institute of Molecular Life Sciences, at the HUN-REN Research Centre for Natural Sciences in Budapest, Hungary. The focus of his laboratory is to understand how non-coding elements of the genomic "dark matter" contribute to the molecular fine-tuning of gene expression, mainly investigating RNA interference pathways and their interplay with DNA transposons. He is especially interested in the molecular details of miRNA maturation, currently investigating the specific intron-derived miRNAs called mirtrons, and the specific regulation of miRNA clusters. In addition to deciphering the molecular details of small RNA pathways, he is also involved in developing potential future therapeutic applications, and in collaboration with other research groups from the institute, he has a patent granted on the molecular identification of cardiomyocytes differentiated from human pluripotent stem cells. Additionally, as a newly appointed associate professor, as of this year (2025) he has started a parallel research group at the Department of Genetics, Cell- and Immunobiology at the Semmelweis University, Faculty of Medicine in Budapest, Hungary.

Preface

During the last decades of the 20th century, seemingly unrelated scientific results on different genetic model organisms lead to the discovery of RNA interference (RNAi), or "gene silencing", which is now considered to be the collective name for various regulatory pathways in eukaryotic organisms. The common effectors in these pathways are the specific ribonucleoprotein complexes containing short, single-stranded RNA molecules, and members of the Argonaute protein family. At least three major pathways are considered to be RNAi-related: the short interfering RNA (siRNA) pathway, representing a molecular immune system against invasive genetic elements; the microRNA (miRNA) pathway, which evolved to be a complex, posttranscriptional regulatory network in eukaryotes; and the Piwi-associated RNA (piRNA) pathway, which was originally described as a defense system against transposable elements, particularly in the germline cells of animals.

The scientific significance of these small RNA-based regulatory pathways was recognized by the awarding of the Nobel Prize in 2006 for the discovery of RNAi and by the recently awarded 2024 Nobel Prize for the discovery of miRNAs. This reprint is mainly a tribute to the diversity of the miRNA pathway. The selected articles provide an overview not only of some molecular aspects of the maturation of these tiny regulators but also provide evidence of how miRNAs control cell proliferation and differentiation, and how their misregulation could lead to the formation of various types of cancer. As certain gene therapy applications also involve miRNA-based strategies, the importance of understanding the molecular details of this pathway clearly extends beyond pure scientific interest.

As Guest Editor, it was my pleasure to organize this collection of articles on miRNAs. Gathering the most recent results in the field and presenting scientific discoveries using cutting-edge technologies in several model systems, my aim was to demonstrate how modern molecular genetics, as well as molecular medicine, can benefit from an in-depth understanding of this endogenous small RNA pathway. I hope that all readers enjoy these articles just as much as I enjoyed compiling this collection.

Tamas I. Orban
Guest Editor

genes

MDPI

Article

CRISPR-Induced Expression of N-Terminally Truncated Dicer in Mouse Cells

Radek Malik and Petr Svoboda *

Institute of Molecular Genetics of the Czech Academy of Sciences, Videnska 1083, 142 20 Prague 4, Czech Republic; malikr@img.cas.cz
* Correspondence: svobodap@img.cas.cz; Tel.: +420-241-063-147

Abstract: RNA interference (RNAi) designates sequence-specific mRNA degradation mediated by small RNAs generated from long double-stranded RNA (dsRNA) by RNase III Dicer. RNAi appears inactive in mammalian cells except for mouse oocytes, where high RNAi activity exists because of an N-terminally truncated Dicer isoform, denoted Dicer^O. Dicer^O processes dsRNA into small RNAs more efficiently than the full-length Dicer expressed in somatic cells. Dicer^O is expressed from an oocyte-specific promoter of retrotransposon origin, which is silenced in other cell types. In this work, we evaluated CRISPR-based strategies for epigenetic targeting of the endogenous Dicer gene to restore Dicer^O expression and, consequently, RNAi. We show that reactivation of Dicer^O expression can be achieved in mouse embryonic stem cells, but it is not sufficient to establish a robust canonical RNAi response.

Keywords: RNAi; Dicer; CRISPR; dCas9; VP64; MS2; sgRNA

1. Introduction

Canonical RNA interference (RNAi) has been defined as sequence-specific RNA degradation induced by long double-stranded RNA (dsRNA) [1]. RNAi is initiated by processing long dsRNA by RNase III Dicer into ~22 nt small interfering RNAs (siRNAs), which guide recognition and endonucleolytic cleavage of complementary mRNA molecules (reviewed in [2]). The mammalian canonical endogenous RNAi pathway is generally weak, if active at all (reviewed in [3]). Although mammalian genomes encode protein factors necessary and sufficient for reconstituting canonical RNAi in vitro [4] or in the yeast [5,6], these protein factors primarily support a gene-regulating microRNA pathway where small RNAs are produced from genome-encoded small hairpin precursors and typically guide translational repression coupled with transcript destabilization (reviewed in [7]). Furthermore, the primary mammalian mechanism responding to dsRNA in somatic cells is not RNAi but a sequence-independent interferon pathway (reviewed in [8]). The interferon pathway is an essential component of mammalian innate immunity and one of the factors impeding RNAi [9–12]. Another key factor limiting mammalian RNAi is low Dicer activity, which can be enhanced by truncating the Dicer's N-terminal helicase domain [9,12–14]. Mouse oocytes, the only known mammalian cell type where RNAi is highly active and functionally important, express a unique, naturally N-terminally truncated Dicer isoform, denoted Dicer^O (Figure 1A) [14]. This truncated isoform arose upon intronic insertion of a long terminal repeat (LTR) from the MTC retrotransposon subfamily, which provides an oocyte-specific promoter and the first exon of Dicer^O (Figure 1B). This isoform is expressed only in mouse oocytes, and it was not observed in transcriptomes of somatic cells [14]. Previous studies showed that functional canonical RNAi could be restored in mammalian somatic cells under specific conditions [9,10,12,14], but it was unclear whether Dicer^O reexpression from the endogenous *Dicer* locus could achieve such an effect as well. Accordingly, we examined whether Dicer^O expression could be induced from the endogenous *Dicer* locus in mouse cells and whether Dicer^{O} reactivation would be sufficient to restore robust endogenous canonical RNAi.

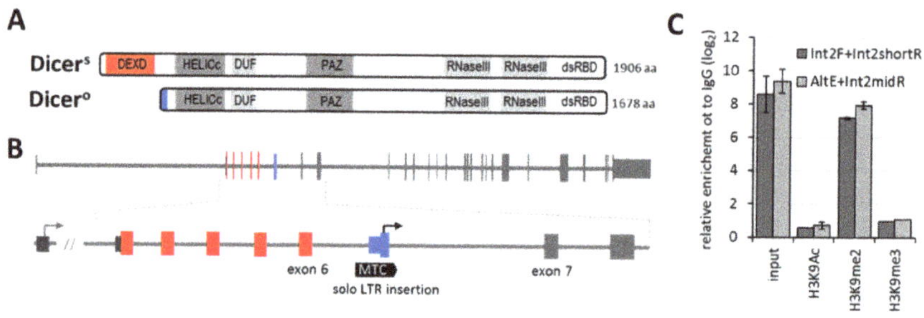

Figure 1. (**A**) Schemes of murine oocyte-specific truncated Dicer protein (Dicer^{O}) and full-length Dicer (Dicer^{S}, S for somatic). (**B**) Genomic organization of the *Dicer1* gene indicating the position of the alternative promoter and the first exon encoding Dicer^{O}. The alternative promoter and the first exon reside in MTC LTR insertion between exons 6 and 7, and its structure and evolutionary history have been described elsewhere [15]. (**C**) Native chromatin immunoprecipitation analysis of H3K9 modifications in the MTC Dicer^{O} promoter region in NIH 3T3 cells. Relative enrichment analyzed by two primer pairs is depicted as $\log_2$-fold enrichment calculated to IgG.

As Dicer^{O} expression is restricted to oocytes and transcription factor(s) controlling it are unknown, we opted for an artificially designed transcriptional activator. Advances in the development of guided nucleases (reviewed in [16]) brought an opportunity to modify these sequence-specific enzymes as platforms for modulating mammalian transcription by fusing them with transcription activation or repression domains [17,18]. Among the guided nucleases, the clustered regularly interspaced short palindromic repeat (CRISPR) Cas9 nuclease gained major popularity because of simple programming of its sequence specificity with a single-guide RNA (sgRNA), which mediates recognition of a specific DNA sequence [19]. RNA-guided nuclease Cas9 was, therefore, converted to a sequence-specific DNA-binding platform, "deactivated Cas9" (dCas9), by inactivation of its two nuclease domains [19]. dCas9 can be fused to a trans-activation domain, such as VP64, to function as a transcriptional activator [20,21]. Further development brought more complex systems where the basic dCas9-VP64 module was extended by modifying the sgRNA structure such that sgRNA loops protruding from the ribonucleoprotein complex could carry sequences serving as binding platforms for additional transcription modulating proteins [22,23].

To test whether dCas9-mediated transcriptional activation could induce Dicer^{O} expression from the *Dicer* gene in mouse embryonic stem cells and fibroblasts, we examined several different dCas9-transcriptional activation designs and obtained the best Dicer^{O} activation with a system consisting of dCas9-VP64 enhanced with an MS2-p65-HSF1 module originally developed by Konermann et al. [23]. In this system, sgRNA loops protruding from the ribonucleoprotein complex carry a minimal hairpin aptamer, which is bound by the MS2 protein that can be fused with an additional factor enhancing transcription [22,23]. We report that we successfully induced robust Dicer^{O} protein expression and observed limited RNAi activity in mouse embryonic stem cells, but not in NIH 3T3 fibroblasts. Our results suggest that induction of RNAi in mouse cells may be possible by inducing Dicer^{O} expression, but achieving robust RNAi activity requires further optimization of Dicer^{O} induction, presumably combined with a strategy reducing the inhibitory effects of innate immunity factors.

2. Materials and Methods

2.1. Plasmids

Catalytically deactivated Cas9 (dCas) was constructed from a pSpCas9n^{D10A} plasmid (PX460; Addgene #48873) by mutating His to Ala at position 840 using Q5 site-directed mutagenesis kit (NEB) according to the manufacturer's instructions. dCas9$^{D10A/H480A}$ was transferred into the pPuro backbone containing the puromycin resistance gene. VP64 was PCR amplified from a pTALE-VP64-EGFP plasmid [14] and inserted into pPuro_dCas9 together with a flexible linker.

Different dCas9 plasmid variants were constructed by stepwise cloning of the following components: (1) Capsid assembly-defective MS2 coat protein variant dlFG lacking the FG loop [24] cloned in the form of a linked dimer [25] to augment dimerization. (2) Mouse p65 activator was PCR amplified from mouse genomic DNA (primers Fwd: 5′-CCATCAGGGCAGATCTCAAACCAGG and Rev: 5′- GGAGCTGATCTGACTCAAAA-GAGC). (3) Human HSF1 was PCR amplified from HeLa cell cDNA (primers Fwd: 5′-GGCTTCAGCGTGGACACCAGTGCC and Rev: 5′-TCAGGAGACAGTGGGGTCCTTG-GCTTTGG). (4) p300 HAT activation domain was PCR amplified from HeLa cell cDNA (primers Fwd: 5′-CCATTTTCAAACCAGAAGAACTACGAC and Rev: 5′-GTCCTGGCT-CTGCGTGTGCAGCTC). When a single plasmid construct was employed, dCas9-VP64 was linked with the enhancing protein via the T2A self-cleavage peptide [23].

The sgRNA backbone containing MS2-binding sites (sgRNA$^{2.0}$) was designed according to [23] and cloned under the U6 promoter. Desired targets were cloned by BsmBI restriction enzyme: Dcr_MT1 (-372): 5′-caccGAACAAATGGCTGCTGAA; Dcr_MT2 (-188): 5′-caccGTCAGTCATCTGAGGGAA; Dcr_MT3 (-78): 5′-caccGGCCCAACCCACTGTGGG; Dcr_MT4 (-162R): 5′-caccGAAGTACGTTCTCTATTG; Dcr_MT5 (-249R): 5′-caccGAGCATC-ACCCTCACTGA; Dcr_MT6 (-34R): 5′-caccGCTTTCTTAATAGAACCC; Dcr_MT7 (-59): 5′-caccGGGGCCATCCCTGGACTG; Dcr_MT8 (-269R): 5′-caccGTGGCAGTAACCCATTTG; Oct4-1 (-534): 5′-caccGGTCTCTGGGGACATATC ; Oct4-2 (-453): 5′-caccGCTGTCTTGTCC-TGGCCT ; Oct4-3 (-216): 5′-caccGAGGTGTCCGGTGACCCA ; Oct4-6 (-170): 5′-caccGAAA-ATGAAGGCCTCCTG; Oct4-7 (-50): 5′-caccGCTCCTCCACCCACCCAG ; Oct4-4 (-275R): 5′-caccGTTGGCACTGCACCCTCT; Oct4-5 (-405R): 5′-caccGTCTAGAGTCCTAGATAT; Oct4-8 (-114R): 5′-caccGTCTTCCAGACGGAGGTT. The mouse *Dicer1* intron 6 region containing the MTC element was PCR amplified from mouse genomic DNA (Fwd: 5′-AAGCTTCTCGAGCCACCTTCAGTGAGGGTG and Rev: 5′-AAGCTTGTATGTCCTTTAC-ACTGATTAAGC) and cloned into pGL4.10 plasmid (Promega; for simplicity referred to as FL) digested with HindIII. The mouse *Oct-4* (*Pou5f1*) promoter was PCR amplified from mouse genomic DNA (Fwd: 5′-CCATGGTGTAGAGCCTCTAAACTCTGGAGG and Rev: 5′-CCATGGGGAAGGTGGGCACCCCGAGCCGG) and cloned into pGL4.10 plasmid digested with NcoI. *Renilla* luciferase-expressing plasmid (Promega; for simplicity referred to as RL) was used for normalization.

An overview of different construct combinations used in the five characterized versions of the transcription-acting systems is provided in Table 1.

Table 1. Overview of versions of transcription activation systems used in the study.

System Version	dCas9 Version	sgRNA Version	Modulating Co-Factor	Comment
Version 1	HA-dCas9$^{D10A/H480A}$-VP64 (HA-tagged)	U6-driven sgRNA	none	
Version 2	HA-dCas9$^{D10A/H480A}$-VP64 (HA-tagged)	U6-driven sgRNA with two MS2-binding sites	MS2^{dlFG}-p65-HSF1	dCas9 and modulator in a single construct (linked with T2A)
Version 3	HA-dCas9$^{D10A/H480A}$-VP64 (HA-tagged)	U6-driven sgRNA with two MS2-binding sites	dimMS2^{dlFG}-p65-HSF1	dCas9 and modulator in a single construct (linked with T2A)
Version 4	HA-dCas9$^{D10A/H480A}$-VP64 (HA-tagged)	U6-driven sgRNA with two MS2-binding sites	dimMS2^{dlFG}-p65-HSF1-HA (HA-tagged)	dCas9 and modulator in separate plasmids
Version 5	HA-dCas9$^{D10A/H480A}$-VP64 (HA-tagged)	U6-driven sgRNA with two MS2-binding sites	dimMS2^{dlFG}-p300-HA (HA-tagged)	dCas9 and modulator in separate plasmids

2.2. Cell Culture and Transfection

Mouse 3T3 cells were maintained in DMEM (Sigma-Aldrich, USA) supplemented with 10% fetal calf serum (Sigma-Aldrich, USA), penicillin (100 U/mL; Invitrogen, USA), and streptomycin (100 µg/mL; Invitrogen, USA) at 37 °C and 5% CO_2 atmosphere. Mouse ESCs were cultured in 2i-LIF media: DMEM supplemented with 15% fetal calf serum, 1× L-glutamine (Invitrogen, USA), 1× nonessential amino acids (Invitrogen, USA), 50 µM β-mercaptoethanol (Gibco, Thermo Fisher Scientific, USA), 1000 U/mL LIF (Merck Millipore, Germany), 1 µM PD0325901, 3 µM CHIR99021, penicillin (100 U/mL), and streptomycin (100 µg/mL). For transfection, the cells were plated on a 24-well plate, grown to 50% density, and transfected using the TurboFect in vitro transfection reagent or lipofectamine 3000 (Thermo Fisher Scientific, USA) according to the manufacturer's protocol. The cells were co-transfected with 50 ng per well of each FL and RL reporter plasmids and 250 ng per well of a dsRNA-expressing plasmid and, eventually, 250 ng per well of a plasmid expressing the tested factor. The total amount of transfected DNA was kept constant (600 ng/well) using a pBluescript plasmid. The cells were collected for analysis 48 h post-transfection.

2.3. Luciferase Assay

Dual-luciferase activity was measured according to [26] with some modifications as described previously [12]. Briefly, the cells were washed with PBS and lysed in PPTB lysis buffer (0.2% vol/vol Triton X-100 in 100 mM potassium phosphate buffer, pH 7.8). Aliquots of 3–5-µL were used for measurement in 96-well plates using a Modulus microplate multimode reader (Turner Biosystems, USA). First, firefly luciferase activity was measured by adding 50 µL substrate (20 mM tricine, 1.07 mM $(MgCO_3)_4 \cdot Mg(OH)_2 \cdot 5$ H_2O, 2.67 mM $MgSO_4$, 0.1 mM EDTA, 33.3 mM DTT, 0.27 mM Coenzyme A, 0.53 mM ATP, and 0.47 mM D-luciferin, pH 7.8) and the signal was integrated for 10 s after a 2 s delay. The signal was quenched by adding 50 µL *Renilla* substrate (25 mM Na_4PP_i, 10 mM sodium acetate, 15 mM EDTA, 500 mM Na_2SO_4, 500 mM NaCl, 1.3 mM NaN_3, and 4 µM coelenterazine, pH 5.0) and *Renilla* luciferase activity was measured for 10 s after a 2 s delay.

2.4. Western Blotting

Cells were grown in 6-well plates. Before collection, the cells were washed with PBS and lysed in RIPA buffer (50 mM Tris, pH 7.5, 150 mM NaCl, 1 mM EDTA, 1 mM EGTA, 1% NP-40 (Igepal CA-630), 0.5% Na-deoxycholate, 0.1% SDS) supplemented with 2× protease inhibitor cocktail set (Merck Millipore, Germany). Proteins were separated in 5% (for Dicer detection) or 10% (for Tubulin detection) polyacrylamide gel and transferred to a PVFD membrane (Merck Millipore, Germany). Anti-Dicer (#349; [27]) and anti-Tubulin (#T6074, 1:5000; Sigma-Aldrich, USA) primary antibodies and HRP-conjugated secondary antibodies (1:50,000) were used for signal detection with SuperSignal West Femto chemiluminescent substrate (Pierce, Thermo Fisher Scientific, USA).

2.5. Native Chromatin Immunoprecipitation

Cells were washed in PBS and resuspended in nuclei preparation buffer I (0.3 M sucrose, 60 mM KCl, 15 mM NaCl, 5 mM $MgCl_2$, 0.1 mM EGTA, 0.5 mM DTT, 0.2% NP-40, 15 mM Tris/HCl, pH 7.5, supplemented with protease inhibitors) and nuclei were released by passing three times through a 21G needle.

The sample was placed over nuclei preparation buffer III (1.2 M sucrose, 60 mM KCl, 15 mM NaCl, 5 mM $MgCl_2$, 0.1 mM EGTA, 0.5 mM DTT, 15 mM Tris/HCl, pH 7.5) and centrifuged at ~2000× *g* for 15–20 min. The pellet (containing purified nuclei) was resuspended in MNase digestion buffer (0.32 M sucrose, 1 mM $CaCl_2$, 4 mM $MgCl_2$, 50 mM Tris/HCl, pH 7.5) to bring final DNA concentration to 1000 mg/mL. Aliquots of 500 µg chromatin DNA were digested by 5 U MNase (micrococcal nuclease) for 5 min at 37 °C. Digestion was stopped by adding EDTA to the final concentration >20 mM.

Supernatant containing mono- and oligo-nucleosomes was used for immunoprecipitation: 30 μg chromatin DNA and 5 μg of primary antibody (anti-H3K9me2 (Abcam #ab1220), anti-H3K9me3 (Abcam #ab8898), anti-H3K9ac (Abcam #ab4441) or control rabbit IgG (Millipore #PP64B)) were used per sample in 700 μL NChIP incubation buffer (50 mM NaCl, 5 mM EDTA, 50 mM Tris/HCl, pH 7.5). Pre-blocked Dynabeads protein G (Invitrogen) or protein A-sepharose 4B, fast flow from *Staphylococcus aureus* (Sigma) were used for pull-down. Beads were washed in wash buffer (10 mM EDTA, 50 mM Tris/HCl, pH 7.5) containing increasing NaCl concentrations (75, 125, 175, and 300 mM; two washes each).

DNA elution was performed in "PCR-friendly" elution buffer (50 mM KCl, 0.45% NP-40, 0.45% Tween-20, 0.01% gelatin, 10 mM Tris/HCl, pH 8.0) supplemented with 1 μL proteinase K (Fermentas) added to the beads pellet. Samples were incubated for 1 h at 55 °C followed by proteinase K inactivation for 10 min at 95 °C. Supernatants were used directly for qPCR analysis using the following primers: Dcr_int2shortR 5′-TGCCCTACAGGTGTGTCTGT; Dcr_int2midR 5′-CTGTGTAGAGGTGTCTGTTTCCA; Dcr_int2F 5′-GCAAAGACCAGCTCCAGCCAT; mDcr_AltE_Fwd 5′-CTCTGCCTTCAGGTTCTGACTTCC.

2.6. qPCR Analyses

Total RNA was isolated using RNeasy MINI kit (Qiagen), and 1 μg amount was used for the reverse transcription (RT) reaction using RevertAid premium (Fermentas) in a 30 μL volume. A 0.5 μL aliquot was used per qPCR reaction. Maxima SYBR Green qPCR master mix (Thermo Scientific) was ge C_T values of the technical replicates were normalized toused for qPCR. qPCR was performed in technical triplicates for each biological sample. Avera housekeeping genes *Hprt* and *B2MG* using the $\Delta\Delta C_T$ method [28]. Primers used for qPCR were as follows: *Pou5f1*: mPou5f1_qPCR_Fwd 5′-GTTGGAGAAGGTGGAACCAA and mPou5f1_qPCR_Rev 5′-GCAAACTGTTCTAGCTCCTTCTG; Dicer_MT (1): mDcr_AltE _Fwd 5′-CTCTGCCTTCAGGTTCTGACTTCC and mDcr_E7_Rev 5′-GCAATCTCTCATACAGCCCACTTC; Dicer_MT (2): mDcr_AltE_Fwd 5′-CTCTGCCTTCAGGTTCTGACTTCC and mDcr_E7_Rev2 5′-CAGTCTACCACAATCTCACAAGGC; *Hprt1*: mHPRT.1.457_Fwd 5′-GCTACTGTAATGATCAGTCAACGG and mHPRT.1.670_Rev 5′-CTGTATCCAACACTTCGAGAGGTC; *B2m*: mβ2-MG.1.342_Fwd 5′- GCAGAGTTAAGCATGCCAGTATGG and mβ2-MG.1.514_Rev 5′-CATTGCTATTTCTTTCTGCGTGC.

3. Results and Discussion

As mentioned above, $Dicer^O$ is not expressed in other mouse cells than oocytes. Using chromatin immunoprecipitation, we examined histone modifications of the *Dicer*^{O} promoter in NIH 3T3 cells to test whether *Dicer*^{O} silencing involves more reversible facultative heterochromatin or more stable constitutive heterochromatin, which may be more difficult to reverse into euchromatin. Chromatin immunoprecipitation showed the presence of facultative heterochromatin mark H3K9me2, but not the constitutive heterochromatin mark H3K9me3 at the MTC LTR insertion controlling *Dicer*^{O} expression (Figure 1C). Regarding DNA methylation, the MTC LTR promoter is a CpG poor promoter where the presence of DNA methylation may not be a significant silencing factor [29].

To develop transcriptional reactivation of $Dicer^O$ in somatic cells, we tested several different designs built on the CRISPR dCas9 system recruiting transcriptional transactivator(s) into the *Dicer*^{O} promoter locus. We started with a basic transcriptional activating system where dCas9 is fused with the VP64 transcriptional transactivator [20,21] and is guided by unmodified sgRNA (Figure 2A).

For initial testing of the CRISPR-driven transcriptional reactivation, we produced stable NIH 3T3 cell lines expressing dCas9-VP64 or dCas9-VP64-EGFP [14] transcriptional activators. This allowed us to reduce the number of transfected plasmids and ensure that all cells would have a constant amount of dCas9-VP64 expression. Cells stably expressing dCas9-VP64 were then co-transfected with an expression vector for sgRNA to target dCas9-VP64 into the promoter of interest cloned into a firefly luciferase reporter and an SV40 promoter-driven *Renilla* luciferase reporter allowing for normalization of the luciferase activity (Figure 2B). Two different firefly reporters were produced to examine the efficiency of the dCas9 activation system, *Pou5f1* (*Oct-4*) promoter and MTC-derived promoter for Dicer^{O}; for each promoter, we designed and tested several sgRNAs.

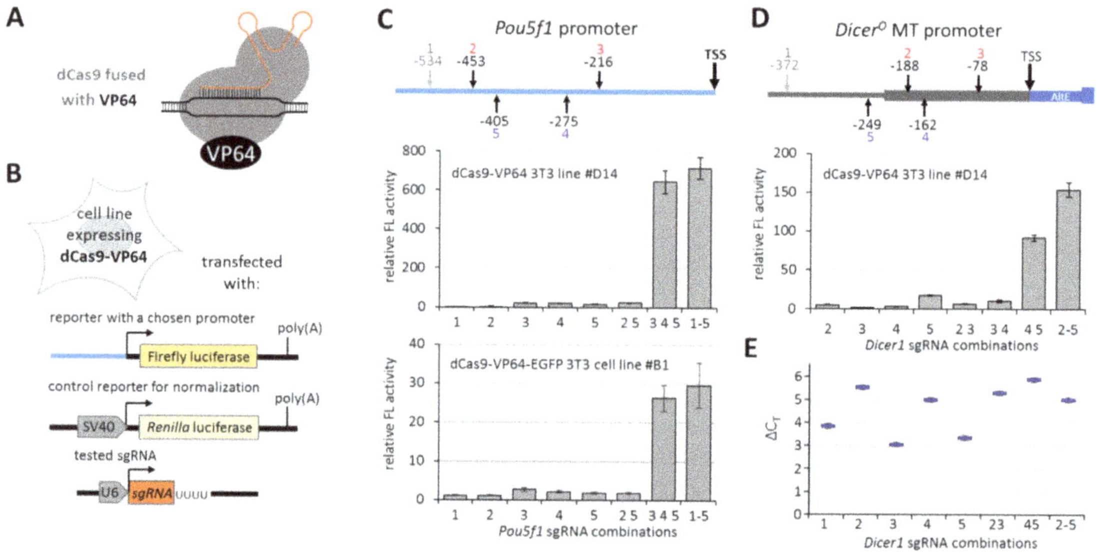

Figure 2. Transcriptional activation using a basic dCas9-VP64 system. (**A**) Schematic depiction of the basic dCas9-VP64 system version 1 composted of dCas9 fused with VP64 and a sgRNA. (**B**) Schematic depiction of the testing strategy built on NIH 3T3 cell lines stably expressing dCas9-VP64 fusions and transfected with plasmids expressing sgRNAs, a firefly reporter driven by a tested promoter, and a *Renilla* luciferase reporter as a control for normalization of transcriptional activation of the tested promoter. (**C**) Transcriptional activation of *Pou5f1* promoter-reporter. The scheme depicts positions of sgRNAs relative to the *Pou5f1* transcription start. sgRNA 1 was outside of the promoter region cloned into the firefly reporter, hence could serve as a negative control. The upper graph depicts results with dCas9-VP64, the lower one with dCas9-VP64-EGFP. Shown is the firefly luciferase activity normalized to *Renilla* luciferase; reporter activity in the absence of sgRNA was set to 1. (**D**) Transcriptional activation of the MTC promoter-reporter. As in (**C**), the scheme depicts the position of sgRNAs relative to the transcription start site, and sgRNA 1 was outside of the cloned promoter region. (**E**) Transcriptional activation of the endogenous MTC promoter in NIH 3T3 cells. The graph depicts relative expression calculated by the $\Delta\Delta C_T$ method; negligible basal expression detected in NIH 3T3 cells was set to 1. All error bars = standard deviation.

Experiments with the *Pou5f1* promoter-reporter and a set of sgRNAs have shown that the system can induce reporter expression (Figure 2C). However, individual sgRNAs had just marginal effects. At the same time, strong transcriptional activation was observed when several sgRNAs were combined (Figure 2C). Because the NIH 3T3 line expressing the dCas-VP64-EGFP fusion yielded an order of magnitude lower transcriptional activation estimated by the luciferase assay, we omitted the dCas-VP64-EGFP fusion from subsequent experiments and continued with dCas-VP64 (denoted version 1). When testing MTC promoter activation, we also observed that individual sgRNAs had a much lower effect than some of their combinations. The most powerful combination was observed with a mixture of all four tested sgRNAs (2–5) that could bind the promoter cloned into the firefly luciferase reporter (Figure 2D). When examining transcriptional activation of the endogenous MTC promoter of *Dicer*^{O} in 3T3 cells, the system yielded up to 90-fold transcriptional activation of *Dicer*^{O} over the background level (Figure 2E). Of note is that qPCR analysis of the endogenous promoter and luciferase reporter gave results differing for some sgRNAs and their combinations (e.g., sgRNA 4 vs. sgRNA 5 in Figure 2D,E). This may reflect different accessibility of the plasmid reporter and the genomic locus.

To test whether we could further enhance transcriptional activation effects, we employed a published sgRNA modification with MS2 binding sites, which allows binding of exposed sgRNA loops to the MS2 domain fused with additional transcriptional activators [23]. We tested three variants of the MS2 system: (1) an MS2-p65-HSF1 fusion construct with enhanced dimerization MS2 variant ($MS2^{dlFG}$ [24]), which was encoded in a single construct together with dCas9-VP64 via T2A self-cleavage peptide as described previously [23]. We denoted this system as of version 2 (Figure 3A); (2) modification of the MS2 component by using a covalently linked MS2-dimer ($dimMS2^{dlFG}$-p65-HSF1 [25]) in a single construct with dCas9-VP64, denoted version 3; and (3) use of separate constructs expressing dCas9-VP64 and $dimMS2^{dlFG}$-p65-HSF1, denoted version 4 (Figure 3B). For each of the versions, we produced stable 3T3 lines and examined *Pou5f1* and MTC reporter expression as described previously (Figure 2B). A detailed overview of the variants of dCas9-VP64 systems is provided in Table 1. Expression of dCas9 and enhancing factors in stable cell lines is provided in Appendix A.

Pou5f1 reporters (Figure 3C) revealed the best enhancing effects of version 4 of the transcriptional activation system; otherwise, the effects were comparable to the Cas9-VP64 version without MS2-based enhancement (Figure 2C). Importantly, transcriptional activation of the endogenous MTC promoter of $Dicer^{O}$ in 3T3 cells with version 4 was much stronger than with version 3 (Figure 3D). When we used transient transfection and compared version 1 and version 4, version 4 was clearly superior to version 1, suggesting that modified sgRNA and $dimMS2^{dlFG}$-p65-HSF1 indeed have an enhancing effect (Figure 3E). In the same experiment, we also examined alternative MS2-based enhancement employing $dimMS2^{dlFG}$-p300 acetyltransferase fusion [22], denoted version 5, but version 4 using separate constructs expressing dCas9-VP64 and $dimMS2^{dlFG}$-p65-HSF1 was clearly superior (Figure 3E). We thus selected version 4 for analysis of the production of the $Dicer^{O}$ protein.

To analyze $Dicer^{O}$ production's activation, we used stable cell lines expressing version 4 (Figure 4A) and transient transfection of sgRNA-expressing constructs (Figure 4B). When deploying the system into embryonic stem cells (ESCs), we were able to detect robust transcriptional activation and production of $Dicer^{O}$ protein, whose abundance in the lysate of transiently transfected cells was comparable to the endogenous Dicer protein (Figure 4C). Although we achieved clear transcriptional activation of *Dicer*^{O} in NIH 3T3 cells, the *Dicer*^{O} transcript levels were approximately a hundredfold less induced than in ESCs, and the truncated protein could not be detected by Western blotting (Figure 4D). Because of this result, we tested the induction of RNAi activity in ESCs.

To test the RNAi activity (Figure 4E), we co-transfected ESCs stably expressing version 4 with a combination of four sgRNAs that yielded the most of the $Dicer^{O}$ protein and our established RNAi assay system described in detail previously [12]. Briefly, the RNAi assay system has three plasmid components: (1) a targeted *Renilla* luciferase reporter with a cognate *Mos* sequence in the 3′ UTR, (2) a non-targeted firefly luciferase reporter, and (3) a plasmid expressing long dsRNA with the *Mos* sequence in the form of a long hairpin (MosIR). As a control for nonspecific dsRNA effects, we used unrelated dsRNA-expressing plasmids (Lin28IR and Elavl2IR) instead of MosIR. As negative control to MosIR, we used a MosMos construct, where *Mos* sequences are oriented head-to-tail. Hence the plasmid has the same sequence as MosIR but does not produce dsRNA. Western blot analysis again showed good induction of $Dicer^{O}$ expression and a marked reduction of the targeted *Renilla* reporter in the presence of MosIR (Figure 4E). The average *Renilla* reporter reduction was ~30% (p-value > 0.05), which was comparable to our previous experiments with cells that have an intact protein kinase R gene encoding one of the dsRNA-sensing components of the interferon response [12,14].

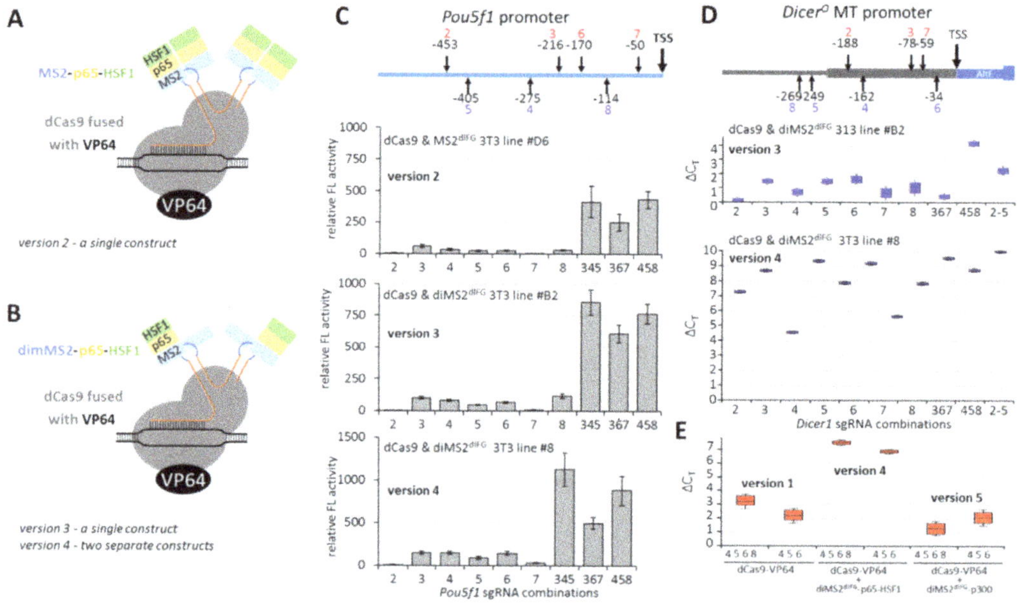

Figure 3. Transcriptional activation using dCas9-VP64 enhanced with MS2 systems. (**A**) Schematic depiction of version 2 based on dCas9-VP64, modified sgRNA with MS2-binding sequence (blue) and MS2 domain fused with p65 and HSF1. The scheme indicates that the MS2 fusion construct is dimerizing on the RNA sequence. (**B**) Schematic depiction of versions 3 and 4. The difference from (A) lies in MS2 being encoded as a covalently linked duplex where one MS2 is fused with p65 and HSF1 (dimMS2-p65-HSF1). Versions 3 and 4 differ in the organization of expression. Version 3 expresses dCas9-VP64 and dimMS2-p65-HSF1 from a single construct, version 4 from separate constructs. (**C**) Transcriptional activation of *Pou5f1* promoter-reporter with versions 2, 3, 4. As in Figure 1C, shown are relative firefly luciferase activities normalized to the *Renilla* control reporter; firefly reporter activity in the absence of sgRNA was set to 1. (**D**) Transcriptional activation of the endogenous MTC promoter in NIH 3T3 cells. The graph depicts relative expression calculated by the $\Delta\Delta C_T$ method. (**E**) Comparison of transcriptional activation of the endogenous MTC promoter in NIH 3T3 cells of version 4 with version 1 (Figure 2) and version 5, which uses dimMS2-p300 instead of dimMS2-p65-HSF1. The graph depicts relative expression calculated by the $\Delta\Delta C_T$ method. All error bars = standard deviation.

To sum up, we tested five different versions of the dCas9-VP64 transcriptional activation system, out of which we selected dCas9-VP64 combined with sgRNA carrying MS2-binding sites and a dimMS2^{dlFG}-p65-HSF1 fusion protein. Our work shows the extent of optimization needed to induce robust transcriptional activation (even if only in ESCs). It underscores the importance of setting up a good testing system for multiple sgRNAs and their combinations. We opted for producing stable lines expressing dCas9 and the enhancing factor. These cell lines represent one of the valuable contributions of this work as they can be used to program dCas9 with sgRNAs to influence gene expression in ESCs and 3T3 cells.

Using this three-component system and a combination of four sgRNAs targeting the MTC promoter of *Dicer*^{O} isoform, we achieved robust Dicer^O protein expression in ESCs comparable to full-length Dicer expression. In fact, Dicer^O protein expression in cells transfected with sgRNAs likely exceeded that of endogenous full-length Dicer in those cells because in Western blots in Figure 4C,E, the Dicer^O signal comes from fewer cells than the full-length Dicer protein detected above it. However, despite the level of Dicer^O expression, the RNAi activity assessed by a reporter assay was minor, if any, suggesting that the sole expression of Dicer^O is not sufficient to bring robust RNAi. On the other hand, this observation is consistent with other data showing that RNAi activity is low in the presence of specific innate immunity factors [9–12].

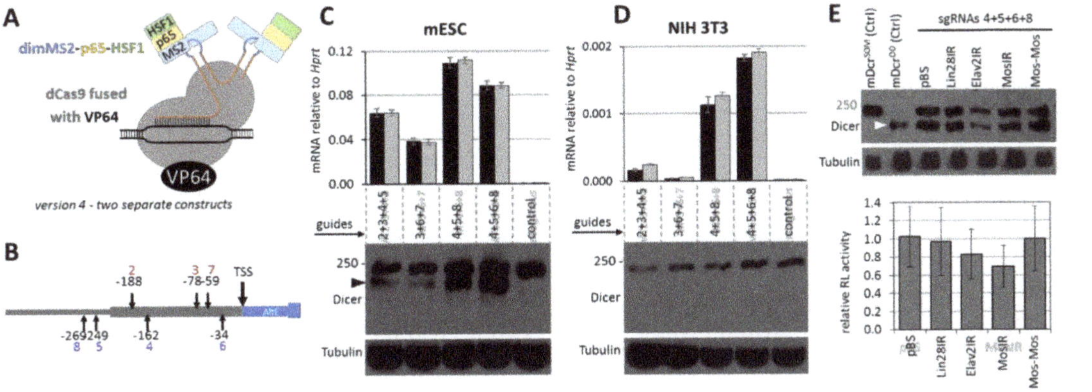

Figure 4. Transcriptional activation of Dicer^{O}. (**A**) Schematic depiction of the finally chosen version 4. (**B**) Schematic depiction of the position of sgRNAs targeting the MTC promoter. (**C**) Activation of Dicer^{O} with shRNA combinations in ESCs analyzed by qPCR and Western blot. Dicer^{O} is visible as the lower band detected by the anti-Dicer antibody (depicted by black arrowhead). (**D**) Activation of Dicer^{O} with shRNA combinations in NIH 3T3 cells analyzed by qPCR and Western blot. Dicer^{O} was not detected. (**E**) Analysis of RNAi effects in ESCs where Dicer^{O} expression was induced with sgRNAs. Western blot depicts activation of Dicer^{O} in ESCs (depicted by white arrowhead). The first two Western blot lanes show control ESCs expressing either the full-length Dicer^{S} or truncated Dicer^{O} [14]. The graph depicts the relative activity of a *Renilla* luciferase reporter carrying a cognate *Mos* sequence as described previously [12]. Briefly, the *Renilla-Mos* reporter is co-transfected with a non-targeted firefly luciferase reporter and a reporter expressing dsRNA from an inverted repeat (IR). pBS and MosMos are controls not expressing dsRNA. All error bars = standard deviation.

Unfortunately, we did not succeed in inducing a similar expression of Dicer^{O} in NIH 3T3 cells. This result seems to be caused by much lower transcriptional activation than in ESCs and could be influenced by different chromatin organization in ESCs and NIH 3T3 cells, which are derived from fibroblasts [30]. ESCs have a relatively open chromatin structure supporting higher transcription and reduced heterochromatin signature [31,32]. Thus, ESCs may better respond to transcriptional reactivation of *Dicer*^{O} than NIH 3T3 cells. The system with histone acetyltransferase p300 (version 5) did not have an additive effect on top of VP64 (Figure 3E), suggesting that targeting acetylation to the locus does not sufficiently enhance transcriptional activation of the MTC promoter. Perhaps a histone lysine demethylase targeted to the locus that would reduce H3K9me2 could make the transcriptional activation of *Dicer*^{O} more efficient.

In conclusion, we showed that reactivation of Dicer^{O} via a dCas9 system is not a viable strategy to induce a robust canonical RNAi pathway in cultured mouse cells. While rather negative, our results are useful for understanding the functional limits of the endogenous RNAi and long dsRNA processing capacity furnished by the expression of Dicer^{O} from its endogenous locus. Although this time we did not observe robust RNAi, it is likely that more sensitive methods, such as RNA sequencing of small RNAs, would reveal the impact of expressed truncated Dicer on the long dsRNA metabolism. In our previous work in 3T3 and ESCs, we observed a robust increase in siRNA biogenesis with ectopic expression of Dicer^{O} but rarely robust RNAi [12]. Our work offers another tool for studying the consequences of Dicer^{O} expression in mouse cells and investigating whether low-activity RNAi would have any measurable impact on cell physiology, especially dsRNA metabolism, miRNA pathway, the sensitivity of innate immunity pathways to dsRNA, and viral resistance.

Author Contributions: Conceptualization, R.M. and P.S.; methodology, R.M.; validation, R.M.; formal analysis, R.M.; investigation, R.M.; resources, P.S.; data curation, R.M. and P.S.; writing—original draft preparation, R.M. and P.S.; writing—review and editing, R.M. and P.S.; project administration, P.S.; funding acquisition, P.S. Both authors have read and agreed to the published version of the manuscript.

Funding: This work was funded by the European Research Council under the European Union's Horizon 2020 research and innovation programme (grant 647403, D-FENS) and Czech Science Foundation EXPRO grant 20-03950X.

Institutional Review Board Statement: Not applicable.

Informed Consent Statement: Not applicable.

Data Availability Statement: Data is contained within the article or supplementary material.

Acknowledgments: We thank Matyas Flemr for their help with the analysis of chromatin at the MTC promoter of Dicer^{O}.

Conflicts of Interest: The authors declare no conflict of interest.

Appendix A

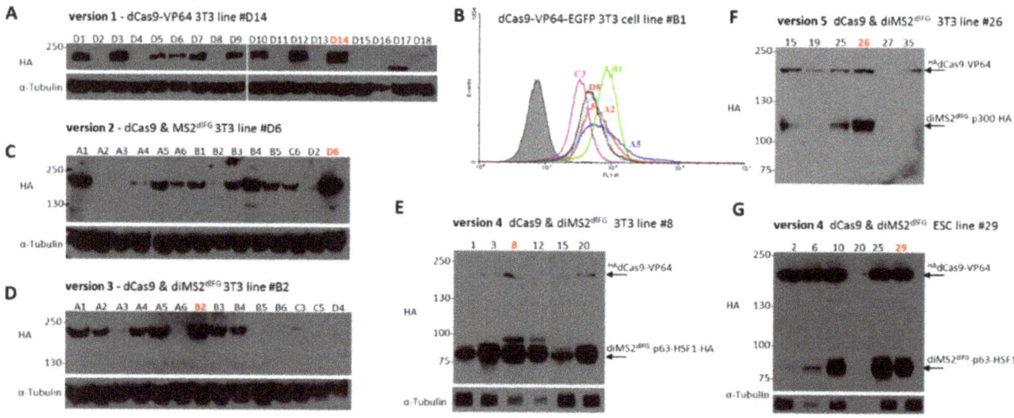

Figure A1. Overview of selected cell lines used in the work: (**A**) dCas9-VP64 3T3 line #D14 (version 1), used in Figure 2C,D, and Figure 3E, (**B**) dCas9-VP64-EGFP 3T3 cell line #B1 used in Figure 2C, (**C**) dCas9 and MS2^{dlFG} 3T3 line #D6 (version 2) used in Figure 3C, (**D**) dCas9 and diMS2^{dlFG} 3T3 line #B2 (version 3) used in Figure 3C,D, (**E**) dCas9 and diMS2^{dlFG} 3T3 line #8 (version 4) used in Figure 3C,E, and Figure 4D, (**F**) dCas9 and diMS2^{dlFG} 3T3 line #26 (version 5) used in Figure 3E, (**G**) dCas9 and diMS2^{dlFG} ESC line #29 (version 4) used in Figure 4C,E. Note that in single construct versions 2 and 3 (panels **C**,**D**), the MS2-p65-HSF1 fusion protein construct is linked to HA-tagged dCas9-VP64 via a T2A self-cleavage peptide [23]. Thus, anti-HA antibody detects only dCas9. When dCa9 and MS2-p65-HSF1 are expressed separately, both proteins carry HA-tag and can be detected (panels **E**–**G**).

References

1. Fire, A.; Xu, S.; Montgomery, M.K.; Kostas, S.A.; Driver, S.E.; Mello, C.C. Potent and specific genetic interference by double-stranded RNA in Caenorhabditis elegans. *Nature* **1998**, *391*, 806–811. [CrossRef] [PubMed]
2. Jinek, M.; Doudna, J.A. A three-dimensional view of the molecular machinery of RNA interference. *Nat. Cell Biol.* **2008**, *457*, 405–412. [CrossRef] [PubMed]
3. Svoboda, P. Renaissance of mammalian endogenous RNAi. *FEBS Lett.* **2014**, *588*, 2550–2556. [CrossRef] [PubMed]
4. Macrae, I.J.; Ma, E.; Zhou, M.; Robinson, C.V.; Doudna, J.A. In vitro reconstitution of the human RISC-loading complex. *Proc. Natl. Acad. Sci. USA* **2008**, *105*, 512–517. [CrossRef]
5. Suk, K.; Choi, J.; Suzuki, Y.; Ozturk, S.B.; Mellor, J.C.; Wong, K.H.; Mackay, J.L.; Gregory, R.I.; Roth, F.P. Reconstitution of human RNA interference in budding yeast. *Nucleic Acids Res.* **2011**, *39*, e43. [CrossRef]
6. Wang, Y.; Mercier, R.; Hobman, T.C.; Lapointe, P. Regulation of RNA interference by Hsp90 is an evolutionarily conserved process. *Biochim. Biophys. Acta (BBA) Bioenerg.* **2013**, *1833*, 2673–2681. [CrossRef]
7. Bartel, D.P. Metazoan MicroRNAs. *Cell* **2018**, *173*, 20–51. [CrossRef]

8. Gantier, M.P.; Williams, B.R. The response of mammalian cells to double-stranded RNA. *Cytokine Growth Factor Rev.* **2007**, *18*, 363–371. [CrossRef]
9. Kennedy, E.M.; Whisnant, A.W.; Kornepati, A.V.R.; Marshall, J.B.; Bogerd, H.P.; Cullen, B.R. Production of functional small interfering RNAs by an amino-terminal deletion mutant of human Dicer. *Proc. Natl. Acad. Sci. USA* **2015**, *112*, E6945–E6954. [CrossRef]
10. Maillard, P.V.; Van Der Veen, A.G.; Deddouche-Grass, S.; Rogers, N.C.; Merits, A.; Sousa, C.R.E. Inactivation of the type I interferon pathway reveals long double-stranded RNA -mediated RNA interference in mammalian cells. *EMBO J.* **2016**, *35*, 2505–2518. [CrossRef]
11. Van Der Veen, A.G.; Maillard, P.V.; Schmidt, J.M.; Lee, A.S.; Deddouche-Grass, S.; Borg, A.; Kjær, S.; Snijders, A.P.; Sousa, C.R.E. The RIG-I-like receptor LGP2 inhibits Dicer-dependent processing of long double-stranded RNA and blocks RNA interference in mammalian cells. *EMBO J.* **2018**, *37*, e97479. [CrossRef]
12. Demeter, T.; Vaskovicova, M.; Malik, R.; Horvat, F.; Pasulka, J.; Svobodova, E.; Flemr, M.; Svoboda, P. Main constraints for RNAi induced by expressed long dsRNA in mouse cells. *Life Sci. Alliance* **2019**, *2*, e201800289. [CrossRef]
13. Ma, E.; Macrae, I.J.; Kirsch, J.F.; Doudna, J.A. Autoinhibition of human dicer by its internal helicase domain. *J. Mol. Biol.* **2008**, *380*, 237–243. [CrossRef]
14. Flemr, M.; Malik, R.; Franke, V.; Nejepinska, J.; Sedlacek, R.; Vlahovicek, K.; Svoboda, P. A retrotransposon-driven dicer isoform directs endogenous small interfering RNA production in mouse oocytes. *Cell* **2013**, *155*, 807–816. [CrossRef]
15. Franke, V.; Ganesh, S.; Karlic, R.; Malik, R.; Pasulka, J.; Horvat, F.; Kuzman, M.; Fulka, H.; Cernohorska, M.; Urbanova, J.; et al. Long terminal repeats power evolution of genes and gene expression programs in mammalian oocytes and zygotes. *Genome Res.* **2017**, *27*, 1384–1394. [CrossRef]
16. Carroll, D. Genome engineering with targetable nucleases. *Annu. Rev. Biochem.* **2014**, *83*, 409–439. [CrossRef]
17. Beerli, R.R.; Segal, D.J.; Dreier, B.; Barbas, C.F. Toward controlling gene expression at will: Specific regulation of the erbB-2/HER-2 promoter by using polydactyl zinc finger proteins constructed from modular building blocks. *Proc. Natl. Acad. Sci. USA* **1998**, *95*, 14628–14633. [CrossRef]
18. Zhang, F.; Cong, L.; Lodato, S.; Kosuri, S.; Church, G.M.; Arlotta, P. Efficient construction of sequence-specific TAL effectors for modulating mammalian transcription. *Nat. Biotechnol.* **2011**, *29*, 149–153. [CrossRef]
19. Jinek, M.; Chylinski, K.; Fonfara, I.; Hauer, M.; Doudna, J.A.; Charpentier, E. A Programmable dual-RNA-guided DNA endonuclease in adaptive bacterial immunity. *Science* **2012**, *337*, 816–821. [CrossRef]
20. Gilbert, L.A.; Larson, M.H.; Morsut, L.; Liu, Z.; Brar, G.A.; Torres, S.E.; Stern-Ginossar, N.; Brandman, O.; Whitehead, E.H.; Doudna, J.A.; et al. CRISPR-Mediated modular RNA-guided regulation of transcription in eukaryotes. *Cell* **2013**, *154*, 442–451. [CrossRef]
21. Maeder, M.L.; Linder, S.J.; Cascio, V.M.; Fu, Y.; Ho, Q.H.; Joung, J.K. CRISPR RNA–guided activation of endogenous human genes. *Nat. Methods* **2013**, *10*, 977–979. [CrossRef]
22. Hilton, I.B.; D'Ippolito, A.M.; Vockley, C.M.; Thakore, P.I.; Crawford, G.E.; Reddy, T.E.; Gersbach, C.A. Epigenome editing by a CRISPR-Cas9-based acetyltransferase activates genes from promoters and enhancers. *Nat. Biotechnol.* **2015**, *33*, 510–517. [CrossRef]
23. Konermann, S.; Brigham, M.D.; Trevino, A.E.; Joung, J.; Abudayyeh, O.O.; Barcena, C.; Hsu, P.D.; Habib, N.; Gootenberg, J.S.; Nishimasu, H.; et al. Genome-scale transcriptional activation by an engineered CRISPR-Cas9 complex. *Nat. Cell Biol.* **2015**, *517*, 583–588. [CrossRef]
24. Peabody, D.S.; Ely, K.R. Control of translational repression by protein-protein interactions. *Nucleic Acids Res.* **1992**, *20*, 1649–1655. [CrossRef]
25. Huggins, C.J.; Malik, R.; Lee, S.; Salotti, J.; Thomas, S.; Martin, N.; Quiñones, O.A.; Alvord, W.G.; Olanich, M.E.; Keller, J.R.; et al. C/EBPγ Suppresses senescence and inflammatory gene expression by heterodimerizing with C/EBPβ. *Mol. Cell Biol.* **2013**, *33*, 3242–3258. [CrossRef]
26. Hampf, M.; Gossen, M. A protocol for combined Photinus and Renilla luciferase quantification compatible with protein assays. *Anal. Biochem.* **2006**, *356*, 94–99. [CrossRef]
27. Sinkkonen, L.; Hugenschmidt, T.; Filipowicz, W.; Svoboda, P. Dicer is associated with ribosomal DNA chromatin in mammalian cells. *PLoS ONE* **2010**, *5*, e12175. [CrossRef]
28. Pfaffl, M.W.; Horgan, G.W.; Dempfle, L. Relative expression software tool (REST(C)) for group-wise comparison and statistical analysis of relative expression results in real-time PCR. *Nucleic Acids Res.* **2002**, *30*, e36. [CrossRef]
29. Weber, M.; Hellmann, I.; Stadler, M.B.; Ramos, L.; Pääbo, S.; Rebhan, M.; Schübeler, D. Distribution, silencing potential and evolutionary impact of promoter DNA methylation in the human genome. *Nat. Genet.* **2007**, *39*, 457–466. [CrossRef]
30. Todaro, G.J.; Green, H. Quantitative studies of the growth of mouse embryo cells in culture and their development into established lines. *J. Cell Biol.* **1963**, *17*, 299–313. [CrossRef]
31. A Martens, J.H.; O'Sullivan, R.J.; Braunschweig, U.; Opravil, S.; Radolf, M.; Steinlein, P.; Jenuwein, T. The profile of repeat-associated histone lysine methylation states in the mouse epigenome. *EMBO J.* **2005**, *24*, 800–812. [CrossRef] [PubMed]
32. Efroni, S.; Duttagupta, R.; Cheng, J.; Dehghani, H.; Hoeppner, D.J.; Dash, C.; Bazett-Jones, D.P.; Le Grice, S.; McKay, R.D.; Buetow, K.H.; et al. Global transcription in pluripotent embryonic stem cells. *Cell Stem Cell* **2008**, *2*, 437–447. [CrossRef] [PubMed]

genes

MDPI

Article

Partial Disturbance of Microprocessor Function in Human Stem Cells Carrying a Heterozygous Mutation in the DGCR8 Gene

Dóra Reé [1,2], Ábel Fóthi [1], Nóra Varga [1], Orsolya Kolacsek [1], Tamás I. Orbán [1,*] and Ágota Apáti [1,2,*]

[1] Institute of Enzymology, Research Center for Natural Sciences, 1117 Budapest, Hungary
[2] Doctoral School of Molecular Medicine, Semmelweis University, 1085 Budapest, Hungary
* Correspondence: orban.tamas@ttk.hu (T.I.O.); apati.agota@ttk.hu (Á.A.)

Abstract: Maturation of microRNAs (miRNAs) begins by the "Microprocessor" complex, containing the Drosha endonuclease and its partner protein, "DiGeorge Syndrome Critical Region 8" (DGCR8). Although the main function of the two proteins is to coordinate the first step of precursor miRNAs formation, several studies revealed their miRNA-independent functions in other RNA-related pathways (e.g., in snoRNA decay) or, for the DGCR8, the role in tissue development. To investigate the specific roles of DGCR8 in various cellular pathways, we previously established a human embryonic stem-cell (hESC) line carrying a monoallelic DGCR8 mutation by using the CRISPR-Cas9 system. In this study, we genetically characterized single-cell originated progenies of the cell line and showed that DGCR8 heterozygous mutation results in only a modest effect on the mRNA level but a significant decrease at the protein level. Self-renewal and trilineage differentiation capacity of these hESCs were not affected by the mutation. However, partial disturbance of the Microprocessor function could be revealed in pri-miRNA processing along the human chromosome 19 miRNA cluster in several clones. With all these studies, we can demonstrate that the mutant hESC line is a good model to study not only miRNA-related but also other "noncanonical" functions of the DGCR8 protein.

Keywords: miRNA processing; C19MC; miRNA cluster; human pluripotent stem cells; differentiation

1. Introduction

MicroRNAs (miRNAs) are small noncoding RNAs, representing an abundant class of key regulatory molecules that act on the posttranscriptional level of gene expression. They function in cytoplasmic ribonucleoprotein complexes that always contain a member of the Argonaute (Ago) protein family. These RNA-induced silencing complexes (RISCs) constantly scan the RNA population of the cell to find their targets by base pair complementarity between the contained miRNA and other, mainly protein coding mRNAs (in animals, the target sites are mostly located at the 3′ untranslated regions). The effector functions of miRNAs are manifested in either translation inhibition or the destabilization and degradation of the target mRNAs. As miRNAs generally have several target molecules, and mRNAs usually contain numerous miRNA binding sites, this posttranscriptional fine-tuning represents an elaborate regulatory network with a complexity comparable with that of transcription factors [1–3]. This explains why miRNAs play important regulatory roles in most cellular processes, having essential functions especially during embryonic development. Imbalance of the miRNA proportions is widely observed in many diseases, particularly in cancer and in neurodevelopmental disorders, suggesting the importance of precise and carefully regulated miRNA dosage in the cells [4,5].

Maturation of most miRNAs occurs via the canonical pathway: transcription of an endogenous locus results in the primary-miRNA (pri-miRNA) containing one or more stem-loop structures that undergo two consecutive RNA cleavage reactions. The first occurs in the nucleus by the Microprocessor complex, a protein heterotrimer containing an RNase III enzyme, Drosha, and two molecules of its regulatory partner protein, the

DiGeorge Syndrome Critical Region 8 (DGCR8) [6]. The so formed precursor-miRNAs (pre-miRNAs) are transferred to the cytoplasm by the Exportin-5 system, where another RNase III enzyme Dicer cleaves the apical loop, forming a double-stranded short RNA with two-nucleotide overhangs at their 3′ end. This intermediate then undergoes subsequent maturation steps when it is associated with an Ago protein and its auxiliary factors: one strand (the "passenger" strand) of the duplex is eliminated, and the final functional RISC is formed [7]. Certain miRNAs mature via alternative pathways during which one of the above cleavage steps (or both) is substituted by other unrelated processing mechanisms, such as the mRNA splicing for the Microprocessor-independent mirtrons, or other RNase activities for the tRNA-derived miRNAs [8].

Functionally linked miRNAs are often clustered in the human genome [9,10], and in the primate lineage, such clusters play essential roles in stem-cell regulation and in placenta formation [11,12]. The distinct regulation of these clusters goes beyond common transcriptional regulation, and details of the exact mechanisms often remain controversial [13]. The chromosome 19 miRNA cluster (C19MC) is an especially long (46 miRNA genes) cluster predominantly expressed in human embryonic stem cells (hESCs) and in the reproductive system. In hESCs, the position-dependent processing of this cluster is regulated by the Microprocessor recruitment, explaining why the miRNA expression levels in the Drosha knockdown of those cells gradually decrease toward the 3′ end of the cluster [14]. To further explore the molecular mechanisms behind this phenomenon, the recruitment and role of DGCR8 or additional Microprocessor-associated factors also need to be studied.

There are different approaches to uncover the regulation driven by miRNAs; knocking out DGCR8, Drosha, or Dicer in ES cells provides useful information about the biological processes affected by the general loss of miRNAs. For instance, global loss of DGCR8 and canonical miRNAs causes embryonic lethality in mice at d 6.5, and the balance between self-renewal and differentiation of DGCR8 knockout mESCs is severely disturbed [15,16]. In contrast, heterozygous deletion of DGCR8 in mES cells shows a cellular phenotype comparable with the wild type: the miRNA and DGCR8 mRNA levels are not significantly changed due to a feedback control by the Microprocessor complex [17]. Mice heterozygous for DGCR8 show reduced expression of DGCR8 and a subset of miRNAs in the prefrontal cortex. These effects are not observed in neonatal mice but emerge from postnatal days 25–30 [18]. DGCR8 deficiency also alters miRNA biogenesis in adult mouse hippocampus: results of the array-based miRNA analysis show that while most of the examined miRNAs are downregulated, some are increased, suggesting more complicated alterations in miRNA biogenesis than expected. In the subgranular zone of these monoallelic knockout mice, cell proliferation is reduced, which may affect adult neurogenesis in the hippocampal dentate gyrus [19]. In the forebrain, pyramidal neurons of DGCR8 +/− mice show decreased branch complexity and decreased total dendrite length. These neurons have altered electrical properties such as imbalance of spontaneous synaptic transmission and short-term plasticity [18,20]. In a human cellular model system of DiGeorge syndrome, cortical neurons derived from DGCR8 +/− hiPSCs show defected neuronal activity and calcium handling, and the monoallelic mutant cells have altered resting membrane potential and abnormal inactivation of voltage-gated calcium channels [21]. Here, we presented an in vitro model for investigating the heterozygous deletion of DGCR8 in hESCs. We studied the aspects of cell renewal capacity, as well as the differentiation pattern of these mutant stem cells, and addressed the questions of whether any DGCR8-related functional deficiency could be detected, among others, on the position-dependent regulation pattern of the C19MC.

2. Materials and Methods

2.1. Cell Culture and Differentiation

The HuES9 embryonic stem-cell line was kindly provided by Douglas Melton (HHMI). This work was performed according to ethical approvals (HuES9 NIH approval NIHhESC-09-0022 and Health Care Research Council, Human Reproduction Committee in Hungary

(Egészségügyi Tudományos Tanács, Humán Reprodukciós Bizottság, ETT HRB) approval number 6681/2012-EHR. HuES9). The parental HuES9 and DGCR8-deficient HVRDe009-A-1 cells were maintained on hESC qualified Matrigel (Corning, New York, NY, USA, #354277) coated plates in mTeSR1 (STEMCELL Technologies, Vancouver, BC, Canada # 85850) with or without 0.8 μM puromycin (ThermoFisher Scientific, Waltham, MA, USA #A1113803). Media were changed every day. Cells were passaged with StemPro Accutase (ThermoFisher Scientific, Waltham, MA, USA #A1110501) at a 1:10 ratio every 3–4 days and plated onto fresh Matrigel-coated plates in mTeSR1-Y (mTeSR1 supplemented with 10 μM Y27632-2HCl (Selleckchem, Planegg, Germany, #S1049) for 24 h to improve cell survival. The genetic identity and normal karyotype of the cultured cells were confirmed by STR (short tandem repeat) analysis and G-Banding performed by UD-GenoMed Medical Genomic Technologies Ltd. For spontaneous differentiation, the hES cells were dissociated with ReLeSR (STEMCELL Technologies, #05872) and plated onto MEF (Merck, Darmstadt, Germany, #PMEF-CFL) covered tissue culture plates. Cells grown on MEFs were maintained in hESC culture media (KO-DMEM (ThermoFisher Scientific, #10829018) supplemented with 15% knockout serum replacement (ThermoFisher Scientific, Waltham, MA, USA, #10828010), 1 mM L-glutamine (ThermoFisher Scientific, Waltham, MA, USA, #25030024), 0.1 mM β-mercaptoethanol (ThermoFisher Scientific, Waltham, MA, USA, #21985023), 1% nonessential amino acids (ThermoFisher Scientific, Waltham, MA, USA, #11140050), and 4 ng/mL human basic fibroblast growth factor (ThermoFisher Scientific, Waltham, MA, USA, #13256-029)). hESC colonies were then dissociated with Collagenase (ThermoFisher Scientific, Waltham, MA, USA, #17018029) and cultured in suspension on low attachment plates in an EB (Embryoid body) medium (KO-DMEM supplemented with 20% FBS, 1 mM L-GLU, 1% non-essential amino acids, and 0.1 mM ß-mercaptoethanol) for 6 days. Next, EBs were transferred onto 0.1% gelatin coated 24-well tissue culture plates or confocal chamber slides (ThermoFisher Scientific, Waltham, MA, USA, #177402) and allowed to differentiate for another 6 days in DMEM (ThermoFisher Scientific, Waltham, MA, USA, #41965062) supplemented with 10% FBS. During differentiation, the media were changed every other day. The differentiated offspring was characterized by immunocytochemical staining and RT-qPCR.

2.2. Single-Cell Cloning

HVRDe009-A-1 hESCs were cultured without puromycin to enrich those cells not expressing the green fluorescent protein (GFP). Next, cells were dissociated with StemPro Accutase, and single cells were plated onto Matrigel-coated 96-well plates with a BD FACSAria™ Cell Sorter based on the GFP expression in cloning media (mTeSR1-Y supplemented with 1/3 MEF-conditioned hESC media). SSCs were expanded in cloning media for 10–20 days then replated onto 24-well Matrigel-coated plates in mTeSR1-Y. When reaching 80% density, SSCs were plated onto 6-well plates. GFP expression was validated by FACS measurements.

2.3. Trichostatin A Treatment

Cells were plated onto 6-well plates and cultured without the addition of puromycin. On day 3, cells were treated with mTeSR1 supplemented with 30 nM or 60 nM Trichostatin A, respectively. Next day, cells were detached, and GFP expression was measured in the treated and untreated control cells with flow cytometry. In parallel, the expression of the endogenous cancer testis gene GAGE was measured via RT-qPCR to prove effective TSA-induced demethylation.

2.4. Flow Cytometry Measurements

SSEA-4 Flow Cytometry was performed as previously described [22]. Briefly, single-cell suspensions were prepared using Accutase. Subsequently, cells were labeled in PBS supplemented with 0.5% BSA (bovine serum albumin, Sigma, St. Louis, MO, USA, #A9418-50G) with anti-human SSEA-4 PE-conjugated antibody (1:100, R&D Systems, Minneapolis,

MN, USA, #RD-FAB1435A-100) on 37 °C for 30 min. Propidium iodide (ThermoFisher Scientific, Waltham, MA, USA, #P1304MP) staining was employed for gating out the positively labeled dead cells. Control measurements with isotype-matched control were included (1:100, R&D Systems, Minneapolis, MN, USA, #IC007A).

2.5. Immunocytochemistry

For immunostaining, cells were seeded onto eight-well Nunc Lab-Tek II Chambered Coverglass (ThermoFisher Scientific, Waltham, MA, USA, #155411), fixed, and permeabilized with 4% PFA (paraformaldehyde) in DPBS (Dulbecco's modified PBS) for 15 min on RT (room temperature). Next, cells were blocked for 60 min at RT in a blocking solution (DPBS supplemented with 2 mg/mL BSA, 0.1% Triton-X 100, and 1% fish gelatin with or without 5% goat serum). Then, the samples were incubated for 60 min at RT in the blocking solution supplemented with the following primary antibodies: OCT3-4 (1:50, Santa Cruz, CA, USA, #SC-5279) and NANOG (1:100, R&D Systems, Minneapolis, MN, USA, #AF1997) as pluripotency markers; AFP (1:500, Sigma, St. Louis, MO, USA#A8452), SMA (1:500, Abcam, Cambridge, UK, #ab7817), and ß-III-tubulin (1:2000, R&D Systems, Minneapolis, MN, USA, #RD-MAB1195) as markers specific for the three lineages. After washing with DPBS, the cells were incubated for 60 min at RT with Alexa Fluor 647-conjugated goat anti-mouse and Alexa Fluor 594-conjugated donkey anti-goat IgG antibodies (1:250, ThermoFisher Scientific, Waltham, MA, USA, #A11029, #A11012). DAPI (ThermoFisher Scientific, Waltham, MA, USA, # D1306) was used for nuclear staining. The samples were then examined by a Zeiss LSM 710 confocal laser scanning microscope.

2.6. RNA Isolation and Gene Expression Studies

Total RNA from the hES and differentiated cells was isolated using a TRIzol reagent (ThermoFisher Scientific, Waltham, MA, USA, #15596018). RNA integrity was analyzed by agarose gel electrophoresis; total RNA concentrations and sample purity were measured by a Nanodrop spectrophotometer (ThermoFisher Scientific, Waltham, MA, USA).

For pri-miRNA analysis, 1 µg total RNA was reverse-transcribed by random oligomers using a High-Capacity cDNA Reverse Transcription Kit (ThermoFisher Scientific, Waltham, MA, USA). cDNA samples were diluted 1:10 before subsequent amplifications. In the case of C19MC pri-miRNA, RT-PCR was carried out by using a SYBR Green PCR Master Mix with custom-made PCR primers (Supplementary Table S1).

For gene expression assays, cDNA samples were synthesized from 400 ng of total RNA using the Promega Reverse Transcription system according to the manufacturer's instructions. mRNA levels for DGCR8 (Hs00987085_m1), Drosha (Hs00203008_m1), NANOG (Hs02387400_g1), AFP (Hs00173490_m1), TBXT (Hs00610080_m1), and PAX6 (Hs00240871_m1) were determined using TaqMan® gene expression assays (ThermoFisher Scientific, Waltham, MA, USA, #4331182).

Real-time PCR measurements were run and analyzed on the StepOne™ Real-Time PCR System (Applied Biosystems) according to the manufacturer's instructions. Quantitative gene expression data were normalized to endogenous control mRNAs: for Taq-Man® analyses RPLP0 (Hs9999902_m1) or PolR2A (Hs00172187_m1); for SYBR Green assay details, see Supplementary Table S1.

For mature miRNA quantification, the expression analysis was performed using the miRCURY LNA™ Universal RT miRNA PCR Assay (Qiagen, Venlo, Netherlands) according to the manufacturer's instructions. Briefly, RNA samples (5 ng/µL) were reverse-transcribed, and the UniSp6 RNA spike-in template was added to each reaction for controlling the quality of cDNA synthesis. cDNA samples were diluted 1:80 before subsequent amplifications. RT-PCR was performed by using a miRCURY SYBR® Green master mix (Qiagen Venlo, Netherlands), and real-time PCRs were run on a StepOnePlus™ platform (ThermoFisher Scientific, Waltham, MA, USA) according to the manufacturer's protocol. In these cases, the hsa-miR-103a internal control miRNA was used for normalization during the relative quantifications by the ΔΔCt method.

2.7. Protein Analysis by Western blotting

Samples were briefly sonicated and subsequently ran on 8% acrylamide gels followed by electroblotting onto PVDF membranes. Membranes were then blocked with blocking solution (5% Milk/TBS-Tween) and incubated with a monoclonal antibody specific to human DGCR8 (1:1000, Abcam, Cambridge, UK, #ab191875) ON (overnight) at 4 °C. Next, anti-rabbit IgG (1:5000, ThermoFisher Scientific, Waltham, MA, USA, #G-21234) was used as a secondary antibody. Pierce ECL Western blotting substrate ((ThermoFisher Scientific, Waltham, MA, USA, # 32106) was used for signal detection; membranes were exposed to Agfa films. Anti-β actin antibody (1:10,000, Abcam, Cambridge, UK, #ab20272) was used to normalize the DGCR8 expression. Expression levels were determined by densitometry of the scanned images using ImageJ and corrected with background and normalized to β-actin and parental HuES9 levels. Briefly, a region of interest (ROI) for a given protein was chosen to be the smallest rectangle shape that can enclose the largest band of that protein and used in all lanes of a blot. The background for the normalization is measured by the same ROI at close proximity to the target band; aspecific bands (if any) were always avoided.

3. Results

3.1. Establishment of Heterozygous DGCR8 Mutant Clones

The heterozygous DGCR8 mutant HVRDe009-A-1 hESC line was established and characterized as previously described [23]. These mutant cells contain a donor DNA sequence in the third exon of DGCR8, which consists of two CAG-driven selection markers, a puromycin resistance gene, and a GFP. The insertion of the donor plasmid results in monoallelic DGCR8 expression in the mutant cells. The HVRDe009-A-1 hESCs show consistent GFP expression when cultured in a puromycin-containing medium; however, during puromycin deprivation, gradual loss of GFP expression was observed (Supplementary Figure S1a,b). Transgene silencing and mosaicism are known to occur in transfected cells, especially when selection is heavily dependent on antibiotic resistance [24,25]. Treatments with Trichostatin A (TSA), an organic compound interfering with the removal of acetyl groups from histones [26], resulted in elevated GFP expression only in the still GFP-positive cells, which suggests that gene silencing by deacetylation was not responsible for GFP loss (Supplementary Figure S1b). When sorting out GFP-positive and -negative populations, and propagated them without puromycin, GFP-negative cells showed a decreased GFP copy number based on real-time quantitative PCR measurements (Supplementary Figure S1c). These results show that the loss of transgene expression is likely a result of genetic rearrangements and the loss of the transgene copy from the DGCR8 locus.

To investigate the exact genetic background in the GFP-positive and -negative cells, we accomplished single-cell cloning of the HVRDe008-A-1 cells based on the GFP expression. We established 12 GFP-positive (Supplementary Figure S2) and 13 GFP-negative single-cell clones. The clones were propagated without puromycin, and the GFP expression in the clones was continuously monitored by high content screening and FACS measurements. By day 30, the GFP expression in 7 out of 12 GFP-positive single-cell clones decreased below 25% (Figure 1a), and by day 60, it further decreased to 0% in two clones (Supplementary Figure S2); on the other hand, GFP-negative clones maintained to keep the negative phenotype, regardless of the remaining parts of the GFP expression cassette (GFP copy number was simultaneously monitored by quantitative real-time PCR measurements; see Figure 1b).

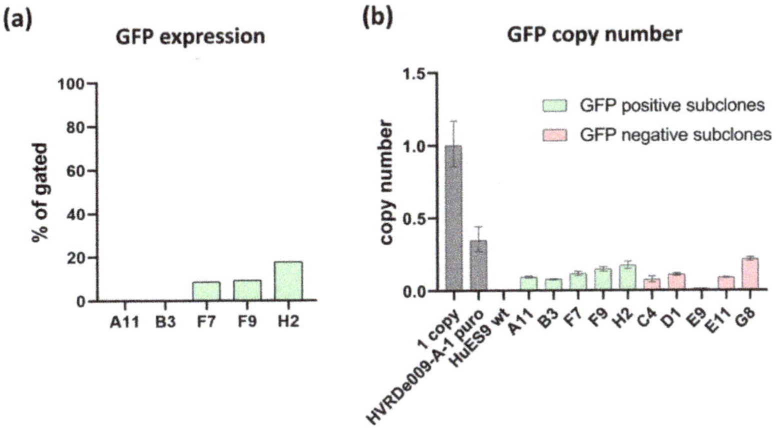

Figure 1. GFP expression in the single-cell clones of HVRDe009-A-1. (**a**) GFP FACS measurements of GFP-positive single-cell-derived clones (passage 10 after single-cell cloning); (**b**) GFP copy number measurements on the GFP-positive and -negative single-cell-derived clones and the HRDVe009-A-1 cells cultured with puromycin. Relative quantitation of copy number values was calculated using RPPH1 as a reference target and 1 copy control gDNA reference sample.

3.2. Genetic Characterization of Selected Single-Cell Clones

Next, we have selected four single-cell clones for further characterization (Supplementary Figure S3), namely HVRDe009-A-1-A11, HVRDe009-A-1-B3, HVRDe009-A-1-C4, and HVRDe009-A-1E9. All the investigated clones were negative for GFP expression (Figure 1a). Moreover, the single-cell clones lost their puromycin resistance and died after 48 h when treated with puromycin.

The results prompted us to analyze the transgene sequence in these clones by diagnostic PCRs. Amplifying different segments of the transgene cassette, we found that significant portions of the GFP expression unit, as well as the puromycin resistance gene, were lost in these cells (Supplementary Figure S4). However, certain regions of the transgene still remained integrated, indicating that at least one DGCR8 allele is disrupted in all of these descendant cells.

To confirm if one intact DGCR8 allele is still present, we sequenced the targeted genomic site in the clones, and Sanger sequencing data provided evidence for the monoallelic disruption of the DGCR8 gene (Supplementary Figure S5).

3.3. DGCR8 +/− hESCs Maintain Pluripotency and Trilineage Differentiation Capacity

The single-cell clones maintained stem-cell-like morphology and normal karyotype after single-cell cloning when cultured in feeder-free conditions (Supplementary Figure S6). To evaluate the effect of the heterozygous DGCR8 mutation on the pluripotency, first we measured the SSEA-4 expression of the single-cell clones, and every clone showed over 90% positivity for this commonly used embryonic stem-cell marker (Figure 2a). Moreover, immunostaining of OCT4 and NANOG pluripotency markers showed homogenous expression in the hESC colonies (Figure 2b).

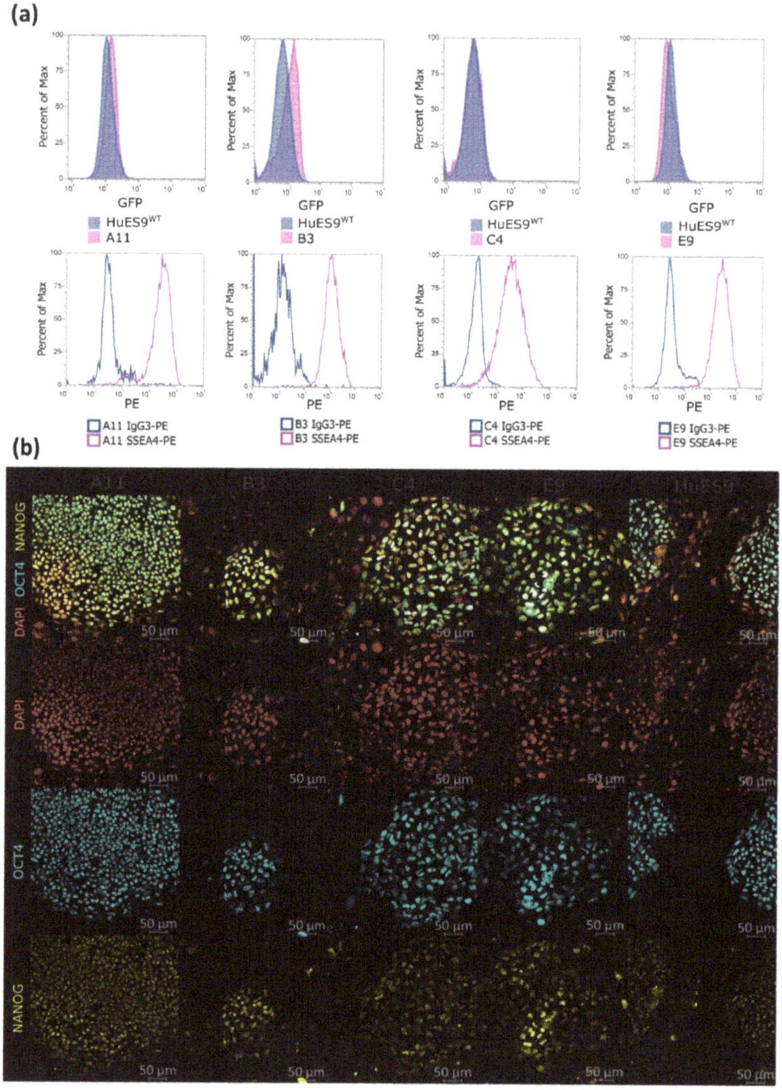

Figure 2. Pluripotency of the monoallelic mutant single-cell clones. (**a**) GFP and SSEA-4 FACS measurements of undifferentiated hESCs. (**b**) Immunostaining of the pluripotency markers OCT4 and NANOG on undifferentiated hESCs.

To assess the differentiation capacity of these mutant hESCs, we performed an in vitro embryoid body (EB) differentiation assay. Immunostaining and real-time quantitative PCR measurements confirmed the continuous decline of the expression of pluripotency marker NANOG and increase in markers specific for the three germ layers (ectoderm: TUJ1/ β-III-Tub, PAX6; mesoderm: TBXT, SMA; endoderm: AFP) (Figure 3a,b). These results indicate that monoallelic deletion of DGCR8 does not cause impairments in the three-lineage differentiation capacity of the hESCs.

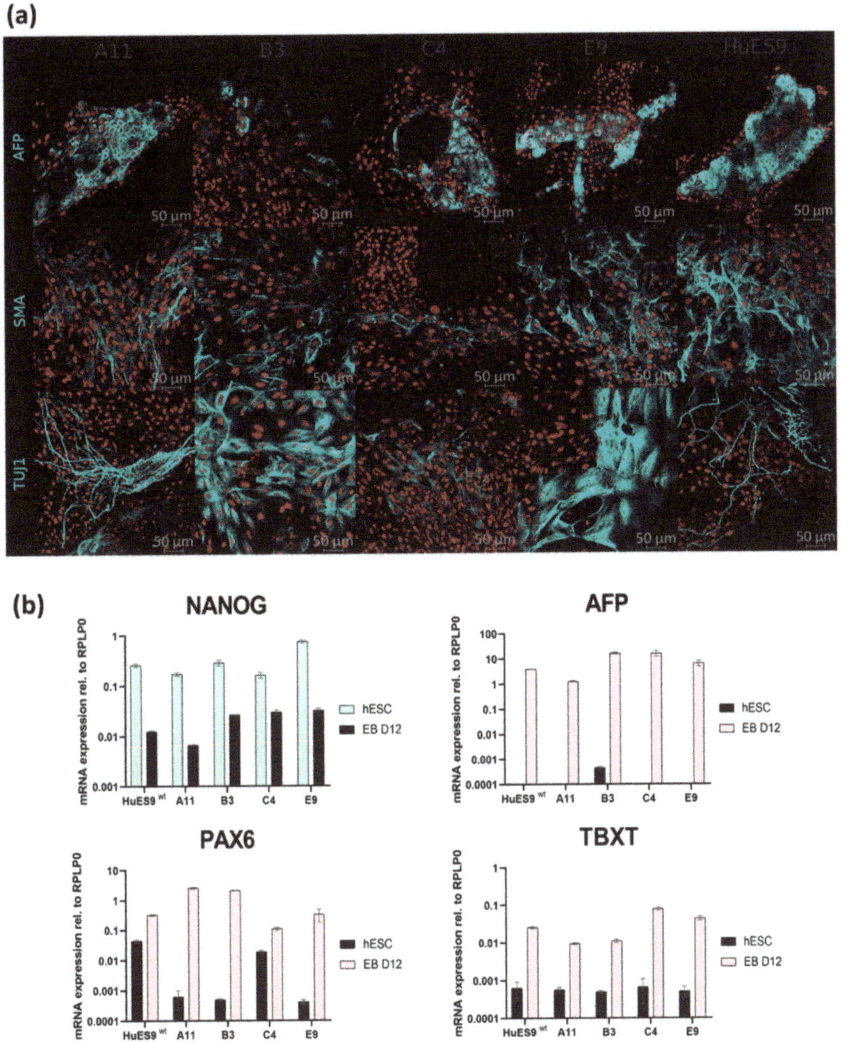

Figure 3. Differentiation capacity of monoallelic mutant single-cell clones. (**a**) Immunostaining of markers specific for the three germ layers (AFP, SMA, TUJ1) on the differentiated offspring (embryoid body day 12). (**b**) mRNA expression levels of pluripotency and lineage-specific markers on undifferentiated hESCs and differentiated offspring (embryoid body day 12).

3.4. Expression of Microprocessor Complex Components DGCR8 and Drosha

When determining the DGCR8 mRNA levels by real-time quantitative PCR, the clones showed slightly fluctuating but not considerably different levels when compared with their HuES9 stem-cell ancestor (homozygous to the wild-type DGCR8 allele). The Drosha mRNA levels also showed some variability among the clones when individually compared with their parental HuES9 cell line (Figure 4a and Supplementary Figure S7a). On the other hand, DGCR8 protein levels were decreased by at least 40–50% based on Western blot detections in each clone (Figure 4b,c and Supplementary Figure S7b). In contrast, Drosha protein levels considerably varied, showing rather clone-specific expression profiles (Figure 4b,c and Supplementary Figure S7c).

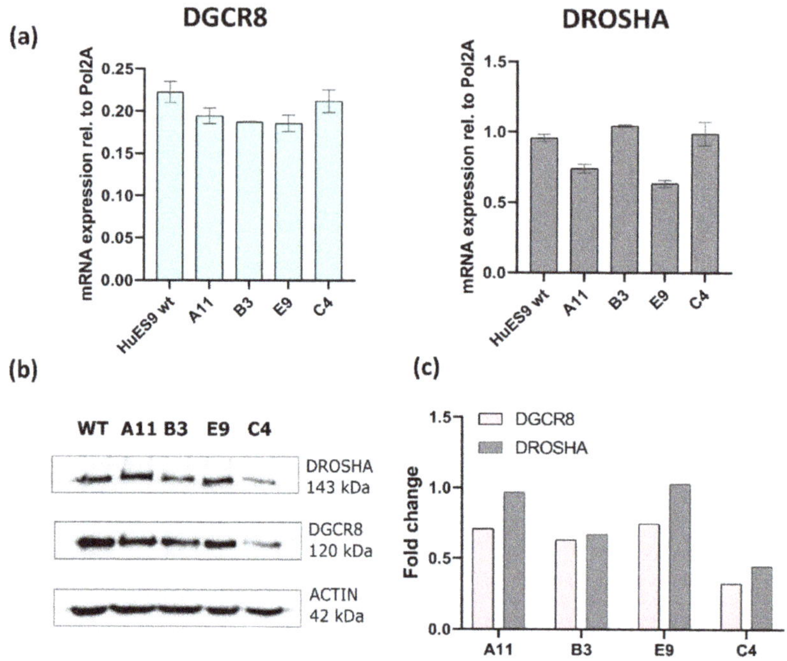

Figure 4. Expression of the Microprocessor components DGCR8 and Drosha in the monoallelic mutant single-cell clones. (**a**) Expression levels of DGCR8 and Drosha mRNAs relative to POL2A endogenous control in a representative experiment. Data are presented as mean ± SD of 3 technical parallels. (**b**) Representative Western blots for DGCR8, Drosha, and β-actin. (**c**) DGCR8 and Drosha protein levels of the representative Western blots normalized to β-actin levels, and to the relative expressions in the parental HuES9 (WT) cell line.

3.5. Pri-miRNA Processing Efficiency of C19MC in DGCR8 Mutant Cells

The considerable decrease in DGCR8 protein levels prompted us to test whether the activity or any functions related to the Microprocessor complex are disturbed in the examined cell clones. In a recent study, we showed that the depletion of Drosha in hESCs caused a gradual decrease in pri-miRNA processing along an extended miRNA cluster, C19MC [14]. We also proposed that depleting DGCR8, the other component of the Microprocessor complex, may result in a similar phenotype; therefore, we tested the processing activity in three selected regions of C19MC in the four DGCR8 heterozygous clones. We detected clone-specific responses: a modest gradual decrease toward the 3′ end of the cluster was revealed in clones E9 and B3, whereas no significant decrease was measured in the other two clones, A11 and C4 (Figure 5). The results did not show a clear correlation with the measured DGCR8 expression levels but rather indicated a potential disturbance of the Microprocessor function among cells where the DGCR8 protein level is reduced due to a heterozygous mutation.

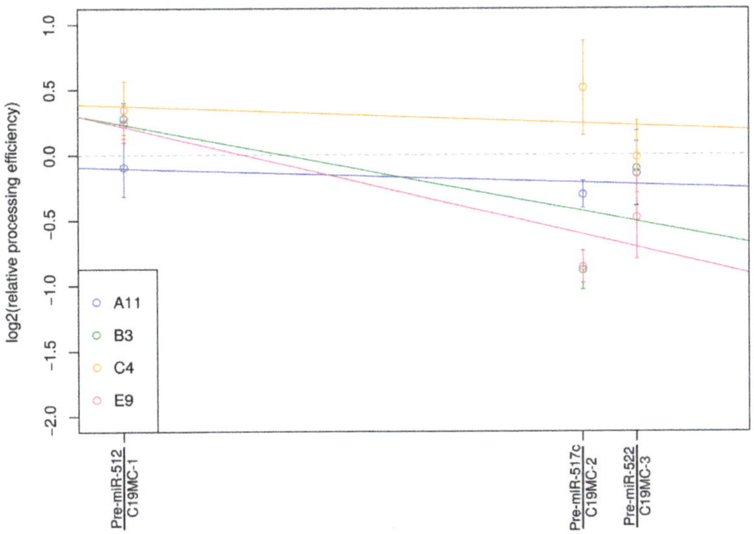

Figure 5. Measuring position-dependent pri-mRNA processing along the C19MC in DGCR8 mutant clones. At 3 selected positions, the ratio of total versus unprocessed pri-miRNAs was determined using distinct primer pairs by real-time PCRs. In this representative experiment, colored circles show mean values of measurement points in different clones; ± SE values of 3 technical parallels are also shown. Colored lines indicate the tendency of decrease in processing efficiency (if any) for a given clone.

4. Discussion

DGCR8 is a central component of the Microprocessor complex, and together with its endonuclease partner Drosha, they play an essential role in the initiating step of canonical miRNA biogenesis [6,27–30]. Disturbance in the level of DGCR8, therefore, significantly impairs miRNA maturation, negatively influencing cell proliferation, differentiation, and apoptosis [31,32], and thereby disturbing important developmental processes, including, among others, cardiovascular and brain development [33,34]. DGCR8 knockout or depletion strategies lead to identifying the functions of key miRNAs [35–37] but also revealed miRNA-unrelated, "noncanonical" functions of this protein, such as controlling the stability of small nucleolar RNAs [38,39] and its role in heterochromatin stabilization [40] or in DNA repair processes [41]. Considering its versatile cellular functions, it does not seem unexpected that the complete loss of DGCR8 is lethal in mice, and such knockout embryonic stem cells have serious differentiation problems [15]. On the other hand, mouse monoallelic DGCR8 mutant PSCs were reported to have no obvious phenotypes or changes in their differentiation capacity, and this was thought to be due to the homeostatic mechanisms affecting the DGCR8 mRNA levels [17,42,43]. However, subsequent studies using these mouse models carrying monoallelic mutations revealed pronounced downstream effects in neurological physiology, manifested in reduced cell proliferation and aberrant neuron morphology in several brain areas [18–20], or in cardiac malfunctioning, resulting in heart failure [44]. Moreover, conditional mouse knockout models in early immune cells resulted in defects in natural killer cell activation and survival, which was connected to the loss of certain miRNA population [35]. In a human cellular model system of DiGeorge syndrome, cortical neurons derived from DGCR8 +/− hiPSCs showed defective neuronal activity and calcium handling, which were in line with the results of the mouse models [21]. However, DGCR8 heterozygous deficiency is still poorly characterized in human ESCs and in differentiated tissue types, mostly due to the lack of suitable knockout or mutant model cell lines.

To investigate the function of the DGCR8 in various human cell types, we previously established a monoallelic mutant hESC-line using the CRISPR/Cas9 system [23]. This is a suitable model to study the molecular and cellular defects not only in stem cells but also in their differentiated isogenic offspring cell types. However, when generating several single-cell clones for further comparative studies, and omitting the antibiotic selection to avoid interference with differentiation, we detected a gradual loss of the inserted transgenes, the GFP marker and the puromycin resistance gene. After the genetic analyses of the clones, we concluded that it is not due to any form of epigenetic silencing but rather to a recombination event initiated by the two CAG promoter sequences present in the transgenic cassette, leading to the disruption of the transgene structure and to the loss of expression (Supplementary Figures S4 and S5). Such transgenic rearrangements and somatic drifts are known problems reported earlier when establishing stable genetic models in several systems [45–48]; however, as we successfully applied similar expression cassettes with two transcription units in earlier studies, even in stem cells [49–51], we did not expect this form of transgene inactivation. Nevertheless, the modified DGCR8 allele still remained mutant in all clones even after the transgene rearrangement, so these cells could be used as DGCR8 monoallelic mutants for further studies.

The effect of DGCR8 haploinsufficiency on cellular phenotypes is controversial: such cells are viable, but since the impairment of this regulatory protein significantly varies for different miRNAs [52], the effect on the functional disturbance is also cell-type-specific. In our investigations at the human stem-cell level, the DGCR8 monoallelic mutation did not disturb either the pluripotency status or the tri-lineage differentiation capacity, the results of which were in line with the mouse models [17,18]. In addition, measuring the mRNA expression levels did not reveal dramatic differences between the normal and heterozygous mutant cell lines, which may have been expected based on the subtle autoregulatory mechanism described earlier for this transcript [17,42,43]. In contrast, studies at the protein level revealed significant decrease in DGCR8 expression as compared with the normal hESCs. This result prompted us to investigate genetic loci where such an imbalance in Microprocessor components is known to disturb molecular functions. One candidate was the C19MC locus, where the local depletion of the Microprocessor complex results in a gradual positional decline of pri-miRNA processing in a long miRNA cluster. This positional effect was more prominent when miRNA positions were compared in a pairwise manner between hESCs and placenta cells, between cell types where this imprinted miRNA cluster is predominantly expressed [14]. When we investigated the single-cell clones carrying the DGCR8 heterozygous mutation, the positional effect was detected in only half of the examined clones, indicating that the significantly lower protein level can result in a partial functional disturbance of the Microprocessor complex in hESCs. However, it is currently unknown why this heterozygous mutation did not show a 100% penetrance in our experiments. One explanation could lie in the naïve versus primed stage of the used human embryonic stem cell line: in a recent study, it was revealed that the C19MC is relatively highly expressed in naïve hESCs, whereas its transcriptional activity is declined in primed stem cells [53]. In our culturing conditions, the HuES9 ESC line rather contains primed cell populations, and the lower expression of the miRNA cluster in those cell types might hinder the detection of the position effect in pri-miRNA processing. In addition, C19MC is also expressed at a much higher level in the trophoblast lineage, a differentiation route that is more effectively initiated from the naïve state of hESCs. It is also conceivable that the effect of the DGCR8 monoallelic mutation may be more pronounced in those cell types. However, further experiments, including RNA sequencing, are needed to uncover the DGCR8 mRNA profile and the potential presence of nontranslatable splicing variants in our established clones, to unequivocally connect all the observed functional changes to the DGCR8 protein itself.

In conclusion, in this study, we could show that the progenies of the hESC line carrying a heterozygous mutation in the DGRC8 gene are good molecular and cellular models to examine the functions of this RNA-binding regulatory protein. Moreover, being the good basis to generate isogenic differentiation lineages, they provide an excellent in vitro platform to study the phenotype of DGCR8 deficiency in several human somatic cell types. Since one consequence of lower DGCR8 expression is the suppressed maturation of canonical miRNAs, the use of our special hESC line can also enhance the studies of mirtrons, a special class of Microprocessor-independent but splicing-dependent miRNAs [54–59]. Taken together, this DGCR8 +/− cell line can well-contribute to the study of the dominant human RNA interference pathway, as well as to the better understanding of the "noncanonical" functions of the DGCR8 protein.

Supplementary Materials: The following supporting information can be downloaded at: https://www.mdpi.com/article/10.3390/genes13111925/s1, Figure S1. GFP expression in HVRDe009-A-1 cell line., Figure S2. Overview of establishment of "GFP-positive" single-cell clones from the HVRDe009-A-1 cell line during puromycin deprivation., Figure S3. Selection of the four single-cell clones for further characterization., Figure S4. Diagnostic PCR results in the parental HVRDe009-A-1 cell line and the derived single-cell clones., Figure S5. Sanger sequencing of DGCR8 mutant allele in HVRDe009-A-1 derived single-cell clones. Figure S6. Representative karyograms of HVRDe009-A-1-derived single-cell clones., Figure S7. Expression of DGCR8 in the monoallelic mutant single-cell clones., Table S1. Primers for SYBR Green assay.

Author Contributions: Conceptualization, D.R., T.I.O. and Á.A.; methodology, D.R., N.V., Á.F. and O.K.; formal analysis, D.R., T.I.O. and Á.A.; writing—original draft preparation, D.R., T.I.O. and Á.A.; writing—review and editing, T.I.O. and Á.A.; funding acquisition, T.I.O. and Á.A. All authors have read and agreed to the published version of the manuscript.

Funding: This research was funded by the National Brain Research Program (NAP) of Hungary (grant numbers: 2017-1.2.1-NKP-2017-00002 to Á.A.), the National Research, Development and Innovation Office (OTKA-K128369 to Á.A. and "HunProtExc" 2018-1.2.1-NKP-2018-00005 to T.I.O.), and the Hungarian Academy of Sciences (PC-II-12/2022 to T.I.O.). The authors acknowledge the financial support received as a Centre of Excellence of the Hungarian Academy of Sciences.

Institutional Review Board Statement: Not applicable.

Informed Consent Statement: Not applicable.

Data Availability Statement: The data presented in this study are available on request from the corresponding authors.

Acknowledgments: The authors thank Beáta Haraszti, Kornélia Némethy, and Zsuzsa Erdei for technical assistance. The HuES9 cell line was kindly provided by Douglas A. Melton, Harvard University.

Conflicts of Interest: The authors declare no conflict of interest.

References

1. Bartel, D.P. Metazoan MicroRNAs. *Cell* **2018**, *173*, 20–51. [CrossRef]
2. Gebert, L.F.R.; MacRae, I.J. Regulation of microRNA function in animals. *Nat. Rev. Mol. Cell Biol.* **2019**, *20*, 21–37. [CrossRef]
3. Huntzinger, E.; Izaurralde, E. Gene silencing by microRNAs: Contributions of translational repression and mRNA decay. *Nat. Rev. Genet.* **2011**, *12*, 99–110. [CrossRef] [PubMed]
4. Beveridge, N.J.; Gardiner, E.; Carroll, A.P.; Tooney, P.A.; Cairns, M.J. Schizophrenia is associated with an increase in cortical microRNA biogenesis. *Mol. Psychiatry* **2010**, *15*, 1176–1189. [CrossRef] [PubMed]
5. Perkins, D.O.; Jeffries, C.D.; Jarskog, L.F.; Thomson, J.M.; Woods, K.; Newman, M.A.; Parker, J.S.; Jin, J.; Hammond, S.M. MicroRNA expression in the prefrontal cortex of individuals with schizophrenia and schizoaffective disorder. *Genome Biol.* **2007**, *8*, R27. [CrossRef]
6. Nguyen, T.A.; Jo, M.H.; Choi, Y.G.; Park, J.; Kwon, S.C.; Hohng, S.; Kim, V.N.; Woo, J.S. Functional Anatomy of the Human Microprocessor. *Cell* **2015**, *161*, 1374–1387. [CrossRef] [PubMed]
7. Iwakawa, H.O.; Tomari, Y. Life of RISC: Formation, action, and degradation of RNA-induced silencing complex. *Mol. Cell* **2022**, *82*, 30–43. [CrossRef] [PubMed]
8. Miyoshi, K.; Miyoshi, T.; Siomi, H. Many ways to generate microRNA-like small RNAs: Non-canonical pathways for microRNA production. *Mol. Genet. Genom. MGG* **2010**, *284*, 95–103. [CrossRef] [PubMed]
9. Altuvia, Y.; Landgraf, P.; Lithwick, G.; Elefant, N.; Pfeffer, S.; Aravin, A.; Brownstein, M.J.; Tuschl, T.; Margalit, H. Clustering and conservation patterns of human microRNAs. *Nucleic Acids Res.* **2005**, *33*, 2697–2706. [CrossRef] [PubMed]

10. Kim, Y.K.; Yu, J.; Han, T.S.; Park, S.Y.; Namkoong, B.; Kim, D.H.; Hur, K.; Yoo, M.W.; Lee, H.J.; Yang, H.K.; et al. Functional links between clustered microRNAs: Suppression of cell-cycle inhibitors by microRNA clusters in gastric cancer. *Nucleic Acids Res.* **2009**, *37*, 1672–1681. [CrossRef]
11. Morales-Prieto, D.M.; Ospina-Prieto, S.; Chaiwangyen, W.; Schoenleben, M.; Markert, U.R. Pregnancy-associated miRNA-clusters. *J. Reprod. Immunol.* **2013**, *97*, 51–61. [CrossRef]
12. Malnou, E.C.; Umlauf, D.; Mouysset, M.; Cavaille, J. Imprinted MicroRNA Gene Clusters in the Evolution, Development, and Functions of Mammalian Placenta. *Front. Genet.* **2018**, *9*, 706. [CrossRef] [PubMed]
13. Michlewski, G.; Caceres, J.F. Post-transcriptional control of miRNA biogenesis. *RNA* **2019**, *25*, 1–16. [CrossRef] [PubMed]
14. Fothi, A.; Biro, O.; Erdei, Z.; Apati, A.; Orban, T.I. Tissue-specific and transcription-dependent mechanisms regulate primary microRNA processing efficiency of the human chromosome 19 MicroRNA cluster. *RNA Biol.* **2021**, *18*, 1170–1180. [CrossRef] [PubMed]
15. Wang, Y.; Medvid, R.; Melton, C.; Jaenisch, R.; Blelloch, R. DGCR8 is essential for microRNA biogenesis and silencing of embryonic stem cell self-renewal. *Nat. Genet.* **2007**, *39*, 380–385. [CrossRef]
16. Cirera-Salinas, D.; Yu, J.; Bodak, M.; Ngondo, R.P.; Herbert, K.M.; Ciaudo, C. Noncanonical function of DGCR8 controls mESC exit from pluripotency. *J. Cell Biol.* **2017**, *216*, 355–366. [CrossRef]
17. Han, J.; Pedersen, J.S.; Kwon, S.C.; Belair, C.D.; Kim, Y.K.; Yeom, K.H.; Yang, W.Y.; Haussler, D.; Blelloch, R.; Kim, V.N. Posttranscriptional crossregulation between Drosha and DGCR8. *Cell* **2009**, *136*, 75–84. [CrossRef]
18. Schofield, C.M.; Hsu, R.; Barker, A.J.; Gertz, C.C.; Blelloch, R.; Ullian, E.M. Monoallelic deletion of the microRNA biogenesis gene Dgcr8 produces deficits in the development of excitatory synaptic transmission in the prefrontal cortex. *Neural Dev.* **2011**, *6*, 11. [CrossRef]
19. Ouchi, Y.; Banno, Y.; Shimizu, Y.; Ando, S.; Hasegawa, H.; Adachi, K.; Iwamoto, T. Reduced adult hippocampal neurogenesis and working memory deficits in the Dgcr8-deficient mouse model of 22q11.2 deletion-associated schizophrenia can be rescued by IGF2. *J. Neurosci.* **2013**, *33*, 9408–9419. [CrossRef]
20. Fenelon, K.; Mukai, J.; Xu, B.; Hsu, P.K.; Drew, L.J.; Karayiorgou, M.; Fischbach, G.D.; Macdermott, A.B.; Gogos, J.A. Deficiency of Dgcr8, a gene disrupted by the 22q11.2 microdeletion, results in altered short-term plasticity in the prefrontal cortex. *Proc. Natl. Acad. Sci. USA* **2011**, *108*, 4447–4452. [CrossRef]
21. Khan, T.A.; Revah, O.; Gordon, A.; Yoon, S.J.; Krawisz, A.K.; Goold, C.; Sun, Y.; Kim, C.H.; Tian, Y.; Li, M.Y.; et al. Neuronal defects in a human cellular model of 22q11.2 deletion syndrome. *Nat. Med.* **2020**, *26*, 1888–1898. [CrossRef] [PubMed]
22. Erdei, Z.; Lorincz, R.; Szebenyi, K.; Pentek, A.; Varga, N.; Liko, I.; Varady, G.; Szakacs, G.; Orban, T.I.; Sarkadi, B.; et al. Expression pattern of the human ABC transporters in pluripotent embryonic stem cells and in their derivatives. *Cytom. Part B Clin. Cytom.* **2014**, *86*, 299–310. [CrossRef]
23. Ree, D.; Borsy, A.; Fothi, A.; Orban, T.I.; Varady, G.; Erdei, Z.; Sarkadi, B.; Rethelyi, J.; Varga, N.; Apati, A. Establishing a human embryonic stem cell clone with a heterozygous mutation in the DGCR8 gene. *Stem Cell Res.* **2020**, *50*, 102134. [CrossRef] [PubMed]
24. Bartoli, A.; Fettucciari, K.; Fetriconi, I.; Rosati, E.; Di Ianni, M.; Tabilio, A.; Delfino, D.V.; Rossi, R.; Marconi, P. Effect of trichostatin a and 5′-azacytidine on transgene reactivation in U937 transduced cells. *Pharmacol. Res.* **2003**, *48*, 111–118. [CrossRef]
25. Kaufman, W.L.; Kocman, I.; Agrawal, V.; Rahn, H.P.; Besser, D.; Gossen, M. Homogeneity and persistence of transgene expression by omitting antibiotic selection in cell line isolation. *Nucleic Acids Res.* **2008**, *36*, e111. [CrossRef] [PubMed]
26. Vanhaecke, T.; Papeleu, P.; Elaut, G.; Rogiers, V. Trichostatin A-like hydroxamate histone deacetylase inhibitors as therapeutic agents: Toxicological point of view. *Curr. Med. Chem.* **2004**, *11*, 1629–1643. [CrossRef]
27. Denli, A.M.; Tops, B.B.; Plasterk, R.H.; Ketting, R.F.; Hannon, G.J. Processing of primary microRNAs by the Microprocessor complex. *Nature* **2004**, *432*, 231–235. [CrossRef]
28. Gregory, R.I.; Yan, K.P.; Amuthan, G.; Chendrimada, T.; Doratotaj, B.; Cooch, N.; Shiekhattar, R. The Microprocessor complex mediates the genesis of microRNAs. *Nature* **2004**, *432*, 235–240. [CrossRef]
29. Han, J.; Lee, Y.; Yeom, K.H.; Kim, Y.K.; Jin, H.; Kim, V.N. The Drosha-DGCR8 complex in primary microRNA processing. *Genes Dev.* **2004**, *18*, 3016–3027. [CrossRef]
30. Landthaler, M.; Yalcin, A.; Tuschl, T. The human DiGeorge syndrome critical region gene 8 and Its D. melanogaster homolog are required for miRNA biogenesis. *Curr. Biol. CB* **2004**, *14*, 2162–2167. [CrossRef]
31. Dexheimer, P.J.; Cochella, L. MicroRNAs: From Mechanism to Organism. *Front. Cell Dev. Biol.* **2020**, *8*, 409. [CrossRef] [PubMed]
32. Ivey, K.N.; Srivastava, D. MicroRNAs as Developmental Regulators. *Cold Spring Harb. Perspect. Biol.* **2015**, *7*, a008144. [CrossRef] [PubMed]
33. Ma, Q.; Zhang, L.; Pearce, W.J. MicroRNAs in brain development and cerebrovascular pathophysiology. *Am. J. Physiol. Cell Physiol.* **2019**, *317*, C3–C19. [CrossRef] [PubMed]
34. Wojciechowska, A.; Braniewska, A.; Kozar-Kaminska, K. MicroRNA in cardiovascular biology and disease. *Adv. Clin. Exp. Med. Off. Organ Wroc. Med. Univ.* **2017**, *26*, 865–874. [CrossRef]
35. Bezman, N.A.; Cedars, E.; Steiner, D.F.; Blelloch, R.; Hesslein, D.G.; Lanier, L.L. Distinct requirements of microRNAs in NK cell activation, survival, and function. *J. Immunol.* **2010**, *185*, 3835–3846. [CrossRef]
36. Melton, C.; Judson, R.L.; Blelloch, R. Opposing microRNA families regulate self-renewal in mouse embryonic stem cells. *Nature* **2010**, *463*, 621–626. [CrossRef]

37. Suh, N.; Baehner, L.; Moltzahn, F.; Melton, C.; Shenoy, A.; Chen, J.; Blelloch, R. MicroRNA function is globally suppressed in mouse oocytes and early embryos. *Curr. Biol. CB* **2010**, *20*, 271–277. [CrossRef]
38. Macias, S.; Cordiner, R.A.; Gautier, P.; Plass, M.; Caceres, J.F. DGCR8 Acts as an Adaptor for the Exosome Complex to Degrade Double-Stranded Structured RNAs. *Mol. Cell* **2015**, *60*, 873–885. [CrossRef]
39. Macias, S.; Plass, M.; Stajuda, A.; Michlewski, G.; Eyras, E.; Caceres, J.F. DGCR8 HITS-CLIP reveals novel functions for the Microprocessor. *Nat. Struct. Mol. Biol.* **2012**, *19*, 760–766. [CrossRef]
40. Deng, L.; Ren, R.; Liu, Z.; Song, M.; Li, J.; Wu, Z.; Ren, X.; Fu, L.; Li, W.; Zhang, W.; et al. Stabilizing heterochromatin by DGCR8 alleviates senescence and osteoarthritis. *Nat. Commun.* **2019**, *10*, 3329. [CrossRef]
41. Hang, Q.; Zeng, L.; Wang, L.; Nie, L.; Yao, F.; Teng, H.; Deng, Y.; Yap, S.; Sun, Y.; Frank, S.J.; et al. Non-canonical function of DGCR8 in DNA double-strand break repair signaling and tumor radioresistance. *Nat. Commun.* **2021**, *12*, 4033. [CrossRef] [PubMed]
42. Shenoy, A.; Blelloch, R. Genomic analysis suggests that mRNA destabilization by the microprocessor is specialized for the auto-regulation of Dgcr8. *PLoS ONE* **2009**, *4*, e6971. [CrossRef] [PubMed]
43. Triboulet, R.; Chang, H.M.; Lapierre, R.J.; Gregory, R.I. Post-transcriptional control of DGCR8 expression by the Microprocessor. *RNA* **2009**, *15*, 1005–1011. [CrossRef] [PubMed]
44. Rao, P.K.; Toyama, Y.; Chiang, H.R.; Gupta, S.; Bauer, M.; Medvid, R.; Reinhardt, F.; Liao, R.; Krieger, M.; Jaenisch, R.; et al. Loss of cardiac microRNA-mediated regulation leads to dilated cardiomyopathy and heart failure. *Circ. Res.* **2009**, *105*, 585–594. [CrossRef] [PubMed]
45. Cottingham, M.G.; Carroll, F.; Morris, S.J.; Turner, A.V.; Vaughan, A.M.; Kapulu, M.C.; Colloca, S.; Siani, L.; Gilbert, S.C.; Hill, A.V. Preventing spontaneous genetic rearrangements in the transgene cassettes of adenovirus vectors. *Biotechnol. Bioeng.* **2012**, *109*, 719–728. [CrossRef]
46. Larson, J.S.; Yin, M.; Fischer, J.M.; Stringer, S.L.; Stringer, J.R. Expression and loss of alleles in cultured mouse embryonic fibroblasts and stem cells carrying allelic fluorescent protein genes. *BMC Mol. Biol.* **2006**, *7*, 36. [CrossRef]
47. Liu, W.; Xiong, Y.; Gossen, M. Stability and homogeneity of transgene expression in isogenic cells. *J. Mol. Med.* **2006**, *84*, 57–64. [CrossRef]
48. Romano, E.; Soares, A.; Proite, K.; Neiva, S.; Grossi, M.; Faria, J.C.; Rech, E.L.; Aragao, F.J. Transgene elimination in genetically modified dry bean and soybean lines. *Genet. Mol. Res. GMR* **2005**, *4*, 177–184.
49. Kolacsek, O.; Krizsik, V.; Schamberger, A.; Erdei, Z.; Apati, A.; Varady, G.; Mates, L.; Izsvak, Z.; Ivics, Z.; Sarkadi, B.; et al. Reliable transgene-independent method for determining Sleeping Beauty transposon copy numbers. *Mob. DNA* **2011**, *2*, 5. [CrossRef]
50. Apati, A.; Paszty, K.; Hegedus, L.; Kolacsek, O.; Orban, T.I.; Erdei, Z.; Szebenyi, K.; Pentek, A.; Enyedi, A.; Sarkadi, B. Characterization of calcium signals in human embryonic stem cells and in their differentiated offspring by a stably integrated calcium indicator protein. *Cell. Signal.* **2013**, *25*, 752–759. [CrossRef]
51. Kolacsek, O.; Erdei, Z.; Apati, A.; Sandor, S.; Izsvak, Z.; Ivics, Z.; Sarkadi, B.; Orban, T.I. Excision efficiency is not strongly coupled to transgenic rate: Cell type-dependent transposition efficiency of sleeping beauty and piggyBac DNA transposons. *Hum. Gene Ther. Methods* **2014**, *25*, 241–252. [CrossRef] [PubMed]
52. Nogami, M.; Miyamoto, K.; Hayakawa-Yano, Y.; Nakanishi, A.; Yano, M.; Okano, H. DGCR8-dependent efficient pri-miRNA processing of human pri-miR-9-2. *J. Biol. Chem.* **2021**, *296*, 100409. [CrossRef] [PubMed]
53. Kobayashi, N.; Okae, H.; Hiura, H.; Kubota, N.; Kobayashi, E.H.; Shibata, S.; Oike, A.; Hori, T.; Kikutake, C.; Hamada, H.; et al. The microRNA cluster C19MC confers differentiation potential into trophoblast lineages upon human pluripotent stem cells. *Nat. Commun.* **2022**, *13*, 3071. [CrossRef] [PubMed]
54. Okamura, K.; Hagen, J.W.; Duan, H.; Tyler, D.M.; Lai, E.C. The mirtron pathway generates microRNA-class regulatory RNAs in Drosophila. *Cell* **2007**, *130*, 89–100. [CrossRef] [PubMed]
55. Ruby, J.G.; Jan, C.H.; Bartel, D.P. Intronic microRNA precursors that bypass Drosha processing. *Nature* **2007**, *448*, 83–86. [CrossRef]
56. Schamberger, A.; Sarkadi, B.; Orban, T.I. Human mirtrons can express functional microRNAs simultaneously from both arms in a flanking exon-independent manner. *RNA Biol.* **2012**, *9*, 1177–1185. [CrossRef]
57. Schamberger, A.; Varady, G.; Fothi, A.; Orban, T.I. Posttranscriptional Regulation of the Human ABCG2 Multidrug Transporter Protein by Artificial Mirtrons. *Genes* **2021**, *12*, 1068. [CrossRef]
58. Sibley, C.R.; Seow, Y.; Curtis, H.; Weinberg, M.S.; Wood, M.J. Silencing of Parkinson's disease-associated genes with artificial mirtron mimics of miR-1224. *Nucleic Acids Res.* **2012**, *40*, 9863–9875. [CrossRef]
59. Sibley, C.R.; Seow, Y.; Saayman, S.; Dijkstra, K.K.; El Andaloussi, S.; Weinberg, M.S.; Wood, M.J. The biogenesis and characterization of mammalian microRNAs of mirtron origin. *Nucleic Acids Res.* **2012**, *40*, 438–448. [CrossRef]

genes

MDPI

Article

Effect of miR-302b MicroRNA Inhibition on Chicken Primordial Germ Cell Proliferation and Apoptosis Rate

Bence Lázár [1,2,†], Nikolett Tokodyné Szabadi [1,†], Mahek Anand [1], Roland Tóth [1], András Ecker [1], Martin Urbán [1], Maria Teresa Salinas Aponte [1], Ganna Stepanova [3], Zoltán Hegyi [4], László Homolya [4], Eszter Patakiné Várkonyi [2], Bertrand Pain [5] and Elen Gócza [1,*]

1 Animal Biotechnology Department, Institute of Genetics and Biotechnology, Hungarian University of Agriculture and Life Sciences, 2100 Godollo, Hungary; lazar.bence@uni-mate.hu (B.L.); Tokodyne.Szabadi.Nikolett@uni-mate.hu (N.T.S.); mahekanand@yahoo.com (M.A.); Toth.Roland.Imre@uni-mate.hu (R.T.); Ecker.Andras@phd.uni-mate.hu (A.E.); Urban.Martin@uni-mate.hu (M.U.); mtsalinasaponte@gmail.com (M.T.S.A.)

2 Institute for Farm Animal Gene Conservation, National Centre for Biodiversity and Gene Conservation, 2100 Godollo, Hungary; varkonyi.eszter@nbgk.hu

3 Faculty of Medicine, Institute of Translational Medicine, Semmelweis University, 1089 Budapest, Hungary; gannastepanova2016@gmail.com

4 Institute of Enzymology, Research Centre for Natural Sciences, 1117 Budapest, Hungary; zotyahegyi@gmail.com (Z.H.); homolya.laszlo@ttk.hu (L.H.)

5 Stem-Cell and Brain Research Institute, USC1361 INRA, U1208 INSERM, 69675 Bron, France; bertrand.pain@inserm.fr

* Correspondence: Gocza.Elen@uni-mate.hu

† These authors contributed equally to this work.

Abstract: The primordial germ cells (PGCs) are the precursors for both the oocytes and spermatogonia. Recently, a novel culture system was established for chicken PGCs, isolated from embryonic blood. The possibility of PGC long-term cultivation issues a new advance in germ cell preservation, biotechnology, and cell biology. We investigated the consequence of gga-miR-302b-5P (5P), gga-miR-302b-3P (3P) and dual inhibition (5P/3P) in two male and two female chicken PGC lines. In treated and control cell cultures, the cell number was calculated every four hours for three days by the XLS Imaging system. Comparing the cell number of control and treated lines on the first day, we found that male lines had a higher proliferation rate independently from the treatments. Compared to the untreated ones, the proliferation rate and the number of apoptotic cells were considerably reduced at gga-miR-302b-5P inhibition in all PGC lines on the third day of the cultivation. The control PGC lines showed a significantly higher proliferation rate than 3P inhibited lines on Day 3 in all PGC lines. Dual inhibition of gga-miR-302b mature miRNAs caused a slight reduction in proliferation rate, but the number of apoptotic cells increased dramatically. The information gathered by examining the factors affecting cell proliferation of PGCs can lead to new data in stem cell biology.

Keywords: primordial germ cells; gga-miR-302b; miRNA inhibition; chicken; cell proliferation; apoptosis

1. Introduction

Germ cell development generates totipotency throughout genetic as well as epigenetic regulation of the genome function. Primordial germ cells (PGCs) are the first germ cell population established during the development and serve as precursors for both the oocytes and spermatogonia [1]. PGCs, similarly to embryonic stem cells (ESCs) [2] and induced pluripotent stem cells (iPSCs) [3], are essential tools in stem cell biology and research. It is possible to genetically modify the PGCs and create germ cell chimeras from the transgenic PGCs, as was shown by Nakamura and co-workers [4]. Chicken PGCs can be easily isolated from the fertilized eggs at the early stages of embryonic development and can be maintained in vitro utilizing already established protocols [5,6]. This accessibility provides a unique tool to explore better medium formulations and investigate stem cell properties, such as factors governing pluripotency and self-renewal [7].

One of the emerged pluripotency factors in stem cell biology are miRNAs [8–11]. They play a fundamental role in maintaining the pluripotency of PGCs [12,13]. MicroRNAs (miRNAS) are small non-coding RNAs found in the genome that post-transcriptionally regulate gene expression via mRNA degradation or translation inhibition. The miRNAs are known to regulate important physiological and pathological processes [14]. The genes for the miRNAs are dispersed throughout the genome: some are intergenic, while others are located in intronic, or in exonic regions [11]. The biogenesis of miRNAs is a multi-step process in higher vertebrates. The process begins with RNA polymerase II transcribing these miRNA genes into primary-miRNA (pri-miRNA) transcipts. This pri-miRNA is further processed by the microprocessor complex consisting of DROSHA, DGCR8 and spliceosome components into precursor miRNA (pre-miRNA). The Exportin-5 enzyme exports pre-miRNAs out of the nucleus into the cytoplasm. This pre-miRNA is further cleaved by Dicer, another Ribonuclease III type enzyme, with the help of the TRBP into a miRNA duplex complex. The duplex consists of a 5P and a 3P strand, and the process of arm selection results in the guide RNA strand incorporating into the RNA-induced silencing complex (RISC), which mediates the RNAi-related gene silencing; the partial base pairing between the mature miRNA and the target mRNA leads to translation inhibition or mRNA degradation. Apart from this canonical maturation process, there are several Drosha- or Dicer-independent pathways of miRNA biogenesis [15–17]. MiRANs are crucial for gene regulation during pluripotency, self-renewal, and differentiation of ESCs and iPSCs [18]. They are expressed during the earliest embryonic developmental stages [9,19]. During embryonic development [19–22], the quantity of miRNAs is strongly regulated [16,20]. If the expression levels of non-coding RNAs become too high, they can act as cancer-promoting agents, which can cause the cells to escape standard mechanisms of control and become malignant in nature [8,17,23]. MicroRNAs belonging to the miR-302 family are emerging as key players in the control of cell growth. Khodayari and co-workers identified a novel mechanism of ephrin-A1 mediated anti-oncogenic signaling in malignant pleural mesothelioma (MPM) through miR-302b upregulation and inhibition of MM tumor sphere growth by inducing apoptosis [24]. Apoptosis is described by specific morphological and biochemical features in which caspase activation plays a central function as a component of both health and disease [25]. Understanding the mechanisms of apoptosis and other variants of programmed cell death provides deeper insight into various disease processes and may thus influence therapeutic strategy.

Our present study identified a novel effect of the gga-miR-302b on chicken PGC proliferation by inhibiting gga-miR-302b-5P (5P) and gga-miR-302b-3P (3P). We demonstrated that, in the case of gga-miR-302b-5P and gga-miR-302b-3P inhibition, PGCs showed a significantly reduced proliferation rate than the control PGCs. The inhibition of gga-miR-302b-5P reduced the apoptotic rate of all the examined PGC lines. Together, these findings suggest that the cells might undergo an additional pathway of cell degradation, besides apoptosis, that needs to be determined experimentally. The differentiation capacity of inhibited cells needs to be explored also.

2. Materials and Methods

2.1. Experimental Animals and Animal Care

Animals were kept according to the standard rules of the Hungarian Animal Protection Law (1998. XXVIII). Permission for experimental animal research at the National Centre for Biodiversity and Gene Conservation, Institute for Farm Animal Gene Conservation (Gödöllő, Hungary), was provided by the National Food Chain Safety Office, Animal Health and Animal Welfare Directorate (Budapest, Hungary). Fertilized eggs from Black Transylvanian Naked Neck Chicken were provided by the National Centre for Biodiversity and Gene Conservation, Institute for Farm Animal Gene Conservation (Gödöllő, Hungary).

2.2. Isolation, Establishment, and Maintenance of PGC Lines

Eggs were collected and incubated before the experiment. Circulating PGCs were isolated from fertilized eggs (HH stage 14–16) from Black Transylvanian Naked Neck chicken embryos and transferred to 300 μL culture medium in a 48-well plate without feeder cells. The culture medium was prepared as described by Whyte and colleagues [6]. The egg surface was cleaned up with 70% EtOH prior to opening. Next, 1 μL of blood was taken by a glass micro-pipette from the dorsal aorta of the embryo under a stereomicroscope. After 1–2 weeks, red blood cells died and PGCs prevailed. PGCs were cultured and used for the experiment. Half of the medium was changed every other day. When the total cell number reached 1.0×10^5, the cells were divided into two and propagated at $2–4.0 \times 10^5$ cells/mL medium in a 24-well plate [26]. Tissue samples for sex-determination were collected from every isolated embryo and stored at −20 °C until further use. The time of isolation, the exact age of the embryos (HH stages) cells [27], and visual the presence/absence of developmental abnormalities were recorded.

2.3. Cell Counting Using Arthur Fluorescence Cell Counter

The cell counting of the chicken PGCs before preparing the cell proliferation assay was performed using the Arthur Novel Fluorescence Cell Counter (NanoEnTek, Pleasanton, CA, USA). Two separate counts in parallel confirmed the results. For each line, PGCs were collected from 6 wells of a 24-well plate. Cell concentration was calculated from 25 μL PGC suspension by the Arthur Cell Counter. After cell number calculation, 1×10^3 cells were placed into each well of one 96-well plate. Further, the cell number was measured every 4 h, for three days, using High-Content Screening Molecular Device. The measurement using ImageXpress Micro XLS Imaging System with a built-in incubator was performed at the Molecular Cell Biology Research Group (Institute of Enzymology, Research Centre for Natural Sciences): 96-well plate, with 6-6 parallel wells, of four previously established PG cell lines: two male (M1: #508-ZZ; M2: #512-ZZ) and two females (F1: #509-ZW; F2: #513-ZW) PGC line prepared for each condition. At the end of the experiment, the samples were collected for apoptosis measurement and immunostaining (Supplementary Figure S1).

2.4. Apoptosis Rate Measurement Using Arthur Fluorescence Cell Counter

Apoptosis rate was measured on Day 3. 10× Annexin V Binding Buffer was diluted to 1× (AD10—Dojindo Molecular Technologies, Inc., Rockville, MD, USA) with tissue culture grade water (Applied Biosystems, Life Technologies, Carlsbad, CA, USA). Both the Annexin V FITC and the Propidium Iodide components were used at 5 μL/test in a final volume of 100 μL. The apoptotic and necrotic rate of the PGCs was measured using the Arthur Novel Fluorescence Cell Counter (NanoEnTek, Pleasanton, CA, USA).

2.5. Immunostaining of PGCs

PGCs were fixed with 4% PFA for 10 min. After washing with PBS, the fixed cells were blocked for 45 min with a blocking buffer containing 5% (*v*/*v*) BSA, then were incubated with each of the primary antibodies including mouse anti-SSEA-1 (1:10, Developmental Studies Hybridoma Bank, Iowa City, Iowa, USA) and rabbit anti-VASA (CVH) (1:1000; kindly provided by Bertrand Pain, Lyon, France). After overnight incubation in the primary antibody solution in a humid chamber at 4 °C, the cells were washed three times with PBS. Then the cells were incubated for 1 h with the secondary antibodies, Alexa-Fluor488-conjugated donkey anti-rabbit-IgG (Applied Biosystems, Life Technologies, Carlsbad, CA, USA) or Alexa-Fluor-555-conjugated donkey-anti-mouse-IgM (Applied Biosystems, Life Technologies, Carlsbad, CA, USA) in a dark, humid chamber, at 37 °C temperature. After washing once with 1× PBS, the nucleus was stained with TO-PRO®-3 stain for 15 min (1:500, Molecular Probes Inc., Eugene, OR, USA). After three rounds of 1x PBS, wash coverslips were mounted on the slide with the application of 20 μL VECTASHIELD® Mounting Media (Vector Laboratories Inc., Burlingame, CA, USA) and analysed by confocal microscopy

(TCS SP8, Leica Microsystems IR GmbH Wetzlar, Hessen, Germany). Negative controls were stained only with the secondary antibodies.

2.6. RNA Isolation

We collected the cells for RNA isolation in RNA Aqueous Lysis Buffer Micro Kit (Applied Biosystems). The isolated RNA was then used for qRT-PCR analysis for stem cell- and germ cell-specific markers as well as for the miRNAs: gga-miR-302b-5P and gga-miR-302b-3P (Tables 1 and 2). The concentration of the extracted RNAs was determined using the NanoDrop One Spectrophotometer (Thermo Fisher Scientific, Waltham, MA, USA). The isolated RNA samples were stored at −70 °C.

Table 1. Primers used at the mRNA analysis.

Gene Symbol	Gene Full Name (*Organism*)	NCBI Number	Primers		Length of the Product (bp)
GAPDH	Glyceraldehyde-3-phosphate dehydrogenase (*Gallus gallus*)	NM_204305.1	FW	GACGTGCAGCAGGA ACACTA	112
			RV	CTTGGACTTTGCCAG AGAGG	
POUV	POU domain class 5 transcription factor 3 (*Pou5f3*) (*Gallus gallus*)	NM_001110178.1	FW	GAGGCAGAGAACA CGGACAA	109
			RV	TTCCCTTCACGTTGG TCTCG	
CVH	DEAD-box helicase 4 (*DDX4*) (*Gallus gallus*)	NM_204708.1	FW	GAACCTACCATCC ACCAGCA	113
			RV	ATGCTACCGAAGTTG CCACA	

Table 2. Primers used at the miRNA analysis.

Name	Gene	Accession Number	Assay ID	Sequence
miR-92	hsa-miR-92	MI0000719	000430	UAUUGCACUUGUCCCGGCCUG
gga-miR-302b-3P	hsa-miR-302b	MI0000772	000531	UAAGUGCUUCCAUGUUUUAGUAG
gga-miR-302b-5P	gga-miR-302b*	MI0003700	008131_mat	ACUUUAACAUGGAGGUGCUUUCU

2.7. CDNA Writing, qRT-PCR

The extracted RNA samples were reverse transcribed into cDNA with High-Capacity cDNA Reverse Transcription Kit, following the instructions of the manufacturer (Applied Bio systems, Life Technologies, Carlsbad, CA, USA). RT master mix was used for cDNA writing. The cDNA was stored at −20 °C. The synthesized cDNA was then used for qRT-PC. The reaction was performed by Eppendorf MasterCycler Realplex machine. TaqMan PCR master mix was applied for the qPCR as a double-stranded fluorescent DNA-specific dye according to the manufacturer's instructions (Applied Bio systems, Life Technologies, Carlsbad, CA, USA). GAPDH was used as an internal control (housekeeping gene). For each gene examined, three parallels were analysed, fluorescence emission was detected and relative quantification was calculated with the GenEx7 program (MultiD Analyses AB, Göteborg, Sweden).

The qRT-PCR analysis was used for checking the expression of the PGC-specific and stem cell-specific markers in the chicken PGCs. The mRNA analysis was done for the following genes given in Table 1 below.

The qRT-PCR analysis was done to check the expression of miR-302b-3P and miR-302b-5P mature miRNA expression. MiR-92 was used as an internal control (Table 2).

2.8. Inhibition gga-miR-302b-5P and gga-miR-302b-3P Using MicroRNA Inhibitors

Cultured chicken PGCs were transfected with inhibitors against the gga-miR-302b-5P and gga-miR-302b-3P at 100 nM final concentration using the transfection agent siPORT

(Applied Biosystems, Life Technologies, Carlsbad, CA, USA). We used the following inhibitors given in Table 3 below.

Table 3. Inhibitors used at the miRNA inhibition assays.

Name	Gene	Catalog Number	Assay ID	Type of Inhibitor
anti-miR-302b-3p	hsa-miR-302b-3P	AM17000	AM10081	anti-miRTM-miRNA inhibitor
anti-miR-302b-5p	gga-miR-302b-5P	4464084	MH11349	mirVanaTM miRNA Inhibitor

2.9. Statistical Analysis

The expression or repression of the target gene relative to the internal control gene in each sample was calculated with the GenEx 7.0 program (Multiday, SE) using the formula $2^{-\Delta\Delta Ct}$, where ΔCt = Ct target gene—Ct internal control and $\Delta\Delta Ct$ = ΔCt test sample—ΔCt control sample. Statistical differences between the examined groups were assessed by *t*-test using the GenEx 7.0 software. The data are presented as mean $\pm$ SD and p values of less than 0.05 were regarded as statistically significant. Labels on the pictures: $p < 0.05$ *, $p < 0.01$ **, $p < 0.001$ ***.

3. Results

There were two male (M1: #508-ZZ; M2: #512-ZZ) and two female (F1: #509-ZW; F2: #513-ZW) PG cell lines used in this study. The cell number of these cell lines was measured every 4 h for three days, using the XLS Imaging system with a built-in incubator. Our aim was to examine the proliferation rate of the PGCs on Day 1, Day 2, and Day 3 after the inhibition of gga-miR-302b-5P (5P) or gga-miR-302b-3P (3P) or using anti-gga-miR-302b-5P and anti-gga-miR-302b-3P inhibitors together (5P/3P inhibition). For example, the proliferation rate on the third day was calculated by dividing the average cell number counted on the third day (h76) by the second day (h52). We compared the proliferation rate of control and treated lines on the first day (28 h/4 h), on the second day (52 h/28 h) and on the third day (76 h/52 h) (Supplementary Figure S1).

3.1. Characterization of Stem Cell- and Germ Cell-Specific Marker Expression Profile of PG Cell Lines at Day 0

The used PG cell lines were characterized before the experiments.

Expression of CVH and POUV were measured. We found higher expression of CVH (germ cell-specific marker, Figure 1a) in male lines (M1, M2) than in females (F1, F2). POUV (stem cell-specific marker, Figure 1b) expression was the highest in the case of the M1 PGC line. Furthermore, a significant difference was found between the proliferation rate of male and female lines. In the case of males, we found a significantly higher proliferation rate compared to female lines (Figure 1c). Gga-miR-302b-5P (Figure 1d) and gga-miR-302b-3P (Figure 1e) miRNAs were also analysed at Day 0. The expression of both miRNAs was the highest in the case of cell line M2, but the difference was not significant. Interestingly, the 5P/3P ratio was the highest in cell line F2 (Figure 1f); although that was not the case regarding the proliferation rate (Figure 1c).

The immunostaining was done to analyse the expression of stem cell-specific SSEA-1 and germ cell-specific CVH. We found high expression for examined markers in all cell lines. Aggregation of cells in the case of the female lines (F1, F2) was also detectable (Figure 2).

The cell number was increased in all cell lines during the cultivation period (4–76 h, Figure 3B). The M1 and M2 (male, ZZ genotypes) lines showed a significantly decreased proliferation rate on Day 2 and Day 3 compared to Day 1 (Figure 3A(a,b)). The proliferation rate of F1 (female, ZW genotype, Figure 3A(c)) PGCs did not change, while the proliferation rate of F2 (female, ZW genotype, Figure 3A(d)) line increased significantly on Day 2 and Day 3 compared to Day 1. This outcome can explain why we got the highest 5P/3P ratio in the case of F2 PGCs (Figure 1f). This finding fit with our earlier result [28] also. The immunostaining revealed high SSEA-1 and CVH expression in all lines (Figure 3A(e–h)).

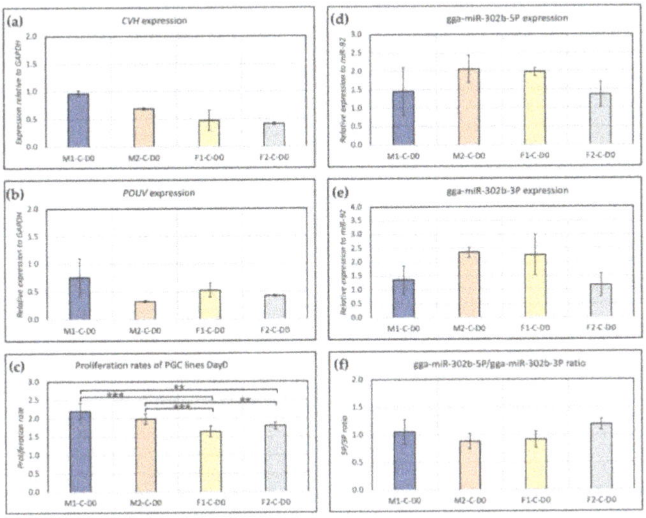

Figure 1. Characterization of PGC lines M1, M2, F1 and F2 on Day 0 of the experiments. Expression of CVH (**a**), POUV (**b**), gga-miR-302b-5P (**d**) and gga-miR-302b-3P (**e**) were analysed. GAPDH (in the case of CVH and POUV) and miR-92 (in the case of microRNAs) were used as housekeeping genes in the experiments. The proliferation rate (**c**) and 5P/3P ratio (**f**) for the PGC lines are also presented ($p < 0.01$ **, $p < 0.001$ ***).

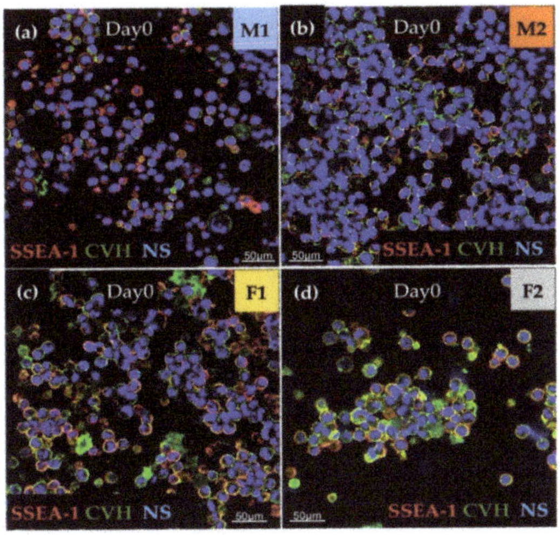

Figure 2. The immunostaining was performed with SSEA-1 (red), CVH (green) and TO-PRO™-3 for nuclear staining (blue). We examined M1 (**a**), M2 (**b**), F1 (**c**) and F2 (**d**) PG cell lines (Scale: 50 µm).

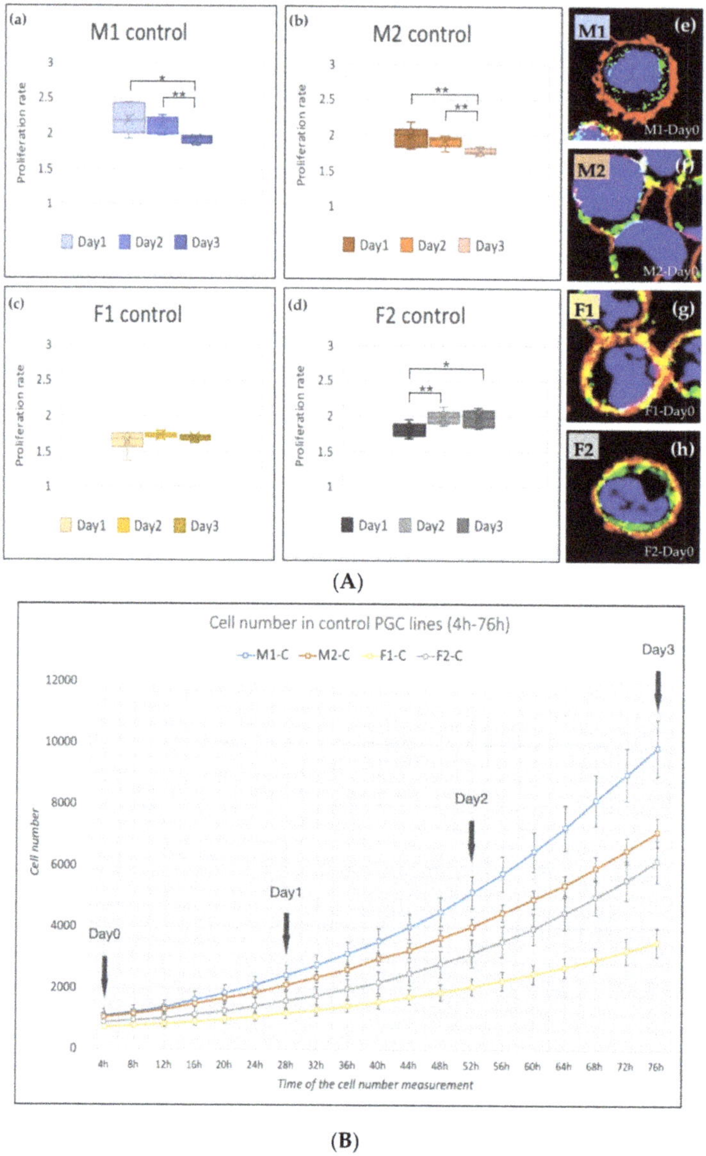

Figure 3. (**A**) The proliferation rate of the control PGC lines on Day 1, Day 2 and Day 3 (**a**–**d**) and immunostaining (**e**–**h**) of the four lines are shown in the figure. The immunostaining was achieved with SSEA-1 (red), CVH (green) and TO-PRO™-3 for nuclear staining (blue). (**B**) Analysis of the cell number. The cell number measurements were performed every 4 h (from 4 h to 76 h). We calculated the proliferation rates on Day 1, Day 2 and Day 3 ($p < 0.05$ *, $p < 0.01$ **).

3.2. *Effect of Anti-gga-miR-302b-5P, -3P and 5P/3P Inhibition on the Proliferation Rate of PGC Lines*

We aimed to examine the proliferation rate of the PGCs on Day 1, Day 2 and Day 3 after the inhibition of gga-miR-302b-5P or gga-miR-302b-3P or the inhibition of both (Supplementary Figure S1).

In the case of the M1 cell line (Figure 4a), on Day 2, the proliferation rate of 5P inhibited cells was significantly lower compared to the control. On Day 3, both 5P and 3P inhibited lines showed lower proliferation rates. The dual inhibition did not cause any detectable differences. We fixed the cells after three days of cultivation. Immunostaining was performed to detect the expression of SSEA-1 (stem cell-specific cell surface marker) and CVH (germ cell-specific marker expressing in the cytoplasm) (Figure 4b–d). In the case of 5P inhibition (Figure 4b), cytoplasm blebbing on most of the cell surface was visible, while in the control condition on Day 0 and Day 3, we could not find membrane deformities (Supplementary Figures S2–S4). The staining of these protrusions showed that they contain only cytoplasm. No sign of nuclear degradation was visible.

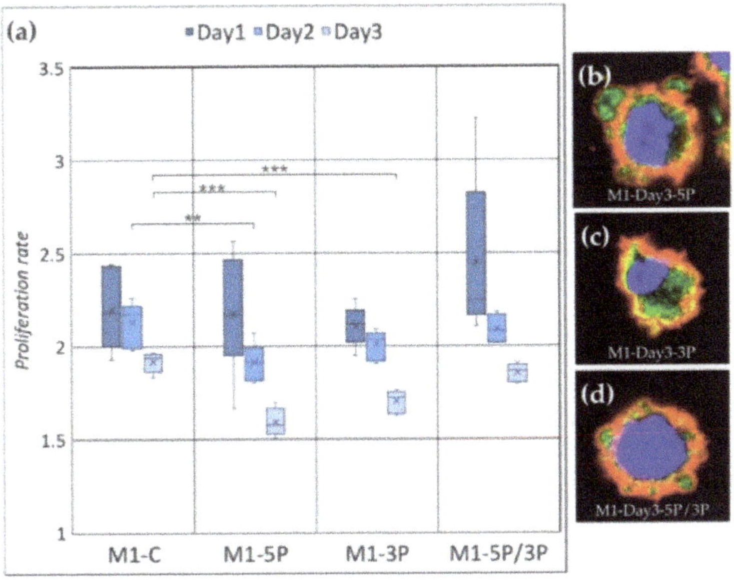

Figure 4. (**a**) Proliferation rate of M1 (ZZ genotype) PGC line on Day 1, Day 2 and Day 3. M1-C: non-inhibited, M1-5P: gga-miR-302b-5P inhibition, M1-3P: gga-miR-302b-3P inhibition, M1-5P/3P: gga-miR-302b-5P and gga-miR-302b-3P inhibition. (**b–d**) Immunostaining of M1 PGCs after three days of cultivation in different culture conditions. The immunostaining was achieved with SSEA-1 (red), CVH (green) and TO-PRO™-3 for nuclear staining (blue) ($p < 0.01$ **, $p < 0.001$ ***).

Cell line M2 (ZZ genotype) (Figure 5a) showed a significant proliferation rate increase compared to the control after inhibition of 3P and 5P/3P on Day 1. On Day 2, no difference was observed between the treated and the control groups. On Day 3, all three inhibitions caused the proliferation to slow down. Small protrusions of the cytoplasm were present only after inhibiting against the 5P (Figure 5b–d).

Cell line F1 (ZW genotype) (Figure 6a) showed a significant proliferation rate increase compared to the control after inhibition of 5P and 3P on Day 1. On Day 2, also significant proliferation rate increase was observed, but in the case of 3P and 5P/3P inhibition. Then, on Day 3, all three inhibitions caused the proliferation to slow down. Small protrusions of the cytoplasm were present after all inhibition treatments (Figure 6b–d).

Cell line F2 (ZW genotype) (Figure 7a) showed a significant proliferation rate increase compared to the control after 3P and 5P/3P inhibition on Day 1. On Day 2, no difference was observed between the treated and the control groups. Then, on Day 3, all three inhibitions caused the proliferation to slow down. Small protrusions of the cytoplasm were present after 3P and 5P/3P inhibition (Figure 7c,d).

Based on our results, we can conclude that, on the third day of the experiment, inhibition of 5P and 3P caused a drastic decrease in the proliferation rate for all cell lines compared to the control (Figure 8); however, the time scale of inhibition was different in the cell lines. Figure 8 shows the proliferation rate comparison on Day 3. It is visible that the control samples had significantly higher proliferation rates in the case of M2, F1, F2 (Figure 8b–d) than the treated ones. M1 PGCs did not show a difference compared to control after 5P/3P inhibition (Figure 8a).

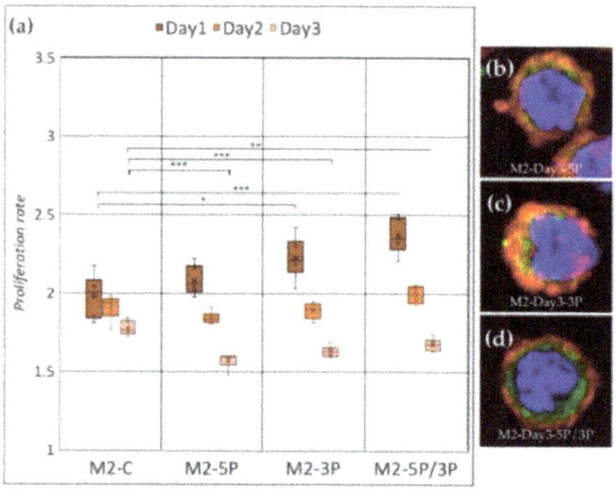

Figure 5. (**a**) Proliferation rate of M2 (ZZ genotype) PGC line on Day 1, Day 2 and Day 3. M2-C: non-inhibited, M2-5P: gga-miR-302b-5P inhibition, M2-3P: gga-miR-302b-3P inhibition, M2-5P/3P: gga-miR-302b-5P and gga-miR-302b-3P inhibition. (**b**–**d**) Immunostaining of M2 PGCs after three days of cultivation with different culture conditions. The immunostaining was achieved with SSEA-1 (red), CVH (green) and TO-PRO™-3 for nuclear staining (blue) ($p < 0.05$ *, $p < 0.01$ **, $p < 0.001$ ***).

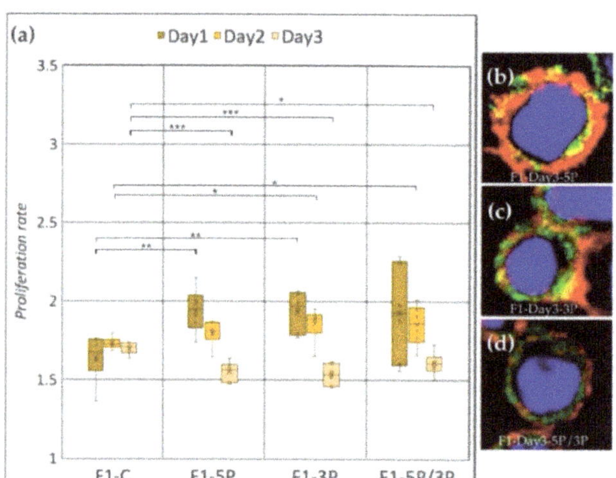

Figure 6. (**a**) Proliferation rate of F1 (ZW genotype) PGC line on Day 1, Day 2 and Day 3. F1-C: non-inhibited, F1-5P: gga-miR-302b-5P inhibition, F1-3P: gga-miR-302b-3P inhibition, F1-5P/3P: gga-miR-302b-5P and gga-miR-302b-3P inhibition. (**b**–**d**) Immunostaining of F1 PGCs after three days of cultivation with different culture conditions. The immunostaining was achieved with SSEA-1 (red), CVH (green) and TO-PRO™-3 for nuclear staining (blue) ($p < 0.05$ *, $p < 0.01$ **, $p < 0.001$ ***).

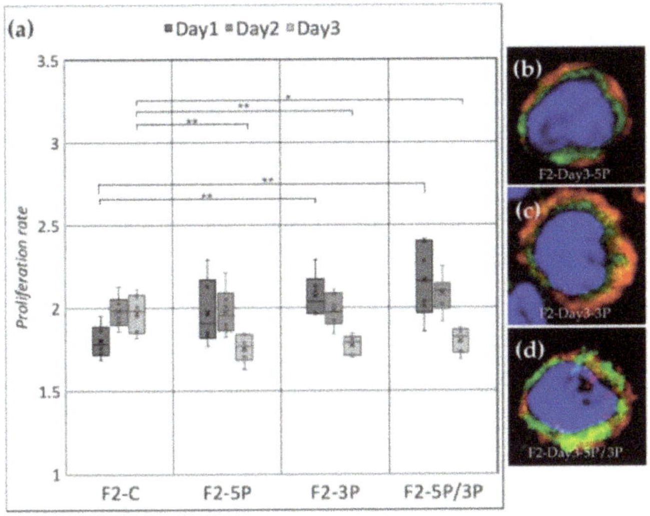

Figure 7. (**a**) Proliferation rate of F2 (ZW genotype) PGC line on Day 1, Day 2 and Day 3. F2-C: non-inhibited, F2-5P: gga-miR-302b-5P inhibition, F2-3P: gga-miR-302b-3P inhibition, F2-5P/3P: gga-miR-302b-5P and gga-miR-302b-3P inhibition. (**b**–**d**) Immunostaining of F2 PGCs after three days of cultivation with different culture conditions. The immunostaining was achieved with SSEA-1 (red), CVH (green) and TO-PRO™-3 for nuclear staining (blue) ($p < 0.05$ *, $p < 0.01$ **).

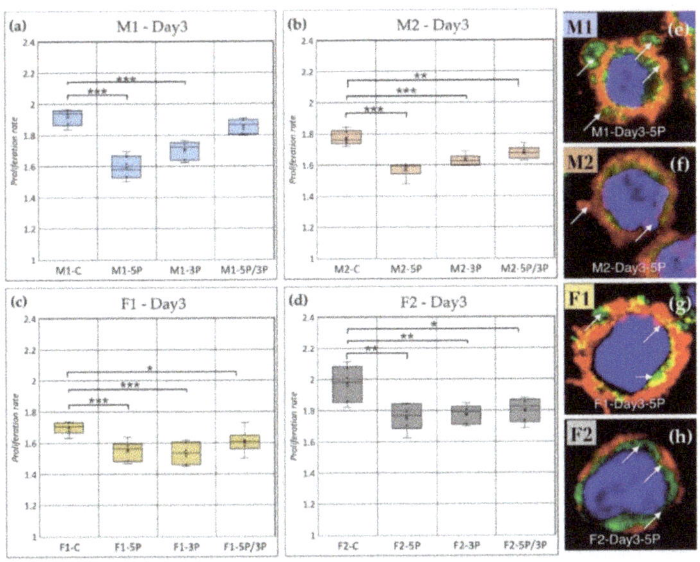

Figure 8. (**a**) Comparison of the proliferation rate of M1 (**a**), M2 (**b**) (ZZ genotype) and F1 (**c**), F2 (**d**) (ZW genotype) PGC lines on Day 3. C: non-inhibited, 5P: gga-miR-302b-5P inhibition, 3P: gga-miR-302b-3P inhibition, 5P/3P: gga-miR-302b-5P and gga-miR-302b-3P inhibition. (**e**–**h**) Immunostaining of M1, M2, F1 and F2 PGCs after three days of cultivation in anti-gga-miR-302b-5P inhibitor-containing medium. The immunostaining was achieved with SSEA-1 (red), CVH (green) and TO-PRO™-3 for nuclear staining (blue). Arrows show the blebbings on the cell surface ($p < 0.05$ *, $p < 0.01$ **, $p < 0.001$ ***).

It is also shown by the immunostaining that the 5P inhibited lines had a higher amount of blebbing cells than the other conditions (Figure 8e–h).

3.3. Determination of the Apoptotic, Late Apoptotic, and Necrotic Cell Ratio of PGC Lines

The apoptotic/necrotic staining was done with the Apoptotic Cell Detection Kit. The cell number calculation was performed using the Arthur Novel Fluorescent Cell Counter. We used two parallel measurements. The average apoptotic, late apoptotic, and necrotic ratios are illustrated in Figure 9. We found that 5P-inhibited PG cells showed a significantly lower apoptotic rate in all cell lines. The 5P/3P inhibition caused the highest percentage of cell death. The highest apoptotic rates were found in the case of the F1 cell line.

Figure 9. Comparison of the apoptotic, late apoptotic, and necrotic cell percentage in M1 (**a**), M2 (**b**) (ZZ genotype) and F1 (**c**), F2 (**d**) (ZW genotype) PGC lines.: C: non-inhibited, 5P: gga-miR-302b-5P inhibition, 3P: gga-miR-302b-3P inhibition, 5P/3P: gga-miR-302b-5P andgga-miR-302b-3P inhibition.

4. Discussion

Chicken Primordial germ cells (PGCs) can be found in the central region of the forming embryo disc. They are the first germ cell population during the development. These cells are precursors of oocytes and spermatogonia. Our study investigated the effect of miR-302b-5P and -3P on cell proliferation and apoptosis via a miRNA inhibition-based assay in PGCs. The information collected by examining the factors affecting cell proliferation of PGCs can lead to new data in stem cell biology [29].

It was published that Lin28 is a negative regulator of let-7 miRNA, and it is essential for PGC development in mouse [30]. The miR-290-295 and miR-17-92 clusters are important in mouse PGCs. MiR-290-295 cluster deficiency in mice leads to embryo lethality and germ cell deficiencies together with PGC migration problems [31]. Overexpression of Lin28 is associated with human germ-cell tumors [30]. In chicken, miR-363 is involved in gonadal development [3]. Moreover, miR-181a inhibits PGC differentiation [32].

The miR-302 cluster is embryonic stem cell-specific and evolutionarily conserved in vertebrates. The miR-302/367 cluster, generally consisting of five members, miR-367, miR-302d, miR-302a, miR-302c and miR-302b, is ubiquitously distributed in vertebrates and occupies an intragenic cluster located in the gene La-related protein 7 (LARP7) [32]. The cluster is cited for playing vital roles in diverse biological processes, such as the pluripotency of human embryonic stem cells (hESCs), self-renewal and reprogramming [28].

The miR-302-367 promoter is known to be transcriptionally regulated by the ESC-specific transcription factors Oct3/4, Sox2 and Nanog, and its activity is restricted to the ESC compartment. Functionally, this cluster regulates the cell cycle in ESCs, promoting self-renewal and pluripotency, therefore representing a master regulator in maintaining hESC stemness. It helps overcome the G1 to S phase transition during the cell cycle.

It was observed that miR-302 is endogenously highly expressed in human embryonic stem cells (hESCs) and human-induced pluripotent stem cells (hiPSCs). Inhibition of miR-302 using antagomirs resulted in downregulation of the self-renewal rate of hESCs, hiPSCs which was observed via cell colony formation assay [33].

We reported earlier a concordant dysregulation between the two arms of gga-miR-302b-5P and 3P [28] In this study, we identify that inhibition of the 3P arm slightly decreased the proliferation, indicating a role of the miR-302b-3P arm in cell proliferation. The cells inhibited with the 5P arm had a lower apoptotic rate than the cells that showed a dual inhibition. This result is consistent with Wu et al. [33], where the 5P acts as a proliferation promoter and oncomiR. Inhibition of miR-302b 5P arm decreased the proliferation rate and lowered the apoptotic rate, indicating a role of the 5P arm in promoting the proliferation of PGCs in vitro and in vivo. Our experimental data have confirmed the proliferation rate reduction after the inhibition of gga-miR-302b-5P and gga-miR-302b-3P (Figure 10).

It has been demonstrated that the members of the miR-302/367 cluster have a critical role in regulating the balance of G1-to-S transition. Three target genes were identified: CDK2, Cyclin D1/D2 and BMI-1 [34].

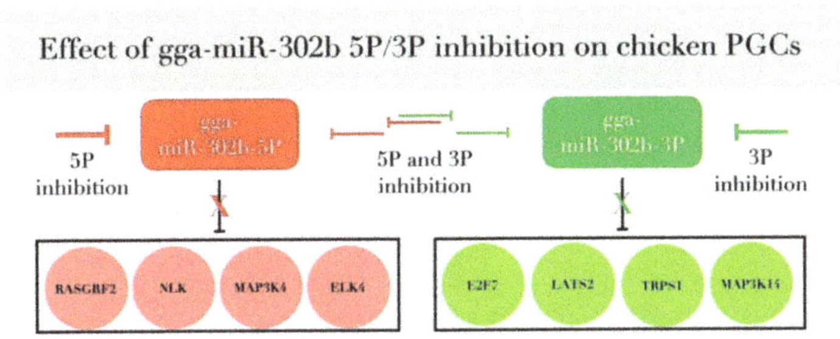

	control	5P inhibition	3P inhibition	5P/3P inhibition
cell proliferation rate	++++	+	++	+++
apoptosis rate	++	+	++	++++
cell membrane abormalities	+	++++	++	+++

Figure 10. This figure summarizes our findings and the most relevant target genes of gga-miR-302b-5P and gga-miR-302b-3P predicted by the bioinformatic analysis published previously [28,35,36].

It is also essential to identify the miR-302b-5P and -3P expression pattern. In nature, both the 5P or 3P miRNA form has been reported, and 5P/3P ratio depends on temporal, spatial, physiological, and pathological conditions. Specific arm selection is supposed to be thermodynamically controlled. Changes in strand selection in cancer cells or developmental stages are possibly associated with the presence of signals. Dicer may affect the 5P/3P strand selection of concurrent expression in cancer cells. In summary, co-regulation of the 5P/3P miRNA in normal and pluripotent tissues and cells is subjected to subtle physiological changes in pre-miRNA processing enzymes and signals [29,37].

It was issued previously that miR-302b has an essential role in cellular glucose metabolism. Glucose energy and metabolism are crucial for embryonic stem cells, induced pluripotent stem cells, and germ cells. A variety of miRNAs regulate glucose metabolism in the pancreas, liver, brain, and muscle adipose tissue. Research papers published by Rengaraj and colleagues [12,32,38,39] described that miR-302b has a significant regulatory effect on glucose phosphate isomerase (GPI), which plays an essential role in glucose metabolism. Alteration of GPI can be associated with the abnormal function of stem and germ cells.

Oct4 and Sox2 are transcription factors essential for pluripotency during early embryogenesis and maintaining embryonic stem cell (ESC) pluripotency. They bind to a conserved promoter region of miR-302 [40]. MiR-302a is predicted to target many cell cycle regulators. Moreover, miR-302a represses the translation of cyclin D1 in hESCs. The transcriptional activation of miR-302 and the translational repression of its targets, such as cyclin D1, may provide a link between Oct4/Sox2 and cell cycle regulation in pluripotent cells. Varying stability and stoichiometry of such complexes offer a means to fine-tune developmental decisions [41].

The apoptotic rate of the cells inhibited against gga-miR-302b-5P showed a negative correlation, while the proliferation rate still dropped. Apoptosis occurs whenever there is an injury to the cell that cannot be repaired. The most common cause of such injury is DNA damage. Whenever cells detect DNA damage, they trigger the response of the p53 gene, which is the most crucial inducer of apoptosis. The second pathological condition under which apoptosis can be observed is whenever misfolded protein accumulates [42]. Multiple types of death can be observed simultaneously in tissues or cell cultures exposed to the same stimulus [43]. The externalization of phosphodiesterase is an early event of apoptosis, occurring while the plasma membrane remains intact. The biological spectrum of cell deaths is much more diverse. The formation of additional "blebbing" on the membrane surface of inhibited cells needs to be addressed. The non-apoptotic membrane blebbing is known to be a cellular migration mechanism [44].

We found that inhibition of miRNA gga-miR-302b-5P has a dual effect. It was reported that the miR-302b acts as an anti-tumor specific miRNA. It has been stated that various miRNAs regulate the intrinsic and the extrinsic pathway for apoptosis in cancer cells. The role of miR-302b-5P in hepatocellular carcinoma (HCC) is still unclear [33]. Guo and co-workers published that upregulation of the miRNAs leads to extensive upregulation of pro-apoptosis genes and pathways, leading to extensive cell death and blebbing [45]. These results indicate the tumor suppressor role of miR-302b-3P in the pathogenesis of gastric cancer. MiR-302b-3p promotes self-renewal properties in Leukemia Inhibitory Factor-Withdrawn Embryonic Stem Cells [46]. miR-302b exhibited anti-tumor activity by reversing EphA2 regulation, which relayed a signalling transduction cascade that attenuated the functions of N-cadherin, β-catenin, and Snail (markers of Wnt/β-catenin and epithelial-mesenchymal transition, EMT). This modulation of EphA2 also had distinct effects on cell proliferation and migration in vivo [47].

MAPK signalling is responsible for maintaining pluripotency and proliferation in mammals. It can be hypothesized that probably the high expression of gga-miR-302b-5P is contributed to controlling the MAPK signaling pathway [21] and TGFβR2. Their targets are downstream targets in other molecular pathways like p53 signalling, FOXO signaling,

TGFβ signalling, and apoptosis. MiR-302b-5P might cause a high proliferation rate in PGC lines through inhibiting the MAPK pathway components.

5. Conclusions

Our findings could help to improve the in vitro conditions for PGC cultivation to gain stable germline competence comparable to in vivo PGCs. Many proteins and ligands are targets of the gga-miR-302b-5P and -3P. We can hypothesize that the inhibition of the 5P arms of miR-302b might lead to an upregulation of these pathways, causing a decrease in proliferation and apoptosis rate. Investigating the cell cycle regulatory function of gga-miR-302b-3P can help to understand its tumor suppressor role of miR-302b-3P in different types of cancer.

Supplementary Materials: The following figures are available online at https://www.mdpi.com/article/10.3390/genes13010082/s1, Figure S1: Description of the experiment, Figure S2: Immunostaining of control M1 PGCs on Day 1, Figure S3: Immunostaining of control M1 PGCs on Day 3, Figure S4: Immunostaining of 5P inhibited M1 PGCs on Day 3.

Author Contributions: Conceptualization, E.G., E.P.V. and B.P.; methodology, B.L., R.T., M.A. and A.E.; validation, E.G., L.H. and E.P.V.; formal analysis, E.G. and B.L.; investigation, B.L., N.T.S., E.G., Z.H., M.U. and A.E; writing—original draft preparation, E.G., A.E., B.L., R.T., G.S. and N.T.S.; writing—review and editing, E.P.V., B.P., R.T., M.T.S.A., M.A. and N.T.S.; visualization, E.G. and Z.H.; supervision, E.G. and L.H. funding acquisition, E.G., E.P.V. and B.P. All authors have read and agreed to the published version of the manuscript.

Funding: This research was funded by VEKOP-2.3.2-16-2016-00012, TKP2020-NKA-24, 2019-2.1.11-TÉT-2019-00036.

Institutional Review Board Statement: All applied methods in NBGK-HGI were approved by the Directorate of Food Safety and Animal Health of the Government Office of Pest County (License number: PE/EA197-4/2016) and by the Institutional Ethical Review Board.

Informed Consent Statement: Not applicable.

Data Availability Statement: The data presented in this study are available on request from the corresponding author.

Acknowledgments: We thank the animal keepers of the National Centre for Biodiversity and Gene Conservation for the egg collection and the housing of poultry.

Conflicts of Interest: The authors declare no conflict of interest.

References

1. Reik, W.; Surani, M.A. Germline and Pluripotent Stem Cells. *Cold Spring Harb. Perspect. Biol.* **2015**, *7*, a019422. [CrossRef]
2. Evans, M.J.; Kaufman, M.H. Establishment in Culture of Pluripotential Cells from Mouse Embryos. *Nature* **1981**, *292*, 154–156. [CrossRef]
3. Takahashi, K.; Yamanaka, S. Induction of Pluripotent Stem Cells from Mouse Embryonic and Adult Fibroblast Cultures by Defined Factors. *Cell* **2006**, *126*, 663–676. [CrossRef] [PubMed]
4. Nakamura, Y.; Kagami, H.; Tagami, T. Development, Differentiation and Manipulation of Chicken Germ Cells. *Dev. Growth Differ.* **2013**, *55*, 20–40. [CrossRef] [PubMed]
5. Nandi, S.; Whyte, J.; Taylor, L.; Sherman, A.; Nair, V.; Kaiser, P.; McGrew, M.J. Cryopreservation of Specialized Chicken Lines Using Cultured Primordial Germ Cells. *Poult. Sci.* **2016**, *95*, 1905–1911. [CrossRef] [PubMed]
6. Whyte, J.; Glover, J.D.; Woodcock, M.; Brzeszczynska, J.; Taylor, L.; Sherman, A.; Kaiser, P.; McGrew, M.J. FGF, Insulin, and SMAD Signaling Cooperate for Avian Primordial Germ Cell Self-Renewal. *Stem Cell Rep.* **2015**, *5*, 1171–1182. [CrossRef] [PubMed]
7. Moradi, S.; Braun, T.; Baharvand, H. MiR-302b-3p Promotes Self-Renewal Properties in Leukemia Inhibitory Factor-Withdrawn Embryonic Stem Cells. *Cell J.* **2018**, *20*, 61–72. [CrossRef] [PubMed]
8. Rad, S.M.A.H.; Bavarsad, M.S.; Arefian, E.; Jaseb, K.; Shahjahani, M.; Saki, N. The Role of MicroRNAs in Stemness of Cancer Stem Cells. *Oncol. Rev.* **2013**, *7*, 8. [CrossRef]
9. Maraghechi, P.; Hiripi, L.; Tóth, G.; Bontovics, B.; Bosze, Z.; Gócza, E. Discovery of Pluripotency-Associated MicroRNAs in Rabbit Preimplantation Embryos and Embryonic Stem-like Cells. *Reproduction* **2013**, *145*, 421–437. [CrossRef] [PubMed]
10. Li, N.; Long, B.; Han, W.; Yuan, S.; Wang, K. MicroRNAs: Important Regulators of Stem Cells. *Stem Cell Res. Ther.* **2017**, *8*, 110. [CrossRef]

11. Houbaviy, H.B.; Murray, M.F.; Sharp, P.A. Embryonic Stem Cell-Specific MicroRNAs. *Dev. Cell* **2003**, *5*, 351–358. [CrossRef]
12. Rengaraj, D.; Lee, B.R.; Lee, S.I.; Seo, H.W.; Han, J.Y. Expression Patterns and Mirna Regulation of DNA Methyltransferases in Chicken Primordial Germ Cells. *PLoS ONE* **2011**, *6*, e19524. [CrossRef]
13. Rengaraj, D.; Park, T.S.; Lee, S.I.; Lee, B.R.; Han, B.K.; Song, G.; Han, J.Y. Regulation of Glucose Phosphate Isomerase by the 3′UTR-Specific MiRNAs MiR-302b and MiR-17-5p in Chicken Primordial Germ Cells. *Biol. Reprod.* **2013**, *89*, 33. [CrossRef]
14. Houbaviy, H.B.; Dennis, L.; Jaenisch, R.; Sharp, P.A. Characterization of a Highly Variable Eutherian MicroRNA Gene Characterization of a Highly Variable Eutherian MicroRNA Gene. *RNA* **2005**, *11*, 1245–1257. [CrossRef] [PubMed]
15. Bartel, D.P. MicroRNAs: Genomics, Biogenesis, Mechanism, and Function. *Cell* **2004**, *116*, 281–297. [CrossRef]
16. Winter, J.; Jung, S.; Keller, S.; Gregory, R.I.; Diederichs, S. Many Roads to Maturity: MicroRNA Biogenesis Pathways and Their Regulation. *Nat. Cell Biol.* **2009**, *11*, 228–234. [CrossRef]
17. Choo, K.B.; Soon, Y.L.; Nguyen, P.N.N.; Hiew, M.S.Y.; Huang, C.J. MicroRNA-5p and -3p Co-Expression and Cross-Targeting in Colon Cancer Cells. *J. Biomed. Sci.* **2014**, *21*, 95. [CrossRef] [PubMed]
18. Hashemzadeh, M.R. Role of Micro RNAs in Stem Cells, Cardiac Differentiation and Cardiovascular Diseases. *Gene Rep.* **2017**, *8*, 11–16. [CrossRef]
19. Darnell, D.K.; Kaur, S.; Stanislaw, S.; Konieczka, J.K.; Yatskievych, T.A.; Antin, P.B. MicroRNA Expression during Chick Embryo Development. *Dev. Dyn.* **2006**, *235*, 3156–3165. [CrossRef] [PubMed]
20. Ha, M.; Kim, V.N. Regulation of MicroRNA Biogenesis. Nature reviews. *Mol. Cell Biol.* **2014**, *15*, 509–524. [CrossRef]
21. Zhan, J.; Jiao, D.; Wang, Y.; Song, J.; Wu, J.; Wu, L.; Chen, Q.; Ma, S. Integrated MicroRNA and Gene Expression Profiling Reveals the Crucial MiRNAs in Curcumin Anti-Lung Cancer Cell Invasion. *Thorac. Cancer* **2017**, *8*, 461–470. [CrossRef] [PubMed]
22. Jiao, D.M.; Yan, L.; Wang, L.S.; Hu, H.Z.; Tang, X.L.; Chen, J.; Wang, J.; Li, Y.; Chen, Q.Y. Exploration of Inhibitory Mechanisms of Curcumin in Lung Cancer Metastasis Using a MiRNA-Transcription Factor-Target Gene Network. *PLoS ONE* **2017**, *12*, e0172470.
23. Picon-Ruiz, M.; Pan, C.; Drews-Elger, K.; Jang, K.; Besser, A.H.; Zhao, D.; Morata-Tarifa, C.; Kim, M.; Ince, T.A.; Azzam, D.J.; et al. Interactions between Adipocytes and Breast Cancer Cells Stimulate Cytokine Production and Drive Src/Sox2/MiR-302b-Mediated Malignant Progression. *Cancer Res.* **2016**, *76*, 491–504. [CrossRef] [PubMed]
24. Khodayari, N.; Mohammed, K.A.; Lee, H.; Kaye, F.; Nasreen, N. MicroRNA-302b Targets Mcl-1 and Inhibits Cell Proliferation and Induces Apoptosis in Malignant Pleural Mesothelioma Cells. *Am. J. Cancer Res.* **2016**, *6*, 1996. [PubMed]
25. Lu, X.X.; Cao, L.Y.; Chen, X.; Xiao, J.; Zou, Y.; Chen, Q. PTEN Inhibits Cell Proliferation, Promotes Cell Apoptosis, and Induces Cell Cycle Arrest via Downregulating the PI3K/AKT/HTERT Pathway in Lung Adenocarcinoma A549 Cells. *BioMed Res. Int.* **2016**, *2016*, 2476842. [CrossRef] [PubMed]
26. Lázár, B.; Molnár, M.; Sztán, N.; Végi, B.; Drobnyák, Á.; Tóth, R.; Tokodyné Szabadi, N.; McGrew, M.J.; Gócza, E.; Patakiné Várkonyi, E. Successful Cryopreservation and Regeneration of a Partridge Colored Hungarian Native Chicken Breed Using Primordial Germ Cells. *Poult. Sci.* **2021**, *100*, 101207. [CrossRef]
27. Hamburger, V.; Hamilton, H.L. A Series of Normal Stages in the Development of the Chick Embryo. *J. Morphol.* **1951**, *88*, 49–92. [CrossRef] [PubMed]
28. Lázár, B.; Anand, M.; Tóth, R.; Várkonyi, E.P.; Liptói, K.; Gócza, E. Comparison of the MicroRNA Expression Profiles of Male and Female Avian Primordial Germ Cell Lines. *Stem Cells Int.* **2018**, *2018*, 1780679. [CrossRef]
29. Cai, D.; He, K.; Chang, S.S.; Tong, D.; Huang, C. MicroRNA-302b Enhances the Sensitivity of Hepatocellular Carcinoma Cell Lines to 5-FU via Targeting Mcl-1 and DPYD. *Int. J. Mol. Sci.* **2015**, *16*, 23668–23682. [CrossRef]
30. West, J.A.; Viswanathan, S.R.; Yabuuchi, A.; Cunniff, K.; Takeuchi, A.; Park, I.H.; Sero, J.E.; Zhu, H.; Perez-Atayde, A.; Frazier, A.L.; et al. A Role for Lin28 in Primordial Germ-Cell Development and Germ-Cell Malignancy. *Nature* **2009**, *460*, 909–913. [CrossRef] [PubMed]
31. Hayashi, K.; Chuva de Sousa Lopes, S.M.; Kaneda, M.; Tang, F.; Hajkova, P.; Lao, K.; O'Carroll, D.; Das, P.P.; Tarakhovsky, A.; Miska, E.A.; et al. MicroRNA Biogenesis Is Required for Mouse Primordial Germ Cell Development and Spermatogenesis. *PLoS ONE* **2008**, *3*, e1738. [CrossRef]
32. Lee, S.I.; Lee, B.R.; Hwang, Y.S.; Lee, H.C.; Rengaraj, D.; Song, G.; Park, T.S.; Han, J.Y. MicroRNA-Mediated Posttranscriptional Regulation Is Required for Maintaining Undifferentiated Properties of Blastoderm and Primordial Germ Cells in Chickens. *Proc. Natl. Acad. Sci. USA* **2011**, *108*, 10426–10431. [CrossRef] [PubMed]
33. Wu, H.-Y.; Cai, K.-T.; Ma, J.; Chen, G.; Dang, Y.-W.; Lu, H.-P.; Pan, S.-L. Evaluation of MiR-302b-5p Expression and Molecular Mechanism in Hepatocellular Carcinoma: Findings Based on RT-qPCR and in Silico Analysis. *Pathol. Res. Pract.* **2019**, *215*, 152424. [CrossRef]
34. Lin, S.L.; Chang, D.C.; Ying, S.Y.; Leu, D.; Wu, D.T.S. MicroRNA MiR-302 Inhibits the Tumorigenecity of Human Pluripotent Stem Cells by Coordinate Suppression of the CDK2 and CDK4/6 Cell Cycle Pathways. *Cancer Res.* **2010**, *70*, 9473–9482. [CrossRef] [PubMed]
35. Cataldo, A.; Cheung, D.G.; Balsari, A.; Tagliabue, E.; Coppola, V.; Iorio, M.; Palmieri, D.; Croce, C.M. MiR-302b Enhances Breast Cancer Cell Sensitivity to Cisplatin by Regulating E2F1 and the Cellular DNA Damage Response. *Oncotarget* **2016**, *7*, 786–797. [CrossRef]
36. Sun, Y.; Liu, W.-Z.; Liu, T.; Feng, X.; Yang, N.; Zhou, H.-F. Signaling Pathway of MAPK/ERK in Cell Proliferation, Differentiation, Migration, Senescence and Apoptosis. *J. Recept. Signal Transduct.* **2015**, *35*, 600–604. [CrossRef] [PubMed]

37. Huang, M.; Li, D.; Huang, Y.; Cui, X.; Liao, S.; Wang, J.; Liu, F.; Li, C.; Gao, M.; Chen, J.; et al. HSF4 Promotes G1/S Arrest in Human Lens Epithelial Cells by Stabilizing P53. *Biochim. Biophys. Acta-Mol. Cell Res.* **2015**, *1853*, 1808–1817. [CrossRef]
38. Rengaraj, D.; Lee, B.R.; Choi, J.W.; Lee, S.I.; Seo, H.W.; Kim, T.H.; Choi, H.J.; Song, G.; Han, J.Y. Gene Pathways and Cell Cycle-Related Genes in Cultured Avian Primordial Germ Cells. *Poult. Sci.* **2012**, *91*, 3167–3177. [CrossRef]
39. Kang, K.S.; Lee, H.C.; Kim, H.J.; Lee, H.G.; Kim, Y.M.; Lee, H.J.; Park, Y.H.; Yang, S.Y.; Rengaraj, D.; Park, T.S.; et al. Spatial and Temporal Action of Chicken Primordial Germ Cells during Initial Migration. *Reproduction* **2015**, *149*, 179–187. [CrossRef] [PubMed]
40. Greer Card, D.A.; Hebbar, P.B.; Li, L.; Trotter, K.W.; Komatsu, Y.; Mishina, Y.; Archer, T.K. Oct4/Sox2-Regulated MiR-302 Targets Cyclin D1 in Human Embryonic Stem Cells. *Mol. Cell. Biol.* **2008**, *28*, 6426–6438. [CrossRef]
41. Orkin, S.H.; Hochedlinger, K. Chromatin Connections to Pluripotency and Cellular Reprogramming. *Cell* **2011**, *145*, 835–850. [CrossRef] [PubMed]
42. Elmore, S. Apoptosis: A Review of Programmed Cell Death. *Toxicol. Pathol.* **2007**, *35*, 495–516. [CrossRef] [PubMed]
43. Fink, S.L.; Cookson, B.T. Apoptosis, Pyroptosis, and Necrosis: Mechanistic Description of Dead and Dying Eukaryotic Cells. *Infect. Immun.* **2005**, *73*, 1907–1916. [CrossRef]
44. Srihawong, T.; Kuwana, T.; Siripattarapravat, K.; Tirawattanawanich, C. Chicken Primordial Germ Cell Motility in Response to Stem Cell Factor Sensing. *Int. J. Dev. Biol.* **2015**, *59*, 453–460. [CrossRef] [PubMed]
45. Guo, M.; Gan, L.; Si, J.; Zhang, J.; Liu, Z.; Zhao, J.; Gou, Z.; Zhang, H. Role of MiR-302/367 Cluster in Human Physiology and Pathophysiology. *Acta Biochim. Biophys. Sin.* **2020**, *52*, 791–800. [CrossRef] [PubMed]
46. Guo, B.; Zhao, Z.; Wang, Z.; Li, Q.; Wang, X.; Wang, W.; Song, T.; Huang, C. MicroRNA-302b-3p Suppresses Cell Proliferation Through AKT Pathway by Targeting IGF-1R in Human Gastric Cancer. *Cell. Physiol. Biochem.* **2017**, *42*, 1701–1711. [CrossRef] [PubMed]
47. Huang, J.; He, Y.; Mcleod, H.L.; Xie, Y.; Xiao, D.; Hu, H.; Chen, P. MiR-302b Inhibits Tumorigenesis by Targeting EphA2 via Wnt/β-Catenin/EMT Signaling Cascade in Gastric Cancer. *BMC Cancer* **2017**, *17*, 886. [CrossRef] [PubMed]

genes

MDPI

Article

Posttranscriptional Regulation of the Human ABCG2 Multidrug Transporter Protein by Artificial Mirtrons

Anita Schamberger *, György Várady, Ábel Fóthi and Tamás I. Orbán *

Institute of Enzymology, ELKH Research Centre for Natural Sciences, H-1117 Budapest, Hungary; varady.gyorgy@ttk.hu (G.V.); fothi.abel@gmail.com (Á.F.)
* Correspondence: schamberger.anita@ttk.hu (A.S.); orban.tamas@ttk.hu (T.I.O.)

Abstract: ABCG2 is a membrane transporter protein that has been associated with multidrug resistance phenotype and tumor development. Additionally, it is expressed in various stem cells, providing cellular protection against endobiotics and xenobiotics. In this study, we designed artificial mirtrons to regulate ABCG2 expression posttranscriptionally. Applying EGFP as a host gene, we could achieve efficient silencing not only in luciferase reporter systems but also at the ABCG2 protein level. Moreover, we observed important new sequential-functional features of the designed mirtrons. Mismatch at the first position of the mirtron-derived small RNA resulted in better silencing than full complementarity, while the investigated middle and 3′ mismatches did not enhance silencing. These latter small RNAs operated most probably via non-seed specific translational inhibition in luciferase assays. Additionally, we found that a mismatch in the first position has not, but a second mismatch in the third position has abolished target mRNA decay. Besides, one nucleotide mismatch in the seed region did not impair efficient silencing at the protein level, providing the possibility to silence targets carrying single nucleotide polymorphisms or mutations. Taken together, we believe that apart from establishing an efficient ABCG2 silencing system, our designing pipeline and results on sequential-functional features are beneficial for developing artificial mirtrons for other targets.

Keywords: mirtron; miRNA; ABCG2; silencing; multidrug transporter

1. Introduction

The human ABCG2 protein is one of the 48 known members of the human ATP-binding-cassette (ABC) protein family. It was originally cloned from the placenta and cells selected for multidrug resistance [1–3], but according to our present knowledge, ABCG2 is also expressed in various differentiated tissues, including ovary, kidney, liver, breast epithelial cells, intestinal epithelia, and the blood–brain barrier [4]. This multidrug transporter protein provides resistance against various endo- and xenobiotics and hypothesized to play a physiological role in the chemoimmunity defense system [5]. The ABCG2 protein has also been identified in many types of tissue-derived stem cells and in human embryonic stem cell lines (hESC), and its role is presumably the protection against different toxins and stress [6,7]. Moreover, its expression was shown to be a reliable marker of the "side-population phenotype" [8]; therefore, investigating its role and function in various stem cells is still an important issue. There are several model systems where ABCG2 is overexpressed or knocked out [9–11]; however, a model where the function of ABCG2 is turned off in a carefully controlled and reversible manner is lacking. MicroRNA (miRNA)-based regulation could present a versatile platform for such purposes, providing a posttranscriptional fine-tuning of gene expression, thereby a careful studying of protein function.

The majority of miRNAs are processed via the canonical miRNA biogenesis pathway. The ~20–24 nucleotide (nt) long, single-stranded mature RNA derives from an imperfect

RNA hairpin structure, which is usually transcribed from the genome by a Pol II polymerase [12,13]. This primary transcript (pri-miRNA) is then cleaved by a nuclear RNase III-like enzyme Drosha (assisted by its partner protein, DGCR8), releasing a ~60–70 nt long hairpin (called pre-miRNA; [14–17]). The pre-miRNA is then transported from the nucleus to the cytoplasm by the Exportin-5 shuttle system [18–21]. In the cytoplasm, Dicer, another RNase III-like enzyme, cleaves the pre-miRNA, liberating the double-stranded miRNA:miRNA* molecule [22,23]. Dicer acts as a molecular ruler, and the cleavage site can be measured either from the 5′- or the 3′-end of the pre-miRNA, depending on the stability of the 5′-end [24–27]. During further processing, one strand (called guide strand) of the liberated small RNA duplex is incorporated into an Argonaute (AGO) protein-containing complex, forming a mature RISC (RNA induced silencing complex) and guiding it to the target transcript. Up to our present knowledge, strand selection is mainly determined by thermodynamic characteristics (strands with low thermodynamic stability at their 5′-end are favorable) and the 5′ nucleotide identity (A and U are favorable [28]). The regulatory effect of miRNAs is usually manifested by the destabilization/degradation and/or translational inhibition of the target mRNA molecule via the partial base pairing of the miRNA and the 3′-untranslated region (3′-UTR) of the mRNA [29,30].

Non-canonical miRNA biogenesis pathways could bypass certain steps of the canonical process, typically one or even both of the two cleavage steps [31–34]. Mirtrons, which are generated in a Drosha-independent pathway, represent the most prominent group of the alternatively processed miRNAs. They reside in short introns, which are essentially equivalent to the precursor form (pre-miRNA) of the given miRNA. Thus, the first step of the mirtronic miRNA processing is different from the canonical one: the pre-miRNA is liberated from the primary transcript by the splicing machinery instead of the Drosha/DGCR8 complex. The mirtron pathway was first described in *Drosophila melanogaster* and *Caenorhabditis elegans* [35,36], and later, it was experimentally demonstrated to be operational also in mammals [37–39].

Mirtrons, owing to their special features, are promising genetic tools for the regulation of genes of interest. They could be expressed by Pol II promoters; therefore, their expression can be spatiotemporally regulated, while their maturation does not interfere in the nucleus with the endogenous canonical miRNA maturation pathway [37,40]. There are several articles investigating the potential of artificially designed mirtrons as silencers and showing additional advantages, such as embedding multiple artificial mirtrons in a gene for delivery and investigating various therapeutic potentials [40–43].

In this study, we present the design of artificial mirtrons for silencing the ABCG2 multidrug transporter protein. Testing several potential candidates, we could successfully silence targets in luciferase reporter assays. Moreover, we could also effectively reduce the protein level of ABCG2. In addition, we observed important sequential-functional features of the designed mirtrons. Changing the complementarity to the target in various positions revealed the importance of the middle and 3′ region in more efficient repression, while one mismatch in the first position or the seed region did not abolish efficient silencing. The various changes also influenced the balance between translation inhibition and mRNA destabilization. As an important aspect, we also point out to consider the presence of nucleotide polymorphisms when designing mirtrons against a particular gene of interest.

2. Materials and Methods

2.1. *Bioinformatics, Statistical Analysis*

During mirtron design, we used several prediction programs to select promising artificial mirtron sequences for further experimental investigations. After designing the different mirtron sequences, we predicted their splicing from the EGFPm coding context. We used the SpliceDB, Softberry program for splicing donor and acceptor site predictions [44,45] and the Human Splice finder for branch point analysis [46]. For structural and delta G (Gibbs free energy) predictions, we used the mFold program [47]. We selected four artificial mirtrons for subsequent investigations, with variable structure and prediction parameters (Supplementary Figures S1–S4). Regarding experimental studies, experiments with three parallels were repeated at least twice. For statistical analysis, a two-sided Student's *t*-test was performed.

2.2. Plasmid Constructs

For the expression of artificial mirtrons, oligonucleotides corresponding to the sense and antisense sequence of the specific mirtrons were hybridized to form a double-stranded DNA, then inserted as an artificial intron into the PvuII site of EGFPm by blunt-end ligation [39,48]. As a control, we used the third intron of the mouse IgCε gene [49] or a canonical intronic miRNA (mir-33b), as we described earlier [39]. For luciferase constructs, hybridized double-stranded oligonucleotides for the wild type and the seed region mutated target sequences of corresponding mirtrons were ligated between the XhoI/NotI restriction sites of *Renilla* luciferase 3′-UTR in the psiCHECK2 vector (Promega, Madison, WI, USA). For ABCG2 experiments, a previously established pCDNA3.1_ABCG2 plasmid was used [50]. All plasmid constructs were verified by Sanger sequencing.

2.3. Cell Cultures and Manipulation

HeLa cell lines were maintained in Dulbecco's modified Eagle's medium (DMEM, cat. #31966047) supplemented with 10% of fetal bovine serum (cat. #10500064), 1% of L-glutamine (cat. #25030081), and 1% of penicillin/streptomycin (cat. # 15070063, all from Thermofisher Scientific) using standard cell culture methodology. Cells were transfected with FuGENE® HD reagent (Roche Applied Science, Penzberg, Germany) in a 6-well or 24-well plate, according to the manufacturer's instruction. To stain the cell nuclei, 10 μM of the Hoechst 33342 dye was used according to the standard protocol. EGFP and Hoechst fluorescence was detected by an IX51 fluorescence microscope (Olympus, Shinjuku City, Tokyo, Japan).

To establish cell lines stably expressing EGFPm-mirtron constructs, we applied the *Sleeping Beauty* transposon-based gene delivery technology as described earlier [51]. Following transfection, cells were sorted for EGFP positivity at day 8 and subsequently at day 15 using a FACS Aria High Speed Cell Sorter (Beckton-Dickinson, Franklin Lake, NJ, USA) to obtain homogenously expressing cell populations. Stable expression was also checked by subsequent FACS analyses. The established cell lines were further used for mRNA level and western blot experiments.

2.4. RNA Analysis

Total RNA was isolated from cultured cells using Trizol reagent (Invitrogen, Waltham, MA, USA). To remove genomic DNA contaminations, RNA samples were treated with DNaseI (New England Biolabs, Ipswich, MA, USA) at 37 °C for 1 h. For cDNA preparations, 1 μg of total RNA was reverse transcribed with random primers using High Capacity cDNA Reverse Transcription Kit (Thermofisher Scientific, Waltham, MA, USA). For splicing experiments, a polymerase chain reaction was performed on cDNA prepared from transiently transfected cells (1000 ng of artificial mirtron expressing plasmids were transfected into cells in a 6-well plate), using the following primers: 5′–TTCTTCAAGTCCGCCATGCC (forward) and 5′–ACTTGTACAGCTCGTCCATGCCG (reverse). To carry out real-time quantitative PCR (qPCR), we used specific TaqMan® assays and reagents. Reactions were performed on StepOne™ or StepOnePlus™ platforms, according to the manufacturer's instructions (Thermofisher Scientific). For relative quantitation, the ΔΔCt method was applied, and we used the RPLP0 mRNA (catalog number: Hs9999902_m1) as endogenous control. For *Renilla* mRNA level experiments, cells were transfected with 500 ng of respective sensor/mutant sensor expressing plasmid in a 6-well plate. For *Renilla* luciferase mRNA detection, a custom-made TaqMan assay was used, containing the following primers: 5′-CGAGTGGCCTGACATCGA (forward),

5′-ACGAAGAAGTTATTCTCAAGCACCAT (reverse) and 5′-CAGGGCGATATCCTC (probe, with 5′-FAM and 3′-MGB labeled). For firefly luciferase mRNA detection, the following custom-made assay was used: 5′-GCTTCGAGG-AGGAGCTGTTC (forward), 5′-CCAGCAGGGCAGACTGAATTT (reverse) and 5′-CAGCC-TGCAAGACTAC (probe, with 5′-FAM and 3′-MGB labeled). For ABCG2 mRNA level measurements, 1000 ng of ABCG2 expressing and 500 ng of psiCHECK2 plasmids were co-transfected into cells, seeded in 6-well plates. For qPCR analysis of ABCG2 mRNA, a pre-developed assay was used (catalog number: Hs01053790_m1).

2.5. *Luciferase Assay*

In each experiment, 300 ng of the mirtron/control expressing plasmids were co-transfected with 15 ng of sensor or mutant sensor luciferase plasmids into cells, seeded on a 24-well plate. Sensors containing two copies of the respective target site were cloned downstream of *Renilla* luciferase in the psiCHECK2 vector. Mutant sensors differ in 3 mismatched nucleotides in the predicted miRNA seed region. Luciferase activity was measured at 48 h posttransfection by a 2030 Multilabel Reader luminometer (PerkinElmer, Waltham, MA, USA) using the Dual-Luciferase Reporter Assay System (Promega). Signal specific for firefly luciferase expressed from the same psiCHECK2 plasmid was used to normalize for transfection efficiency. To fully exclude any non-specific effects, luciferase activities of the sensors were also measured in the presence of an unrelated miRNA (hsa-mir-33b) as non-cognate control.

2.6. *Western Blot (Immunoblot)*

Artificial mirtron and control expressing stable HeLa cell lines were transfected with 500 ng ABCG2 expressing plasmid in a 6-well plate. Cells were lysed and collected 48 h after transfection. After briefly sonicated, samples were run on 7.5% acrylamide gel, then electroblotted onto PVDF membrane (BioRad, Hercules, CA, USA). Membranes were blocked by 5% milk/TBS-Tween and incubated with mouse monoclonal BXP-21 antibody (kindly provided by Dr. George Scheffer) overnight at 4 °C for ABCG2 detection. Next, membranes were incubated in HRP-conjugated Anti-Mouse IgG secondary antibody solution (Jackson's, cat # 715-035-151) for 1 h at room temperature. For signal detection, an ECL reagent (Thermofisher Scientific) was used, and the membranes were exposed to Agfa films. Monoclonal Anti-β-Actin-Peroxidase antibody (Sigma, cat. #A3854) was used for β-actin detection as a control. Experiments were repeated at least three times, and one representative experiment is shown in the figures. Expression levels were quantified by densitometry of the scanned images using the ImageJ software.

3. Results

3.1. *Artificial Mirtron Design*

Although there are numerous advantages of mirtrons as silencers, for the process of artificial mirtron design, some criteria should be considered. First, regarding splicing, a GU 5′-end as 5′ splicing donor site and a (C)AG 3′- end as 3′ acceptor site is advantageous. Then, a functional polypyrimidine tract and a branch point should be placed somewhere in the mirtron sequence (Figure 1A). Additionally, as was mentioned above, there are some features to be considered for proper processing by Dicer and for the loading of the appropriate strand of the small RNA duplex into a functional RISC. Besides, there are some other concerns regarding efficient silencing, such as the complementarity of the small RNA to its target.

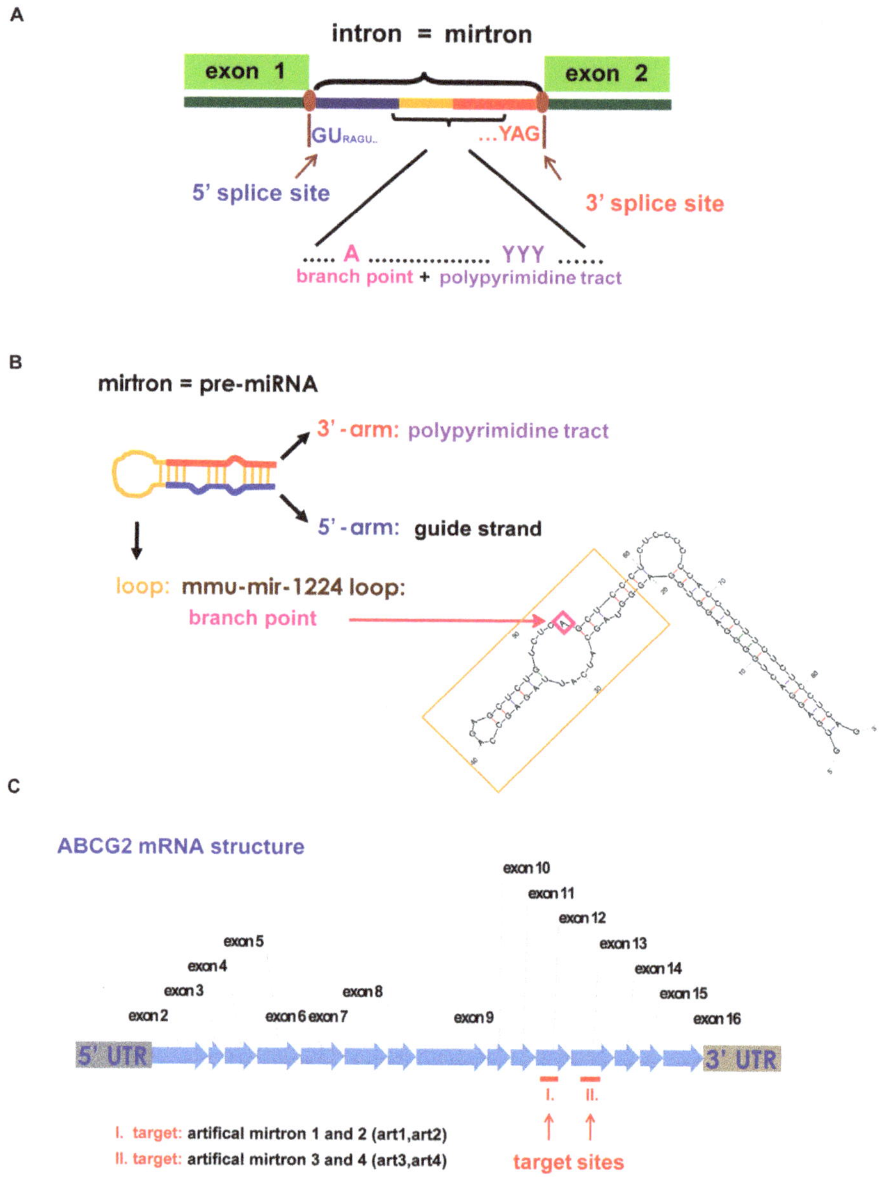

Figure 1. Artificial mirtron design. (**A**) A schematic representation of essential splicing criteria of introns in humans. Since mirtrons are small introns that are liberated by the splicing machinery instead of the Drosha/DGCR8 complex, their sequences should contain the indicated sequence motifs. (**B**) A demonstration of the artificial mirtron design. In the designed mirtronic pre-miRNAs, the 5′-arm gives the potential small guide RNA, and the loop region is originated from *Mus musculus* mir-1224 loop region containing the branch point, while the 3′-arm contains the polypyrimidine tract. The shown putative structure of the mmu-mir-1224 was predicted by the mFold program. (**C**) A schematic representation of the mRNA of ABCG2 membrane transporter protein. Target sites are shown by red, while the experimentally investigated targeting artificial mirtrons (arts) are also indicated.

Theoretically, the guide strand can be placed either in the 5′- or in the 3′-arm of a mirtron. However, in all cases, we chose the 5′-arm because, in this case, the most important part of the potential guide RNA, the 5′-end and therefore the seed region is well defined by splicing, avoiding potential heterogeneous ends, resulted by Dicer processing. Hence, it is easier to plan target specificity and influence strand selection. Concerning the branch point, we analyzed several mammalian mirtrons and found that some of them have their potential branch point in the loop region, while some have it in the 3′-arm region. We decided to position it in the loop region while the polypyrimidine tract was placed in the 3′-arm. We used the mmu-mir-1224 mirtron loop sequence as the loop of our artificial mirtrons since it had the best scores for branch point and splicing prediction analysis during the design process (Figure 1B). Regarding target site selection, we selected target sequences from the coding region of ABCG2 (Figure 1C) since it was previously shown to be applicable [40] and because our earlier effort to target 3′ UTR did not result in efficient silencing (data not shown). During the design process, we selected AC dinucleotides in the ABCG2 cDNA beside pyrimidine-rich sequences in the 5′ neighborhood to be the potential target of a 5′-arm derived mirtronic small RNA. Thus, the 5′-arm of the artificial mirtron (artmir) is complementary to the target site, the loop region contains the branch point, and the 3′-arm has the polypyrimidine tract (Figure 1B). We designed several sequence variants to test complementarity/silencing ability correlations and chose candidates for experimental investigations by bioinformatic analysis. Here we show four artificial mirtron variants targeting two constitutive exons as potential target sites: art1 and art2 for target I (residing in exon 12), and art3 and art4 for target II (residing in exon 13; Figure 1C).

3.2. *Investigating Splicing Ability of the Selected Artificial Mirtrons*

For the expression of artificial mirtrons, we used our earlier established expression system [39]. A modified EGFP sequence (EGFPm) was used, of which the coding region was separated into two exons. The artificial mirtrons were cloned as introns between the two exons. Therefore, EGFP fluorescence indicates accurate splicing, and artificial mirtron expression can be easily monitored (Figure 2A).

In the case of all four artmirs, we observed quite strong EGFP expression in the transfected cells, suggesting proper splicing (Figure 2B and Supplementary Figure S5). Investigation of splicing by RT-PCR indeed revealed successful, very efficient splicing. We detected a small amount of unspliced mRNA form only in the case of art1 (Figure 2C). Splicing accuracy was confirmed by sequencing of the gel-purified PCR products (Supplementary Figure S6). Our experimental results were consistent with our splicing predictions of the design phase. We chose sequences with very high values of splicing donor, acceptor and branch point predictions, and among them, art1 had the lowest values (see Supplementary Figures S1–S4).

Since the expression cassette in our plasmid was located inside a *Sleeping Beauty* transposon (Figure 2A), co-transfection with a transposase expressing plasmid allowed us to make stable cell lines by sorting the cells based on the EGFP signal. We successfully established all four artificial mirtron-expressing stable cell lines for further experiments.

3.3. *Functional Testing of Artificial Mirtrons by Luciferase Reporter Assay*

Since all of the examined artmirs could be effectively spliced out from the host gene, we tested their ability to silence gene expression. We used luciferase sensor assays, for which two copies of the particular target were cloned downstream of the *Renilla* luciferase coding region. Besides the fully complementary seed region containing sensor, we also used a mutant sensor bearing 3 mismatches in the seed region to check seed region specificity (Figure 3A).

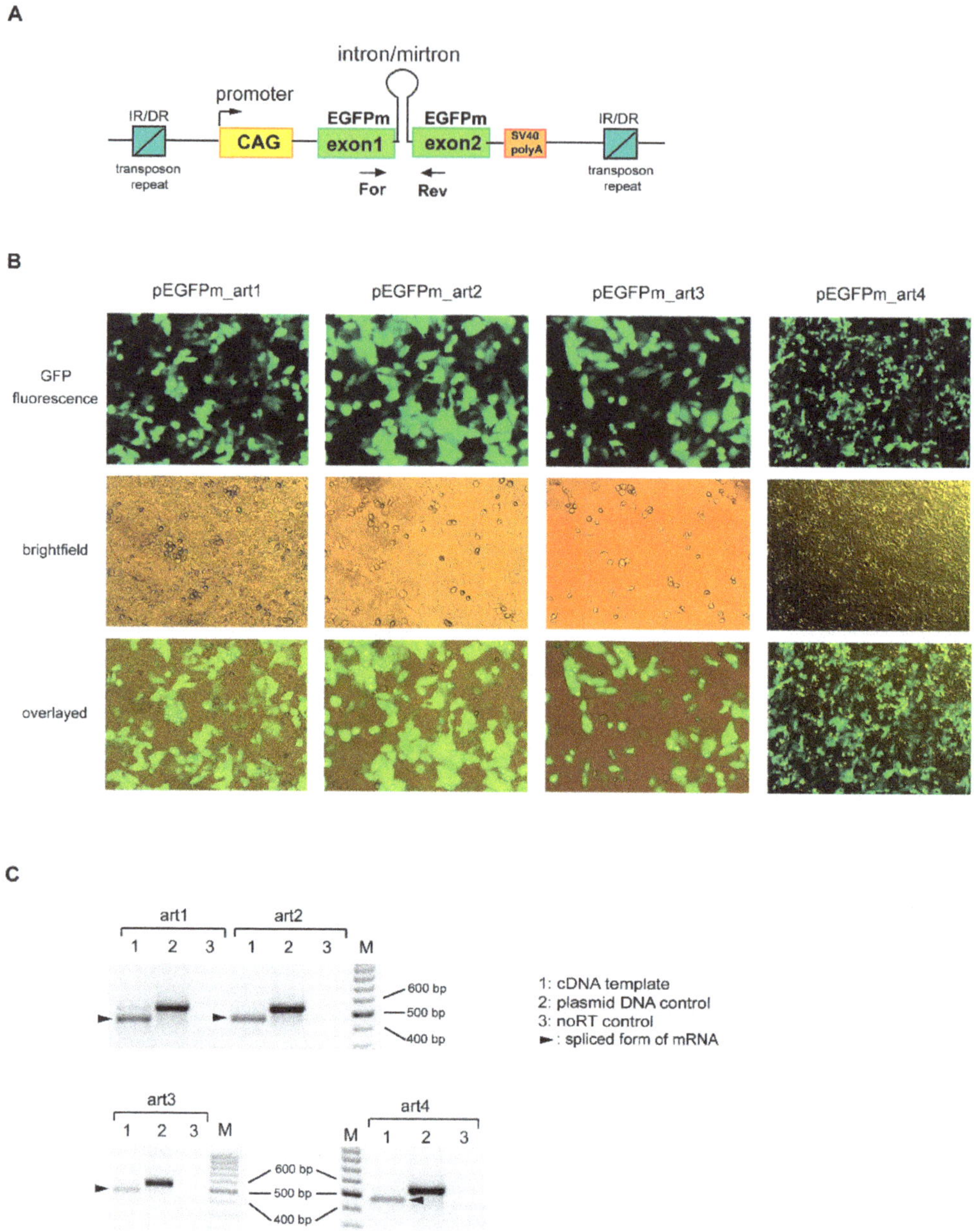

Figure 2. The splicing ability of selected artificial mirtrons (art1–4). (**A**) The *Sleeping Beauty* transposon constructs used for the expression of artificial mirtrons as EGFPm introns. Arrows indicate primers used for RT-PCR. IR/DR: inverted repeat/direct repeat sequences of the transposon. (**B**) Representative fluorescence microscopy images of HeLa cells expressing artmirs. EGFP positive cells indicate successful splicing. (**C**) RT-PCR results for the analysis of splicing. The spliced forms of RNAs are shown by black arrowheads. PCR products from plasmid DNA indicate the size of the unspliced mRNAs. M: DNA ladder as a marker.

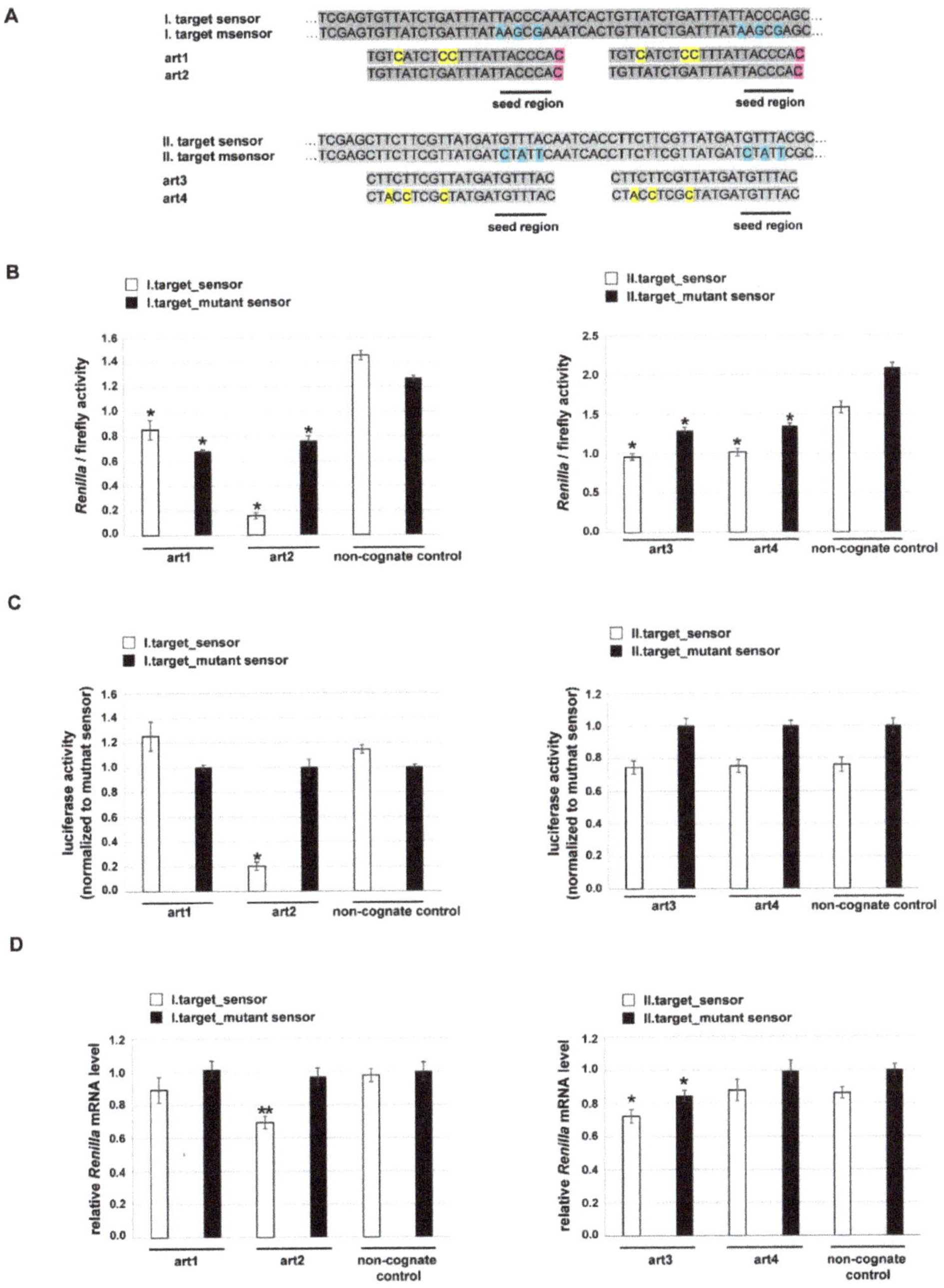

Figure 3. Testing the silencing ability of artificial mirtrons by luciferase assay. **(A)** DNA sequence alignment of luciferase targets and their respective artmirs (sense DNA sequences are shown). The psiCHECK2 vector-based sensor constructs contained two antisense copies of the particular target, cloned downstream of *Renilla* luciferase reporter. Sensors and mutant sensors (msensor) differ in 3 nucleotides in the potential seed region, indicated by the blue background. Art1 and art2 have

a mismatch at their first position to the target since an extra G (its complementary 'C' in DNA is indicated by purple background) was added to their 5′-end to fulfill mirtron criteria. Art3 is fully complementary to its sensor. Art1 and art4 have extra mismatches in their middle and 3′-end regions (indicated by yellow background), compared to their counterparts, art2 and art3, respectively. (**B**) Dual-luciferase assay measurements to assess the silencing ability of the designed artmirs. Mean values of at least 3 parallel experiments are shown, and error bars represent standard deviations; *: $p < 0.001$ relative to respective control. (**C**) To examine 'seed-specific' silencing, luciferase activity (*Renilla*/firefly) of sensor-containing experiments are normalized to the respective mutant sensor values (set to 1 for each mirtron). *: $p < 0.001$, relative to respective normalizing control. (**D**) Luciferase mRNA level measurements by qPCR. Error bars represent standard deviations; *: $p < 0.05$, **: $p < 0.01$, relative to respective control.

For target I., we detected downregulation of both sensor types in the case of both artmirs compared to a non-cognate control (Figure 3B left). Mutant sensors were silenced at a similar extent, by ~30%. Regarding the sensor (having a fully complementary seed region), both artmirs could achieve repression, but art2 had a much higher silencing capacity on it. The extent of the repression was ~43% for art1, whereas ~87% for art2. However, if we compare the downregulation of the sensor to the mutant sensor, we see a significant difference only in the case of art2 (Figure 3C left). The knockdown efficiency, in this case, was also high, more than 80%. For target II., we detected downregulation of both sensor types in the case of the corresponding artmirs compared to a non-cognate control (Figure 3B right). Silencing efficiencies were similar (~30%) among art3 and art4 in the case of both sensor types. However, a comparison of the sensor repression to the mutant sensor repression indicated no differences between artmirs and the non-cognate control (Figure 3C right).

As mentioned above, we designed artmirs to have different sequence complementarity to the target (Figure 3A), and we wanted to test whether there is a difference in their mechanism of silencing: is the observed luciferase repression realized by the cleavage/destabilization of the target mRNA or via translational repression? For this, we measured the mRNA level of *Renilla* luciferase by quantitative PCR. In the case of target I., no significant decrease could be detected in the mutant sensor containing *Renilla* luciferase mRNA. Regarding the sensor containing *Renilla* luciferase mRNA, we detected a significant decrease only in the case of art2, where a ~30% reduction was observed, compared to the control (Figure 3D left). Concerning target II., there was no change in either sensor or mutant sensor containing *Renilla* luciferase mRNA level when art4 was expressed. However, in the case of art3, we observed a slight, ~15% reduction of both sensor types (Figure 3D right).

3.4. Targeting ABCG2 Expression by Artificial Mirtrons

Next, we investigated the ability of the designed mirtrons to silence the expression of the human ABCG2 gene. First, we examined their impact on ABCG2 mRNA expression. Compared to the control, we observed a significant decrease, ~45% only in the case of art3 (Figure 4A).

Based on our previous luciferase experiments, we expected a reduction in the mRNA level also for art2 (see Figure 3D). However, the human ABCG2 gene has various polymorphisms compared to the reference sequence and sequencing the target site I. of our expression construct revealed the presence of one silent polymorphism (CCC > CCA, Pro480Pro). This extra mismatch in the 3rd nucleotide position of the seed region of art1 and art2 compared to the ABCG2 mRNA could well explain the results (Figure 4B).

Finally, we tested if the designed artmirs can influence ABCG2 expression at the protein level. For this, we carried out western blot experiments. We could detect a significant reduction in ABCG2 protein expression by art2 and a much less prominent decrease by art3. However, in the case of art1 and art4, carrying extra mismatches in the 3′ and the middle region of the miRNAs, we observed no significant changes compared to the control (Figure 4C,D).

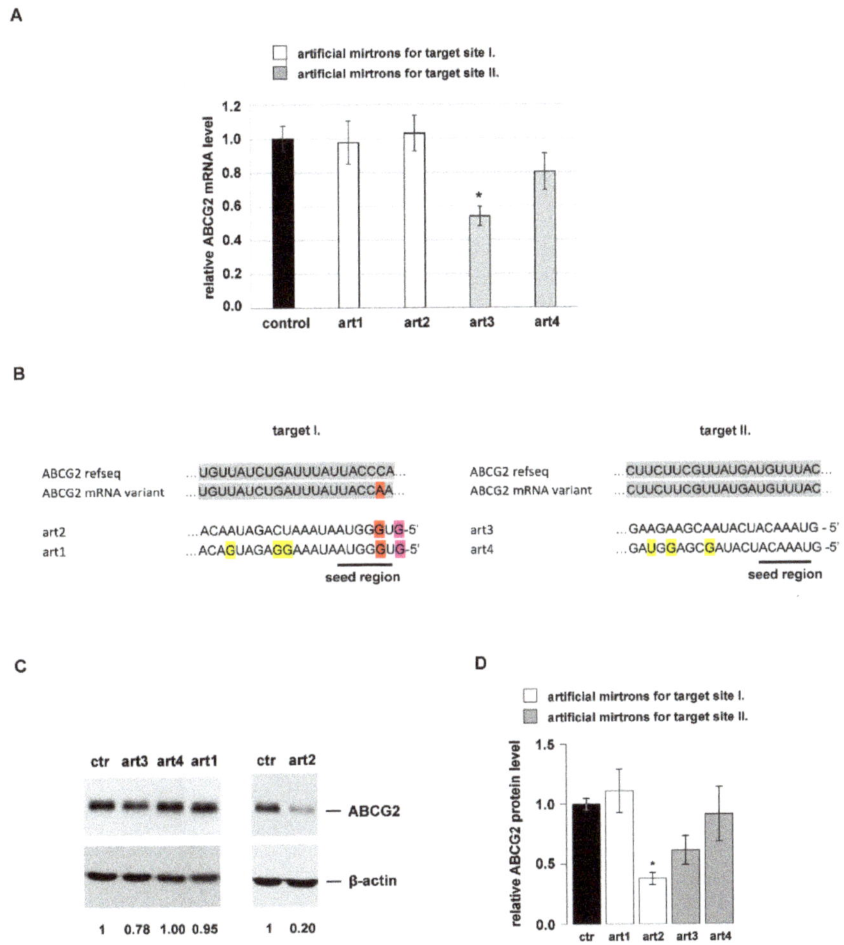

Figure 4. Targeting ABCG2 expression by artificial mirtrons. (**A**) Investigating the mirtron-induced silencing of ABCG2 mRNA by qPCR. Mean values are shown, and error bars represent standard deviations; *: $p < 0.005$ relative to respective control. (**B**) An illustration of artmirs complementary to the respective ABCG2 mRNA target site. Artmirs were designed based on the reference sequence (NM_004827.3); however, experiments were carried out with a common ABCG2 variant, therefore having a mismatch in target I. sequence (indicated by red background). Purple and yellow backgrounds indicate nucleotides, as in Figure 3A. (**C**) Examination of ABCG2 protein repression by artmirs, a representative western blot experiment is shown; normalized values of ABCG2 protein levels are shown under the gel image. (**D**) Relative protein levels measured after ABCG2 silencing by the artificial mirtron constructs. The mean values of 3 independent experiments are shown for artmirs, and the mean value of 5 independent experiments is shown for control (ctr). Error bars represent S.E.M., *: $p < 0.005$.

4. Discussion

In this study, we aimed to design artificial mirtrons to silence ABCG2 expression and investigate some sequential features that could influence efficient silencing. As was mentioned above, mirtrons can serve as useful tools for gene silencing, and they can be exploited in particular when genome editing is not amenable or silencing should be reversible. As artificial introns, they could be placed in various reporter genes, or for therapeutic applications, they may be combined with other genes of interest, achieving more than one genetic effect simultaneously with one expression cassette. Here, we present an artificial mirtron-based approach by which a significant silencing effect can be achieved on the ABCG2 multidrug transporter protein using the EGFP host protein. By developing and combining with appropriate Pol II promoters, it can serve as a useful tool for the investigation of ABCG2 function in various stem cells, including human embryonic stem cells and cells exhibiting the so-called "side-population phenotype" [6,8,52]. To date, there are only a few studies addressing the development of artificial conventional mirtrons to silence gene expression and their potential use in therapeutic applications. In those studies, the silencing effect of artmirs was investigated mostly in luciferase reporter assays and at the target mRNA level [40,41,43]. In a subsequent article, the potential application of 3′-tailed artificial mirtrons was studied, where in addition to the mRNA level, an efficient decrease could also be detected on the indirectly measured protein level of VEGFA [42]. Our data further strengthen the applicability of artificial mirtrons as gene silencers since our careful design could result in mirtrons efficiently reducing the ABCG2 expression when the target protein level was measured directly.

When examining the designed artmirs, interesting sequential features could be observed. We designed artmirs complementary to their targets or having mismatches at various positions. Using mutant sensors in luciferase assays revealed that 3 nucleotide mismatches in the seed region did not abolish silencing at the protein level since all four artmirs could have a silencing effect on the target, compared to the non-cognate control (Figure 3B). The extent of silencing was comparable to that measured on the sensor for art1, art3 and art4, indicating 'non-seed-specific' repression. However, in the case of art2, a strong 'seed-specific' silencing effect was observed (~80% reduction). Regarding mRNA levels, we detected a significant reduction in the *Renilla* mRNA level only in the case of art2 and art3. Art2 reduced its sensor mRNA level by ~30%, while art3 had a smaller effect but surprisingly on both sensor types (Figure 3D). Worth noting, that while art3 is fully complementary to its target, art2 has one mismatch outside the seed region, at the first position, due to the mirtron design rule (having G at the 5′-end). However, despite this mismatch, art2 decreased its sensor mRNA level and more extensively than art3. Nevertheless, when the target is located in the original genomic context, the ability of art2 to reduce the ABCG2 mRNA level was abolished by an additional mismatch positioned in the seed region (3rd position, Figure 4B). In summary, data of the luciferase experiments suggest that art1 and art4 silenced their targets via translational repression, while art2 and art3 could accelerate the degradation of their target mRNA to some extent, even if having mismatches to the target (1st nucleotide of art2 in its sensor, or seed mismatches in mutant sensor of art3).

Concerning ABCG2, only art3 repressed its mRNA level (~45%), but it only resulted in a slight reduction of the amount of protein. Conversely, art2 had no impact on mRNA level but exhibited a quite strong repression at the protein level. In contrast to these, art1 and art4 had no effect on either ABCG2 mRNA or protein level. Considering the sequence environment, ABCG2 mRNA has one, while *Renilla* mRNA has two target sites; however, art3 can regulate the former one more efficiently (~45% versus 15%). Nevertheless, it is worth noting, that in ABCG2 mRNA, the target sequence resides in the cDNA region instead of the 3′ UTR. The results indicate that flanking sequences could strongly influence the miRNA effect. In the natural mRNA context, art3 achieves a slight decrease in ABCG2 protein level, most probably by degrading its mRNA, while art2 operates only by translational repression.

Regarding our data, we noticed that in the case of the 'miRNA-mimic' artmirs (art1 and art4) containing mismatches in the 3′ and the middle region, losing base pairing at the investigated positions did not accelerate silencing either at luciferase or at the ABCG2 protein level, compared to their respective counterparts (art2 and art3). In addition, we also noticed that the presence of rare polymorphisms in the target region should also be considered since they could influence base pairing and thereby efficient silencing of the designed mirtrons (Figure 4B). On the other hand, some other sequential features can be very useful during artificial mirtron design: for example, our data support the possibility of adding a non-complementary G to the 5′-end of a mirtron without decreasing silencing ability, which is very important and useful since it is a strong mirtron criterion [35,36].

Taken together, some of our results using artificial mirtrons are in line with earlier data, such as the reduction of the target mRNA level in the case of full complementarity between the target and the small RNA [29,53]. However, we observed some additional, not expected features, e.g., a reduction in the target mRNA level when the first nucleotide of the small RNA is not complementary. Further experiments are needed to reveal whether these findings are common phenomena or a consequence of the given target sequence and/or its context, which may have different accessibility by the RISC. Another explanation could be an altered RISC assembly when the various small RNA guides are preferentially associated with different Argonaute proteins. Notably, the most successful silencer artmir of LRRK2 was associated with the greatest amount to AGO4 [41].

5. Conclusions

In summary, using our artificial mirtron design and testing scheme, we could successfully establish an efficient silencing system for the ABCG2 multidrug transporter. In addition, we observed important new sequential-functional features of the designed mirtrons. Our silencing system could be directly applied to study the function of this membrane protein in several in vitro or in vivo models. Moreover, combining the artmirs with host proteins other than EGFP, this system would also be suitable for versatile, functional studies in stem cells, where ABCG2 plays an important yet not fully understood role. However, apart from the concrete established model system, we believe that our mirtron design pipeline could also be efficiently applied to target other genes in future studies.

Supplementary Materials: The following are available online at https://www.mdpi.com/2073-4425/12/7/1068/s1, Figure S1: Predicted splicing parameters and structure of artificial mirtron 1 (art1), Figure S2: Predicted splicing parameters and structure of artificial mirtron 2 (art2), Figure S3: Predicted splicing parameters and structure of artificial mirtron 3 (art3), Figure S4: Predicted splicing parameters and structure of artificial mirtron 4 (art4), Figure S5. Microscopy images of untransfected HeLa cells, Figure S6: Sequence alignment of RT-PCR products of spliced mRNAs versus their unspliced form.

Author Contributions: Conceptualization, A.S.; methodology, A.S.; formal analysis, A.S.; investigation, A.S., G.V. and Á.F.; writing—original draft preparation, A.S.; writing—review and editing, T.I.O. and A.S.; funding acquisition, T.I.O. and A.S. All authors have read and agreed to the published version of the manuscript.

Funding: This research was funded by grants from the National Research, Development and Innovation Office, Hungary: NKFIH-OTKA, grant numbers: PD121287 and FK124661 to A.S., and VEKOP-2.1.1-15-2016-00156, VEKOP-2.3.3-15-2017-00014 and project no. 2018-1.2.1-NKP-2018-00005 to T.I.O.

Institutional Review Board Statement: Not applicable.

Informed Consent Statement: Not applicable.

Data Availability Statement: All data are included in the manuscript or in the supplementary materials.

Acknowledgments: The authors would like to thank Kornélia Némethy for excellent technical assistance and Ágota Apáti for the help with microscopy.

Conflicts of Interest: The authors declare no conflict of interest.

References

1. Doyle, L.A.; Yang, W.; Abruzzo, L.V.; Krogmann, T.; Gao, Y.; Rishi, A.K.; Ross, D.D. A multidrug resistance transporter from human MCF-7 breast cancer cells. *Proc. Natl. Acad. Sci. USA* **1998**, *95*, 15665–15670. [CrossRef]

2. Allikmets, R.; Schriml, L.M.; Hutchinson, A.; Romano-Spica, V.; Dean, M. A human placenta-specific ATP-binding cassette gene (ABCP) on chromosome 4q22 that is involved in multidrug resistance. *Cancer Res.* **1998**, *58*, 5337–5339.
3. Miyake, K.; Mickley, L.; Litman, T.; Zhan, Z.; Robey, R.; Cristensen, B.; Brangi, M.; Greenberger, L.; Dean, M.; Fojo, T.; et al Molecular cloning of cDNAs which are highly overexpressed in mitoxantrone-resistant cells: Demonstration of homology to ABC transport genes. *Cancer Res.* **1999**, *59*, 8–13.
4. Krishnamurthy, P.; Schuetz, J.D. Role of ABCG2/BCRP in biology and medicine. *Annu. Rev. Pharmacol. Toxicol.* **2006**, *46*, 381–410 [CrossRef]
5. Sarkadi, B.; Homolya, L.; Szakacs, G.; Varadi, A. Human multidrug resistance ABCB and ABCG transporters: Participation in a chemoimmunity defense system. *Physiol. Rev.* **2006**, *86*, 1179–1236. [CrossRef]
6. Apati, A.; Orban, T.I.; Varga, N.; Nemeth, A.; Schamberger, A.; Krizsik, V.; Erdelyi-Belle, B.; Homolya, L.; Varady, G., Padanyi, R.; et al. High level functional expression of the ABCG2 multidrug transporter in undifferentiated human embryonic stem cells. *Biochim. Biophys. Acta* **2008**, *1778*, 2700–2709. [CrossRef]
7. Apati, A.; Szebenyi, K.; Erdei, Z.; Varady, G.; Orban, T.I.; Sarkadi, B. The importance of drug transporters in human pluripotent stem cells and in early tissue differentiation. *Expert Opin. Drug Metab. Toxicol.* **2016**, *12*, 77–92. [CrossRef] [PubMed]
8. Zhou, S.; Schuetz, J.D.; Bunting, K.D.; Colapietro, A.M.; Sampath, J.; Morris, J.J.; Lagutina, I.; Grosveld, G.C.; Osawa, M.; Nakauchi, H.; et al. The ABC transporter Bcrp1/ABCG2 is expressed in a wide variety of stem cells and is a molecular determinant of the side-population phenotype. *Nat. Med.* **2001**, *7*, 1028–1034. [CrossRef] [PubMed]
9. Hegedus, C.; Szakacs, G.; Homolya, L.; Orban, T.I.; Telbisz, A.; Jani, M.; Sarkadi, B. Ins and outs of the ABCG2 multidrug transporter: An update on in vitro functional assays. *Adv. Drug Deliv. Rev.* **2009**, *61*, 47–56. [CrossRef]
10. Gameiro, M.; Silva, R.; Rocha-Pereira, C.; Carmo, H.; Carvalho, F.; Bastos, M.L.; Remiao, F. Cellular models and in vitro assays for the screening of MODULATORS of P-gp, MRP1 and BCRP. *Molecules* **2017**, *22*, 600. [CrossRef]
11. Kovacsics, D.; Brozik, A.; Tihanyi, B.; Matula, Z.; Borsy, A.; Meszaros, N.; Szabo, E.; Nemeth, E.; Fothi, A.; Zambo, B.; et al. Precision-engineered reporter cell lines reveal ABCG2 regulation in live lung cancer cells. *Biochem. Pharmacol.* **2020**, *175*, 113865. [CrossRef]
12. Lee, Y.; Kim, M.; Han, J.; Yeom, K.H.; Lee, S.; Baek, S.H.; Kim, V.N. MicroRNA genes are transcribed by RNA polymerase II. *EMBO J.* **2004**, *23*, 4051–4060. [CrossRef]
13. Cai, X.; Hagedorn, C.H.; Cullen, B.R. Human microRNAs are processed from capped, polyadenylated transcripts that can also function as mRNAs. *RNA* **2004**, *10*, 1957–1966. [CrossRef] [PubMed]
14. Denli, A.M.; Tops, B.B.; Plasterk, R.H.; Ketting, R.F.; Hannon, G.J. Processing of primary microRNAs by the Microprocessor complex. *Nature* **2004**, *432*, 231–235. [CrossRef] [PubMed]
15. Gregory, R.I.; Yan, K.P.; Amuthan, G.; Chendrimada, T.; Doratotaj, B.; Cooch, N.; Shiekhattar, R. The Microprocessor complex mediates the genesis of microRNAs. *Nature* **2004**, *432*, 235–240. [CrossRef] [PubMed]
16. Han, J.; Lee, Y.; Yeom, K.H.; Kim, Y.K.; Jin, H.; Kim, V.N. The Drosha-DGCR8 complex in primary microRNA processing. *Genes Dev.* **2004**, *18*, 3016–3027. [CrossRef] [PubMed]
17. Landthaler, M.; Yalcin, A.; Tuschl, T. The human DiGeorge syndrome critical region gene 8 and Its D. melanogaster homolog are required for miRNA biogenesis. *Curr. Biol.* **2004**, *14*, 2162–2167. [CrossRef] [PubMed]
18. Yi, R.; Qin, Y.; Macara, I.G.; Cullen, B.R. Exportin-5 mediates the nuclear export of pre-microRNAs and short hairpin RNAs. *Genes Dev.* **2003**, *17*, 3011–3016. [CrossRef] [PubMed]
19. Lund, E.; Guttinger, S.; Calado, A.; Dahlberg, J.E.; Kutay, U. Nuclear export of microRNA precursors. *Science* **2004**, *303*, 95–98. [CrossRef]
20. Bohnsack, M.T.; Czaplinski, K.; Gorlich, D. Exportin 5 is a RanGTP-dependent dsRNA-binding protein that mediates nuclear export of pre-miRNAs. *RNA* **2004**, *10*, 185–191. [CrossRef]
21. Okada, C.; Yamashita, E.; Lee, S.J.; Shibata, S.; Katahira, J.; Nakagawa, A.; Yoneda, Y.; Tsukihara, T. A high-resolution structure of the pre-microRNA nuclear export machinery. *Science* **2009**, *326*, 1275–1279. [CrossRef]
22. Bernstein, E.; Caudy, A.A.; Hammond, S.M.; Hannon, G.J. Role for a bidentate ribonuclease in the initiation step of RNA interference. *Nature* **2001**, *409*, 363–366. [CrossRef]
23. Ketting, R.F.; Fischer, S.E.; Bernstein, E.; Sijen, T.; Hannon, G.J.; Plasterk, R.H. Dicer functions in RNA interference and in synthesis of small RNA involved in developmental timing in *C. elegans*. *Genes Dev.* **2001**, *15*, 2654–2659. [CrossRef]
24. Park, J.E.; Heo, I.; Tian, Y.; Simanshu, D.K.; Chang, H.; Jee, D.; Patel, D.J.; Kim, V.N. Dicer recognizes the 5′ end of RNA for efficient and accurate processing. *Nature* **2011**, *475*, 201–205. [CrossRef]
25. Tian, Y.; Simanshu, D.K.; Ma, J.B.; Park, J.E.; Heo, I.; Kim, V.N.; Patel, D.J. A phosphate-binding pocket within the platform-PAZ-connector helix cassette of human Dicer. *Mol. Cell* **2014**, *53*, 606–616. [CrossRef] [PubMed]
26. Macrae, I.J.; Zhou, K.; Li, F.; Repic, A.; Brooks, A.N.; Cande, W.Z.; Adams, P.D.; Doudna, J.A. Structural basis for double-stranded RNA processing by Dicer. *Science* **2006**, *311*, 195–198. [CrossRef] [PubMed]
27. MacRae, I.J.; Zhou, K.; Doudna, J.A. Structural determinants of RNA recognition and cleavage by Dicer. *Nat. Struct. Mol. Biol.* **2007**, *14*, 934–940. [CrossRef] [PubMed]
28. Suzuki, H.I.; Katsura, A.; Yasuda, T.; Ueno, T.; Mano, H.; Sugimoto, K.; Miyazono, K. Small-RNA asymmetry is directly driven by mammalian Argonautes. *Nat. Struct. Mol. Biol.* **2015**, *22*, 512–521. [CrossRef] [PubMed]

29. Huntzinger, E.; Izaurralde, E. Gene silencing by microRNAs: Contributions of translational repression and mRNA decay. *Nat. Rev. Genet.* **2011**, *12*, 99–110. [CrossRef]
30. Pasquinelli, A.E. MicroRNAs and their targets: Recognition, regulation and an emerging reciprocal relationship. *Nat. Rev. Genet.* **2012**, *13*, 271–282. [CrossRef]
31. Miyoshi, K.; Miyoshi, T.; Siomi, H. Many ways to generate microRNA-like small RNAs: Non-canonical pathways for microRNA production. *Mol. Genet. Genom.* **2010**, *284*, 95–103. [CrossRef]
32. Yang, J.S.; Lai, E.C. Alternative miRNA biogenesis pathways and the interpretation of core miRNA pathway mutants. *Mol. Cell* **2011**, *43*, 892–903. [CrossRef] [PubMed]
33. Curtis, H.J.; Sibley, C.R.; Wood, M.J. Mirtrons, an emerging class of atypical miRNA. *Wiley Interdiscip. Rev. RNA* **2012**, *3*, 617–632. [CrossRef] [PubMed]
34. Daugaard, I.; Hansen, T.B. Biogenesis and function of ago-associated RNAs. *Trends Genet.* **2017**, *33*, 208–219. [CrossRef]
35. Ruby, J.G.; Jan, C.H.; Bartel, D.P. Intronic microRNA precursors that bypass Drosha processing. *Nature* **2007**, *448*, 83–86. [CrossRef]
36. Okamura, K.; Hagen, J.W.; Duan, H.; Tyler, D.M.; Lai, E.C. The mirtron pathway generates microRNA-class regulatory RNAs in Drosophila. *Cell* **2007**, *130*, 89–100. [CrossRef]
37. Sibley, C.R.; Seow, Y.; Saayman, S.; Dijkstra, K.K.; El Andaloussi, S.; Weinberg, M.S.; Wood, M.J. The biogenesis and characterization of mammalian microRNAs of mirtron origin. *Nucleic Acids Res.* **2012**, *40*, 438–448. [CrossRef]
38. Havens, M.A.; Reich, A.A.; Duelli, D.M.; Hastings, M.L. Biogenesis of mammalian microRNAs by a non-canonical processing pathway. *Nucleic Acids Res.* **2012**, *40*, 4626–4640. [CrossRef]
39. Schamberger, A.; Sarkadi, B.; Orban, T.I. Human mirtrons can express functional microRNAs simultaneously from both arms in a flanking exon-independent manner. *RNA Biol.* **2012**, *9*, 1177–1185. [CrossRef]
40. Seow, Y.; Sibley, C.R.; Wood, M.J. Artificial mirtron-mediated gene knockdown: Functional DMPK silencing in mammalian cells. *RNA* **2012**, *18*, 1328–1337. [CrossRef] [PubMed]
41. Sibley, C.R.; Seow, Y.; Curtis, H.; Weinberg, M.S.; Wood, M.J. Silencing of Parkinson's disease-associated genes with artificial mirtron mimics of miR-1224. *Nucleic Acids Res.* **2012**, *40*, 9863–9875. [CrossRef]
42. Kock, K.H.; Kong, K.W.; Hoon, S.; Seow, Y. Functional VEGFA knockdown with artificial 3′-tailed mirtrons defined by 5′ splice site and branch point. *Nucleic Acids Res.* **2015**, *43*, 6568–6578. [CrossRef]
43. Curtis, H.J.; Seow, Y.; Wood, M.J.A.; Varela, M.A. Knockdown and replacement therapy mediated by artificial mirtrons in spinocerebellar ataxia 7. *Nucleic Acids Res.* **2017**, *45*, 7870–7885. [CrossRef]
44. Burset, M.; Seledtsov, I.A.; Solovyev, V.V. Analysis of canonical and non-canonical splice sites in mammalian genomes. *Nucleic Acids Res.* **2000**, *28*, 4364–4375. [CrossRef]
45. Burset, M.; Seledtsov, I.A.; Solovyev, V.V. SpliceDB: Database of canonical and non-canonical mammalian splice sites. *Nucleic Acids Res.* **2001**, *29*, 255–259. [CrossRef]
46. Desmet, F.O.; Hamroun, D.; Lalande, M.; Collod-Beroud, G.; Claustres, M.; Beroud, C. Human Splicing Finder: An online bioinformatics tool to predict splicing signals. *Nucleic Acids Res.* **2009**, *37*, e67. [CrossRef] [PubMed]
47. Zuker, M. Mfold web server for nucleic acid folding and hybridization prediction. *Nucleic Acids Res.* **2003**, *31*, 3406–3415. [CrossRef] [PubMed]
48. Schamberger, A.; Orban, T.I. Experimental validation of predicted mammalian microRNAs of mirtron origin. *Methods Mol. Biol.* **2014**, *1182*, 245–263. [CrossRef]
49. Lacy-Hulbert, A.; Thomas, R.; Li, X.P.; Lilley, C.E.; Coffin, R.S.; Roes, J. Interruption of coding sequences by heterologous introns can enhance the functional expression of recombinant genes. *Gene Ther.* **2001**, *8*, 649–653. [CrossRef] [PubMed]
50. Orban, T.I.; Seres, L.; Ozvegy-Laczka, C.; Elkind, N.B.; Sarkadi, B.; Homolya, L. Combined localization and real-time functional studies using a GFP-tagged ABCG2 multidrug transporter. *Biochem. Biophys. Res. Commun.* **2008**, *367*, 667–673. [CrossRef]
51. Orban, T.I.; Apati, A.; Nemeth, A.; Varga, N.; Krizsik, V.; Schamberger, A.; Szebenyi, K.; Erdei, Z.; Varady, G.; Karaszi, E.; et al. Applying a "double-feature" promoter to identify cardiomyocytes differentiated from human embryonic stem cells following transposon-based gene delivery. *Stem Cells* **2009**, *27*, 1077–1087. [CrossRef] [PubMed]
52. Sarkadi, B.; Orban, T.I.; Szakacs, G.; Varady, G.; Schamberger, A.; Erdei, Z.; Szebenyi, K.; Homolya, L.; Apati, A. Evaluation of ABCG2 expression in human embryonic stem cells: Crossing the same river twice? *Stem Cells* **2010**, *28*, 174–176. [CrossRef] [PubMed]
53. Elbashir, S.M.; Harborth, J.; Lendeckel, W.; Yalcin, A.; Weber, K.; Tuschl, T. Duplexes of 21-nucleotide RNAs mediate RNA interference in cultured mammalian cells. *Nature* **2001**, *411*, 494–498. [CrossRef] [PubMed]

genes

MDPI

Article

miRBind: A Deep Learning Method for miRNA Binding Classification

Eva Klimentová [1,†], Václav Hejret [1,2,†], Ján Krčmář [3], Katarína Grešová [1,2], Ilektra-Chara Giassa [1,*] and Panagiotis Alexiou [1]

1 Central European Institute of Technology (CEITEC), Masaryk University, 60177 Brno, Czech Republic
2 Faculty of Science, National Centre for Biomolecular Research, Masaryk University, 61137 Brno, Czech Republic
3 Faculty of Informatics, Masaryk University, 60200 Brno, Czech Republic
* Correspondence: igiassa@mail.muni.cz
† These authors contributed equally to this work.

Abstract: The binding of microRNAs (miRNAs) to their target sites is a complex process, mediated by the Argonaute (Ago) family of proteins. The prediction of miRNA:target site binding is an important first step for any miRNA target prediction algorithm. To date, the potential for miRNA:target site binding is evaluated using either co-folding free energy measures or heuristic approaches, based on the identification of binding 'seeds', i.e., continuous stretches of binding corresponding to specific parts of the miRNA. The limitations of both these families of methods have produced generations of miRNA target prediction algorithms that are primarily focused on 'canonical' seed targets, even though unbiased experimental methods have shown that only approximately half of in vivo miRNA targets are 'canonical'. Herein, we present miRBind, a deep learning method and web server that can be used to accurately predict the potential of miRNA:target site binding. We trained our method using seed-agnostic experimental data and show that our method outperforms both seed-based approaches and co-fold free energy approaches. The full code for the development of miRBind and a freely accessible web server are freely available.

Keywords: miRNA:target prediction; miRNA binding; CLASH; convolutional neural network

1. Introduction

miRNAs are endogenous small (~17–25 nucleotides long) ncRNAs that negatively regulate gene expression at the level of messenger RNA (mRNA) [1]. The first miRNA (lin-4) was discovered in *Caenorhabditis elegans* in 1993 [2,3]; in humans, the first miRNA that was discovered is let-7, first identified in 2000 in *C. elegans* [4]. To date, 2654 mature human miRNAs have been deposited in the miRbase [5] miRNA database. miRNAs are processed from hairpin-containing primary transcripts (pri-miRNAs); they are subsequently processed into precursor miRNAs (pre-miRNAs) [6], exported to the cytoplasm [7], and cleaved into small double-stranded RNAs [8,9]. The mature miRNA duplex is then loaded into an argonaute (AGO) protein to form a miRNA-induced silencing complex (miRISC). Mature miRNAs interact with the AGO proteins and guide them, via base pairing, toward target RNAs. Such targeting may lead to translational repression and deadenylation-induced mRNA degradation [10,11]. Each miRNA can have thousands of binding sites on the transcriptome, and an mRNA can contain dozens of potential miRNA binding sites [12]. Animal miRNAs occasionally show extensive, but more often only partial, complementarity with their target sites [13,14]. The 5′ end of the miRNA, and especially the hexamer-spanning nucleotides 2–8, were identified very early on as being important for miRNA target recognition and were termed the 'seed' region [15]. Target recognition is primarily achieved via base pairing that involves the seed region [16]; however, seed pairing is not always sufficient for functional target interactions, and additional interactions with the miRNA 3′ end may be necessary for specific targeting [17]. Targeting can also be facilitated by additional sequence elements, such as an unpaired adenosine in the 5′-end of the miRNA [18]. An estimated quantification of non-canonical miRNA binding sites calculates that approximately 60% of all identified interactions are based on non-canonical seeds [19,20].

Early approaches to miRNA target prediction implemented additional features, such as the evolutionary conservation levels of targets, positioning of target sites on 3′-UTRs, nucleotide content of targets, and others [16]. Another family of miRNA target prediction methods utilized alignment or co-fold methodologies, ignoring the 'seed' region [21–23]. In these approaches, an idealized structure is calculated, based on the affinity of the miRNA sequence to the putative target sequence; then, measures such as alignment score or the free energy of binding of the two molecules are used to score binding potential. When the first high-throughput miRNA targeting datasets became available [24,25], 'seed'-based approaches appeared to outperform the 'folding'-based methods on all benchmarks of precision and sensitivity [26]. The following years saw a wealth of high-throughput miRNA targeting data being produced, utilizing techniques in the CLIP-Seq (cross-linking immunoprecipitation sequencing) [27] family, which identified thousands of miRNA-target mRNA pairs [28]. CLIP is based on the stabilization of protein-RNA complexes in their cellular environment by UV cross-linking, the immunoprecipitation of ribonucleoprotein complexes (RNPs), and the isolation and sequencing of bound RNAs. An important limitation of such techniques is that they do not produce specific miRNA:target site pairs. Instead, they produce peaks of Ago protein-binding to which miRNAs need to be assigned, often using a miRNA target prediction program that utilizes the 'seed' heuristic. Even though more functional non-canonical 'seed'-binding sites are being continuously discovered, they remain underrepresented by all miRNA prediction programs and databases of validated miRNA targets.

CLASH (cross-linking, ligation, and sequencing of hybrids) is the first reported high-throughput method for the direct identification of RNA–RNA interactions [18] The CLASH method allows for the precise mapping of interactions for which the downstream consequences are unknown and/or difficult to measure. Because cross-linking is performed in living cells, the dynamic state of the RNA interactome can be probed as a function of physiological conditions. It offers two types of information: precise AGO-binding sites on RNAs (similar to CLIP methods), and RNA–RNA hybrids that are formed within the AGO RNA-binding pocket [29]. In 2013, the first unbiased experimental method for the identification of Ago1 miRNA binding sites in cell culture was performed, using the CLASH technique. The study showed that approximately 60% of identified miRNA:target duplexes contain a non-canonical 'seed', while 18% of targets show binding in the 3′ end of the miRNA without any 'seed' binding [20]. In this seminal paper, the authors used a miRNA:target site 'co-fold free energy' approach to predict the type of binding. Subsequent studies utilized CLASH and CLEAR (covalent ligation of endogenous Argonaute-bound RNAs) [30] techniques to identify more miRNA:target site pairs, and solidify the abundance and functionality of 'non-seed' target sites. Despite this well-documented functionality of the 'non-seed' target sites, the vast majority of miRNA target prediction programs today still use the 'seed' heuristic as a first filtering step [31].

As the studies on miRNA:target binding rules present researchers with new challenges, advances in computational approaches, including machine learning (ML), have gained great significance. ML techniques have been applied to predict miRNA targets and there have been several reviews covering advancements in the field [32–34]; however, ML performance largely depends on the user-defined variables that were selected to train the model. Among the early adopters of ML approaches for miRNA target prediction are miRanda-mirSVR [35], DIANA-micro-T-CDS [36–38], mirTarget2 [39], TargetMiner [40], and SVMicrO [41]. miRanda-mirSVR incorporates support vector regression (SVR), while DIANA-micro-T-CDS utilizes generalized linear models (GLM); the rest of the mentioned classifiers are based on support vector machines (SVM). Deep learning (DL), an emerging field of ML, solves the issue of handcrafted features by embedding the computation of these features into the ML model itself [42]. Thus, DL is highly appropriate for uncovering the miRNA binding rules, where clear rules or features are unknown. In the past decade, Deep Neural Networks (DNN) [43] have found extensive use in many scientific fields, including bioinformatics [44]. Convolutional neural networks (CNNs) are a subtype of DNN that utilize several layers of convolutional neurons to learn increasingly complicated representations of input data. Input data is provided in a raw format, allowing the CNN to learn what patterns in the input data are important for a specific task. Selecting an appropriate training dataset and suitable evaluation metrics are of pivotal importance when building an effective DL model.

ResNet [45] is another type of DNN that uses an innovative architecture that enables the training of very deep models. They address the problem of vanishing gradient and help with optimization by utilizing a special kind of residual block that adds a skip connection. ResNet is used as a backbone for many computer vision tasks.

An important factor to be addressed by the miRNA target prediction methods that are based on classifiers is class imbalance. Each mRNA can be regulated by dozens of miRNAs and each miRNA has thousands of potential binding sites on the transcriptome [12]. The imbalance between the number of actual, experimentally verified binding sites (positive class) and all other regions on the transcriptome (negative class) has a significant, deteriorating effect on the performance of the prediction methods.

Since class imbalance is identified as the leading challenge influencing the performance of any prediction model, we propose miRBind, a novel method dealing with this issue, which is based on targeted sample selection and subsequent label smoothing. miRBind is a ResNet-based method trained on unbiased miRNA:target site CLASH data [20] and is shown to consistently outperform both 'seed' and 'co-fold' approaches in a binding-site classification task. We also provide an alternative CNN approach, a six-convolutional-layer network, with comparable performance. For ease of access, we provide a standalone Python program, as well as a freely available web server with a user-friendly interface (GitHub repository https://github.com/ML-Bioinfo-CEITEC/miRBind accessed on 28 November 2022, web server https://ml-bioinfo-ceitec.github.io/miRBind/ accessed on 28 November 2022).

2. Materials and Methods

2.1. Data Preparation

We retrieved the positive miRNA:target interaction dataset from the original CLASH study published by Helwak et al. [20]. We produced the negative dataset by randomly matching the miRNAs and target sites found in the positive dataset. A detailed explanation of the dataset's production follows.

The positive dataset was constructed from the miRNA:target interactions identified by Helwak et al. in 2013, via their CLASH [20] experiment. The interacting miRNA:target pairs were downloaded and processed by standardizing the length of the miRNA sequences to 20 nt, anchored at the 5′ end of the miRNA, and centering and resizing the target coordinates to the window length of 50 bp. The resized target sequences were extracted using bedtools and the hyb reference [46] (https://github.com/gkudla/hyb/tree/master/data/db (accessed on 14 November 2021)). These processed miRNA:target pairs are known as the positive dataset. The positive dataset was divided into training, testing, and validation sets containing 15,392, 2000, and 1000 miRNA:target pairs.

Negative sets were constructed by excluding the interacting miRNA:mRNA partners provided by the 2013 CLASH experiment. More specifically, we created the negative datasets by matching real target sequences with random miRNAs from the same experiment, excluding the original positive set. We have elected to adopt this approach instead of choosing other parts of the genome/transcriptome as 'fake' targets, so as to avoid introducing any nucleotide content or other biases. We believe that shuffling the miRNA:target pairs is a fair and realistic approach that matches the way in which miRNAs may be assigned to Ago-CLIP peaks. To create the final datasets, positive and negative dataset parts were combined for the training and validation sets for the positive:negative ratios of 1:1, 1:10,

1:20, and 1:100. The CNN models were trained on all ratios, while miRBind was trained on the 1:1, 1:10 and 1:20 ratios, and DNABERT was fine-tuned on the 1:1 and 1:10 training sets. Testing sets were constructed for the positive:negative ratios of 1:1, 1:10 and 1:100.

To avoid any potential target sequence or experimental biases, we elected to completely hide the sequences from the convolutional neural network during training, instead using a two-dimensional representation of miRNA and putative target sequence, in which any Watson–Crick binding nucleotide pair is represented by 1, and any non-binding pair by 0. This creates a 20 × 50 two-dimensional matrix of 1 s and 0 s, which is the input for our training method (Figure 1A).

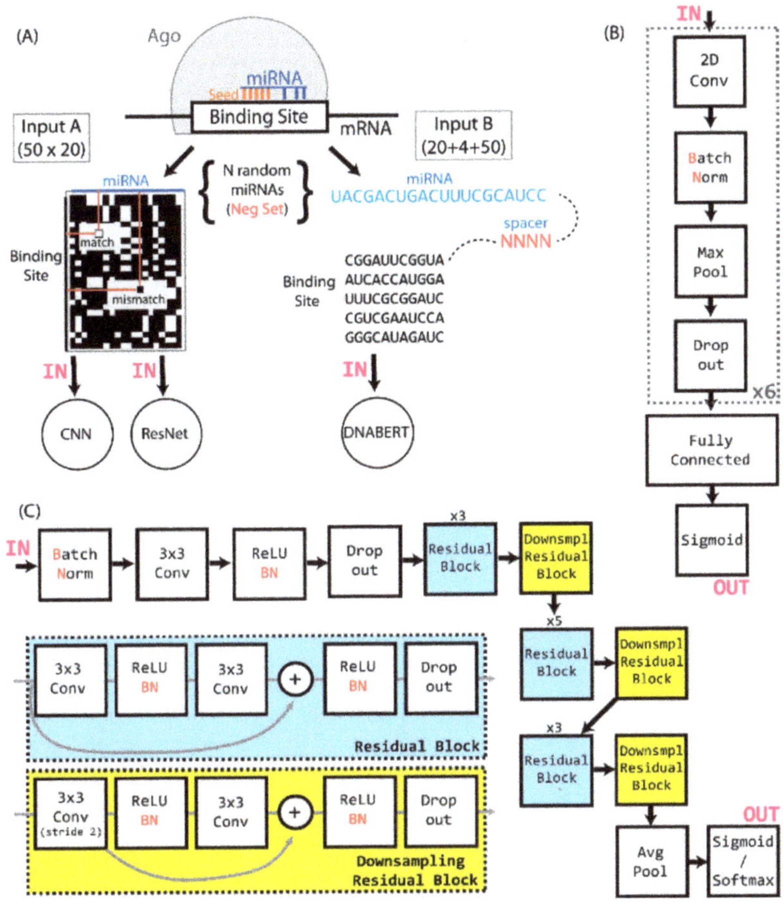

Figure 1. (**A**) Ago binding sites that are identified by CLASH are converted to two-dimensional arrays of matching and mismatching nucleotides, based on the miRNA and target sequence, and then used as input to the CNN and ResNet. The input to the DNABERT is constructed by interlaying the miRNA and target sequences with a 4-nt spacer. (**B**) A compact representation of the CNN network architecture. (**C**) A representation of the miRBind architecture.

2.2. Independent Chimeric Read Dataset (miRNA eCLIP)

We have produced a novel evaluation dataset based on miRNA:target gene interactions, using a novel miR-eCLIP method [47]. The experimental process was performed on our behalf by Eclipse Bioinnovations (miR-eCLIP) and all primary data is deposited on NCBI:GEO with the accession (GSE218466). Briefly, Ago2-associated miRNA:target chimeric pairs were identified by: (a) removing the sequenced reads that fully map (over 85% of read-length) to the reference genome; (b) identifying reads that partially map on miRNA annotated in miRbase on a mature miRNA collection (Release 22.1), but not in databases of rRNAs, tRNAs, yRNAs, and vRNAs (annotations from NCBI [48], Ensembl [49], and the UCSC genome browser [50]—a full list of annotations can be found in the pipeline documentation (see below); (c) unmapped (soft-clipped) parts of these miRNA reads were remapped on the reference genome and annotated on known transcripts from Ensembl. Multiple overlapping genomic alignments mapping to the same miRNA were collapsed into single interactions and were extended to a 50 nt length around the center of the genomic alignment, to make sure that the whole binding site was captured. In all, we produced 477 such high-confidence miRNA:target gene interactions. Negative interactions were again produced by randomly shuffling the miRNA and target sequences found in the positive sample. The fully documented and publicly available pipeline for chimeric interaction detection is available at https://github.com/ML-Bioinfo-CEITEC/HybriDetector/ (accessed on 28 November 2022).

2.3. Benchmarking Approaches

There are certain factors that should be considered when selecting a method for benchmarking comparison: (i) the method must be able to predict binding-site affinity (i.e., to give a score representing the potential of a microRNA to bind to a target site), or, at the minimum, binary classification of the binding site (e.g., seed binding); (ii) the method must work directly on sequences and using sequences only (microRNA, target site); (iii) the method must have an implementation method that is relatively easy to use (standalone program). To this end, we excluded from our study the following prediction methods: (a) target prediction methods that aim to predict microRNA:target gene interactions, (b) methods based on a combination of multiple inputs (e.g., evolutionary conservation of the target), and (c) methods that do not have the full implementation to predict target sites. It is not feasible to re-implement those theoretical methods that lack functional code.

2.3.1. CNN Approach

We utilized a convolutional neural network consisting of multiple layered blocks, composed of a convolutional layer, leaky ReLU, batch normalization, pooling, and a dropout layer. The output of the last dropout layer is flattened and connected to the layered blocks of dense, leaky ReLU, with batch normalization and a dropout layer. The last layer is formed of a single neuron with a sigmoid activation function that outputs the probability of input miRNA:target site binding. A schematic illustration of the network architecture can be found in Figure 1B. The network was compiled with the Adam optimizer and utilized the binary cross-entropy loss function. The models were trained over 10 epochs, with a batch size of 32. To find the best set of parameters to use, a hyperparameter search was performed separately for all positive:negative ratios (1:1, 1:10, 1:20, and 1:100), using the training set for model training and the evaluation set for comparison. The best model consisted of 6 blocks with convolutional layers, followed by 2 blocks with dense layers (Figure 1B). Convolutional layers had kernels sized 5×5, the dropout rate in the dropout layers was 0.3, and the learning rate was 0.00152. Through a subsequent evaluation, we concluded that the best-performing model is the one trained on the 1:10 training set.

2.3.2. DNABERT

DNABERT [51] is a previously published transformer-based model that has achieved superior performance across various downstream DNA sequence prediction tasks. DNABERT uses tokenized k-mer sequences as its input, which also contains 3 additional tokens, and it can be fine-tuned for multiple purposes. Since the input to DNABERT is a set of sequences, we converted each miRNA:target pair into a single sequence, in which miRNA and target sequences were interlaid with 4 N nucleotides, as depicted in Figure 1A. We used a

DNABERT model that was pretrained on the 6-mers and finetuned it on the 1:1 and 1:10 training sets, using a batch size of 64, 4 gradient accumulation steps, a learning rate of 2×10^{-4}, a weight decay of 0.01, and early stopping with patience of 5.

2.3.3. miRBind

The architecture of miRBind is a modified version of ResNet [45], in which the initial 7×7 convolution and pooling layers are removed and the number of residual blocks is optimized, as shown in Figure 1C. Since our input size is a 50×20 matrix, the initial layer and pooling, if left in place, would reduce the input for the subsequent layers to 12×5, thus not allowing the network to learn from the data. The miRBind architecture is illustrated in Figure 1C.

To address the imbalance between positive and negative classes in our 1:100 dataset, we developed a novel approach called instance hardness-based label smoothing. The approach is inspired by techniques presented in previous studies [52,53], which keep important, discriminative samples in the final training dataset and discard easily classifiable samples to rebalance the skewed ratio. Since the aim is to reduce the majority (negative) class, a sample is considered important (or 'hard') if it was incorrectly classified as a false positive. In order to 'punish' the false labeling of an instance, we utilized the probability, p, that the model misclassified a sample, and we introduced a novel instance hardness label smoothing approach, which changes the labels of samples to prevent models from producing overconfident predictions. Our approach first calculates the estimates of instance hardness (IH) on a model, or a committee thereof, to provide a better estimation of IH [54]. Subsequently, IH is smoothed out by being mapped from the range of [0,1] to the range of [0,0.5]. This ensures that the new, soft labels are not flipped and that for those samples that were the hardest to classify (an IH close to 1), the model is forced to be unsure rather than learn a new label for the given samples. miRBind was trained on the 1:1, 1:10, and 1:20 training sets, with the 1:20 model exhibiting the best performance.

2.3.4. RNAhybrid

RNAhybrid [55] is, at its core, a variation of classic RNA secondary structure prediction. It determines the most favorable hybridization site between two sequences (a short and a long RNA) in a kind of domain mode. That means that the short sequence is hybridized to the best-fitting part of the long sequence. The method offers a web-service interface, as well as a standalone version. We evaluated the performance of the RNAhybrid by utilizing its standalone version with the default parameters.

2.3.5. RNACofold

RNACofold [56] (henceforth mentioned as 'cofold') is a tool offered by the ViennaRNA package [57]. It computes the hybridization energy and base-pairing pattern of an input pair of interacting RNA molecule sequences. To simplify the performance comparison, minimum free energy scores were normalized to a range from 0 to 1, where 1 represents the strongest binding.

2.3.6. RNA22

RNA22 [58] is a pattern-based approach for the discovery of microRNA binding sites and their corresponding microRNA/mRNA complexes that relies only on the sequences of miRNAs and their targets. The algorithm is based on a Markov chain that finds recurring patterns in miRNA sequences. Potential targets are then searched with the identified patterns, and areas with accumulated hits are paired with miRNAs, based on the nucleotide pairing and free energy. The standalone version of the RNA22 program was used and run with default parameters, apart from the 'maximum folding energy for heteroduplex', which was set to the maximum value of −5 kcal/mol. The score of the RNA22 was calculated as (1—p value), where p value is the output of RNA22 that characterizes each miRNA:target pair with which the method is predicted to interact. Each score was subsequently normalized by taking into account the minimum and maximum values of the scores.

2.3.7. Seed

The 'seed' approach identifies a perfectly complementary match of the 2–7 miRNA hexamer on the target sequence. Since this is a binary decision, no area under the curve may be calculated.

2.3.8. Web Interface

We built a publicly accessible, user-friendly web interface (https://ml-bioinfo-ceitec.github.io/miRBind/ (accessed on 28 November 2022) that allows for the prediction of the score (probability) of the binding between a user-submitted pair (or pairs) of miRNA and the target site. The web interface is implemented in HTML/CSS and JavaScript. The default model is trained on the CLASH dataset with a 1:1 positive:negative ratio. All relevant files and the default model are available on the miRBind GitHub repository.

2.3.9. Evaluation Measures

For the assessment of classification tasks, a set of useful metrics are used. As has been demonstrated, the area under the precision-recall curve (auPRC) is the most informative visual analysis tool for highly imbalanced binary classification [59]. Sensitivity or recall is the proportion of true positive observations. Precision (Pre) is the ratio of true positive observations to the total number of predicted positive observations.

$$\text{Recall} = \frac{\text{TP}}{\text{TP} + \text{FN}}, \quad (1)$$

$$\text{Precision} = \frac{\text{TP}}{\text{TP} + \text{FP}}, \quad (2)$$

TP and FP are the numbers of true positive and false positive assessments, respectively. Additionally, TN is the number of true negative assessments.

3. Results

To evaluate our models, we followed a two-step process: first, we evaluated the performance of each of the DNN (miRBind, CNN, and DNABERT) for several positive:negative ratios of the validation sets, and selected the model that performs best among them. A complete overview of the evaluation of the models on the three ratios of the test sets is presented in Table 1. The models with the highest area under the precision-recall curve (auPRC) were the miRbind, CNN, and DNABERT models, trained on 1:20 and 1:10 and fine-tuned on 1:1 training sets, respectively.

Subsequently, we evaluated the performance of these best-performing models with the rest of the methods. To this end, we plotted the precision-recall curves (Figure 1) and we calculated the areas under the curves (Figure 2 and Table 2). As we can see in Table 2, miRBind outperforms 'cofold' with an AUPRC of 0.9689, vs. 0.7784 for 'cofold' with the 1:1 dataset. The difference is more pronounced in the more realistic 1:100 dataset, where miRBind100 shows an AUPRC of 0.5372 vs. 0.0413 for 'cofold'. RNAhybrid exhibits better performance than the 'cofold' method, but this is still significantly lower than miRBind across all test sets, with its performance rapidly deteriorating with the increasing ratio of negatives. The 'seed' measure performs similarly to the co-fold method, showing a high precision score on the 1:1 balanced dataset, which has previously offered a promising method for miRNA target prediction programs, as well as for assigning miRNAs to CLIP-Seq peaks. Based on the trade-off between precision and recall at different prediction score thresholds (Figure 3), we have selected two score threshold cutoffs at 0.1 ('normal') and 0.5 ('strict') for general use.

Table 1. The area under the precision-recall curve (AUPRC) for the miRBind, CNN, and DNABERT models, tested against 1:1, 1:10, and 1:100 left-out test sets. The number used in the naming of each model indicates the positive:negative ratio on which the model was trained. We selected the best-performing miRBind and CNN models (miRBind20 and CNN10, henceforth called miRBind and CNN, respectively), as the proposed methods of our work.

AUPRC	Test Set 1:1	Test Set 1:10	Test Set 1:100
miRBind1	0.9495	0.7447	0.3079
miRBind10	0.9614	0.8092	0.4531
miRBind20	0.9689	0.8410	0.5372
CNN1	0.9602	0.7862	0.4095
CNN10	0.9634	0.7969	0.4464
CNN20	0.9590	0.7880	0.4365
CNN100	0.9599	0.8005	0.4466
DNABERT1	0.9267	0.6300	0.1923
DNABERT10	0.9250	0.6440	0.2286

Table 2. The area under the precision-recall curve for the miRBind, CNN, and DNABERT models, and the RNAhybrid, 'cofold', RNA22, and 'seed' approaches, tested against 1:1, 1:10, and 1:100 left-out test sets. The 'seed' method is evaluated based on its sensitivity and precision.

AUPRC	Test Set 1:1	Test Set 1:10	Test Set 1:100
miRbind	0.9689	0.8410	0.5372
CNN	0.9634	0.7969	0.4464
DNABERT	0.9267	0.6300	0.1923
RNAhybrid	0.8439	0.4539	0.0924
Cofold	0.7784	0.2842	0.0413
RNA22	0.6203	0.1507	0.0265
Seed	Sens: 0.1425 Prec: 0.8796	Sens: 0.1425 Prec: 0.4612	Sens: 0.1425 Prec: 0.0824

The performance of the methods was also evaluated based on the area under the receiver operator characteristics curve (AUROC), as shown in Table 3.

Table 3. Area under the receiver operator characteristics curve (AUROC) for the miRBind, CNN, and DNABERT models, as well as the RNAhybrid, 'cofold', RNA22, and 'seed' approaches, tested against 1:1, 1:10, and 1:100 left-out test sets. The 'seed' method is evaluated based on the false positive rate (fpr) and true positive rate (tpr).

AUROC	Test Set 1:1	Test Set 1:10	Test Set 1:100
miRBind	0.9643	0.9654	0.9652
CNN	0.9612	0.9626	0.9628
DNABERT	0.9293	0.9310	0.9310
RNAhybrid	0.8351	0.8406	0.8381
Cofold	0.7839	0.7839	0.7812
RNA22	0.5343	0.5342	0.5375
Seed	fpr: 0.0195 tpr: 0.1425	fpr: 0.0167 tpr: 0.1425	fpr: 0.0159 tpr: 0.1425

To further investigate the predictive power of our approach, we evaluated its performance on the miRNA eCLIP dataset (Table 4). The precision-recall curves are presented in Figure 4. We validate that the miRBind model outperforms other methods in all imbalance categories. Our CNN model follows miRBind as the second-best method in all categories. It is interesting that the simple seed measure performs with much higher precision in this dataset than in the original CLASH data. Notably, in the 1:100 imbalanced dataset, the seed for the miRNA eCLIP dataset has double the precision and double the sensitivity than that for the CLASH dataset. This variation could point to differences between the seed-mediated binding affinities of Ago1 (CLASH) and Ago2 (miRNA eCLIP) proteins, or other experimental variations between the two experiments. In contrast, all other methods, including ours, seem to have a drop in performance between the two experiments.

Table 4. The area under the precision-recall curve (AUPRC) for miRBind, CNN, and DNABERT models, as well as RNAhybrid, 'cofold', RNA22, and 'seed' approaches, tested against the 1:1, 1:10, and 1:100 Ago eCLIP test sets.

AUPRC	Test Set 1:1	Test Set 1:10	Test Set 1:100
miRbind	0.8413	0.4668	0.1545
CNN	0.8223	0.4268	0.1147
DNABERT	0.6787	0.1904	0.0238
RNAhybrid	0.7615	0.2932	0.0469
Cofold	0.6862	0.1946	0.0246
RNA22	0.7116	0.2628	0.0392
Seed	Sens: 0.3774 Prec: 0.9278	Sens: 0.3774 Prec: 0.6020	Sens: 0.3774 Prec: 0.1586

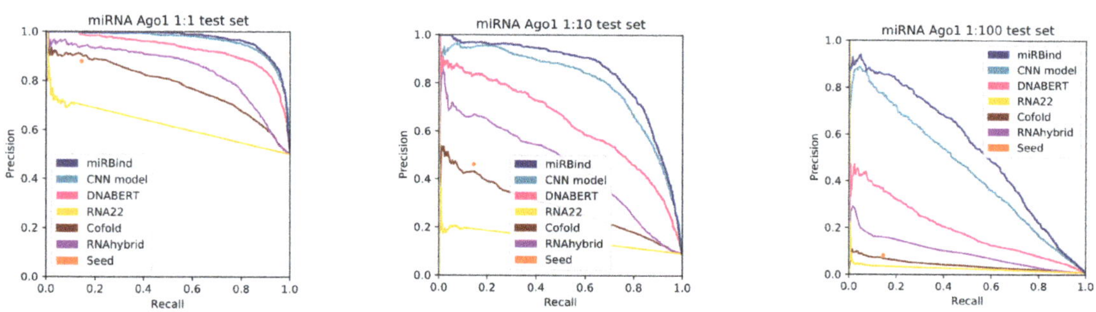

(a) (b) (c)

Figure 2. Precision-recall curves for all the methods, tested against (**a**) 1:1, (**b**) 1:10, and (**c**) 1:100 left-out test sets.

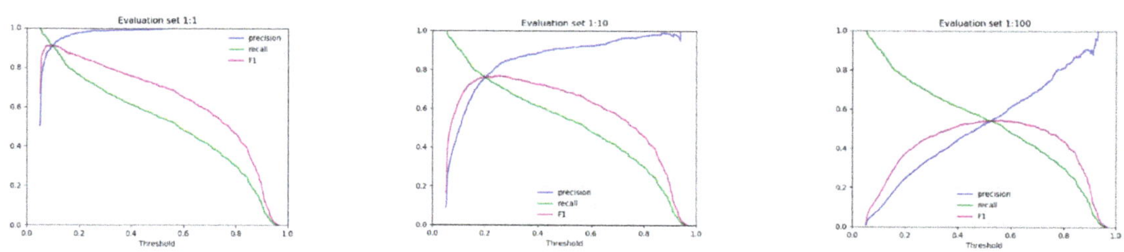

(a) (b) (c)

Figure 3. The precision, recall, and F1 score (the harmonic mean of precision and recall) against prediction score threshold for (**a**) 1:1, (**b**) 1:10, and (**c**) 1:100 datasets. The 'normal' (0.1) and 'strict' (0.5) score thresholds are suggested to users of miRBind.

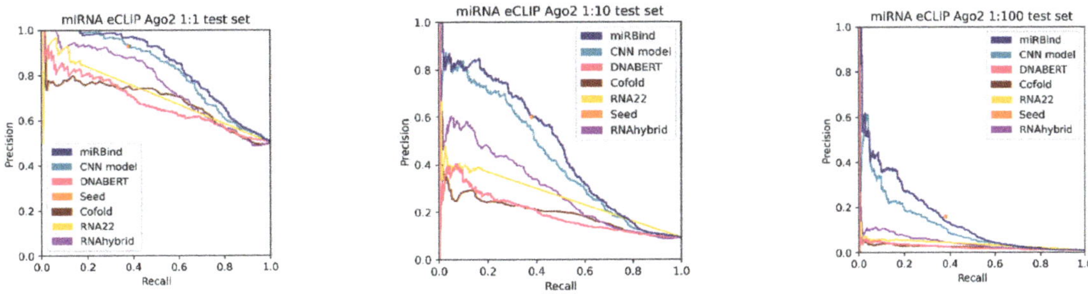

(a) (b) (c)

Figure 4. Precision-recall curves for all the prediction methods, tested against the (**a**) 1:1, (**b**) 1:10, and (**c**) 1:100 Ago eCLIP test sets.

4. Discussion

We have shown that the assignment of targets to miRNAs based on the 'seed' or 'co-fold' methods is unreliable in the highly imbalanced 1:100 scenario. Although these methods show over 90% precision in the balanced 1:1 scenario for a sensitivity/recall of 15–20%, in the 1:100 scenario, they produce nine false positives for each true positive they identify, yielding a precision of approximately 10%. In contrast, miRBind shows an almost perfect precision of up to 50% sensitivity/recall with the 1:1 balanced dataset and is more robust with the imbalanced 1:100 dataset. For a sensitivity of 50%, miRBind retrieves an approximately equal number of true positives and false positives.

We have additionally validated our method on a completely new dataset of miR eCLIP, which was produced via a similar technique to the CLASH dataset used for training. The fact that our method outperforms the state of the art in such a different dataset further reinforces the theory that it has learned some rules of interaction between miRNAs and their targets and that this is not some experiment-specific bias. Our method was trained on CLASH data from the Ago1 protein and was tested on data from the miR eCLIP on the Ago2 protein. Even though the two proteins are considered to have similar modes of utilizing miRNAs to bind the targets, we see that the Ago2 dataset responds better to a simple seed prediction than the Ago1 dataset. We can infer that the Ago2 dataset is enriched in the canonical seed sequences, but we may not speculate if that represents a real difference between the Ago1 and Ago2 proteins, or just a difference between the CLASH and miR eCLIP methodologies. In principle, neither of these methods should enrich seed-based binding over other modes of binding. However, we cannot be confident as to whether some secondary effect of the experimental variation is in play.

To conclude, we presented two deep learning approaches for the prediction of miRNA:target binding, comprising a CNN that consists of six convolutional layers and a ResNet-based neural network. To avoid any potential target sequence or experimental biases, we elected to completely hide the sequences from the networks during training. To that end, we used a two-dimensional 50×20 representation of the miRNA and putative target sequence, in which any Watson–Crick binding nucleotide pair is represented by 1, and any non-binding pair by 0. To evaluate our approaches, we utilized three commonly used methods, namely, 'seed', RNAcofold, and RNA22. We also explored the capabilities of a pre-trained and fine-tuned DNABERT for the given task. We showed that our CNN and ResNet-based approaches outperform the state-of-the-art methods for miRNA:target site prediction.

We expect miRBind to be used by bioinformaticians interested in miRNA target prediction, as part of a larger pipeline, and by researchers interested in miRNA binding, for example, as a way to allocate miRNAs to CLIP-Seq peaks. For the first group of users that may want to run large numbers of pairs, we provide a standalone Python script with the miRBind method, which can be used locally on a CPU or GPU or may even be run on the freely available Google Colaboratory CPU and GPU. For the second group, who may not have the programming expertise that is needed, we provide a free web server at https://ml-bioinfo-ceitec.github.io/miRBind/ (accessed on 28 November 2022), which users can utilize to predict the potential binding between any miRNA-like sequence and any target sequence.

We believe that miRBind can be used to predict the pairing of miRNAs to their targets, improving on more basic methods. The fact that it has been trained without theoretical preconceptions beyond the Watson–Crick pairing makes it unbiased toward seed and non-seed bindings, an important feature that is needed to explain the increasing number of non-canonical binding sites that are being experimentally identified. One caveat of miRBind is that it was trained on Ago1 CLASH data, while Ago2 is the dominant Ago family protein involved with miRNAs. To date, no other Ago2 CLASH dataset has been published; as such, our validation dataset will be extremely important for future miRNA target-prediction methods. Additionally, the field is now open for further exploration of the differences between the Ago1 and Ago2 binding rules.

Author Contributions: Conceptualization, E.K., V.H., I.-C.G. and P.A.; data curation, E.K. and V.H.; formal analysis, E.K., V.H. and K.G.; funding acquisition, P.A.; investigation, E.K., V.H., J.K., K.G., I.-C.G. and P.A.; methodology, E.K., V.H., J.K. and P.A.; project administration, I.-C.G. and P.A.; resources, P.A.; software, E.K., V.H., J.K., K.G. and I.-C.G.; supervision, I.-C.G. and P.A.; validation, E.K., V.H., J.K. and K.G.; visualization, E.K., V.H., J.K. and K.G.; writing—original draft, E.K., V.H., J.K., K.G. and I.-C.G.; writing—review and editing, E.K., K.G., I.-C.G. and P.A. All authors have read and agreed to the published version of the manuscript.

Funding: This research has been supported by Grantová Agentura České Republiky, 19-10976Y. Grant to P.A.

Institutional Review Board Statement: Not applicable.

Informed Consent Statement: Not applicable.

Data Availability Statement: All datasets and the full code for miRBind can be found at https://github.com/ML-Bioinfo-CEITEC/miRBind. The fully documented and publicly available pipeline for chimeric interaction detection is available at https://github.com/ML-Bioinfo-CEITEC/HybriDetector/. The miR eCLIP data are deposited with accession GSE218466.

Acknowledgments: The Bioinformatics Core Facility of CEITEC Masaryk University is gratefully acknowledged for the obtaining of the scientific data presented in this paper.

Conflicts of Interest: The authors declare no conflict of interest.

References

1. Bartel, D.P. Metazoan MicroRNAs. *Cell* **2018**, *173*, 20–51. [CrossRef]
2. Lee, R.C.; Feinbaum, R.L.; Ambros, V. The C. Elegans Heterochronic Gene Lin-4 Encodes Small RNAs with Antisense Complementarity to Lin-14. *Cell* **1993**, *75*, 843–854. [CrossRef] [PubMed]
3. Wightman, B.; Ha, I.; Ruvkun, G. Posttranscriptional Regulation of the Heterochronic Gene Lin-14 by Lin-4 Mediates Temporal Pattern Formation in C. Elegans. *Cell* **1993**, *75*, 855–862. [CrossRef] [PubMed]
4. Pasquinelli, A.E.; Reinhart, B.J.; Slack, F.; Martindale, M.Q.; Kuroda, M.I.; Maller, B.; Hayward, D.C.; Ball, E.E.; Degnan, B.; Müller, P.; et al. Conservation of the Sequence and Temporal Expression of Let-7 Heterochronic Regulatory RNA. *Nature* **2000**, *408*, 86–89. [CrossRef] [PubMed]
5. Kozomara, A.; Griffiths-Jones, S. MiRBase: Integrating MicroRNA Annotation and Deep-Sequencing Data. *Nucleic Acids Res.* **2011**, *39*, D152–D157. [CrossRef] [PubMed]
6. Adams, L. Pri-MiRNA Processing: Structure Is Key. *Nat. Rev. Genet.* **2017**, *18*, 145. [CrossRef] [PubMed]
7. Lund, E.; Güttinger, S.; Calado, A.; Dahlberg, J.E.; Kutay, U. Nuclear Export of MicroRNA Precursors. *Science* **2004**, *303*, 95–98. [CrossRef] [PubMed]
8. O'Brien, J.; Hayder, H.; Zayed, Y.; Peng, C. Overview of MicroRNA Biogenesis, Mechanisms of Actions, and Circulation. *Front. Endocrinol.* **2018**, *9*, 402. [CrossRef]
9. Saliminejad, K.; Khorram Khorshid, H.R.; Soleymani Fard, S.; Ghaffari, S.H. An Overview of MicroRNAs: Biology, Functions, Therapeutics, and Analysis Methods. *J. Cell. Physiol.* **2019**, *234*, 5451–5465. [CrossRef]

10. Filipowicz, W.; Bhattacharyya, S.N.; Sonenberg, N. Mechanisms of Post-Transcriptional Regulation by MicroRNAs: Are the Answers in Sight? *Nat. Rev. Genet.* **2008**, *9*, 102–114. [CrossRef]
11. Dueck, A.; Ziegler, C.; Eichner, A.; Berezikov, E.; Meister, G. MicroRNAs Associated with the Different Human Argonaute Proteins. *Nucleic Acids Res.* **2012**, *40*, 9850–9862. [CrossRef] [PubMed]
12. Pasquinelli, A.E. MicroRNAs and Their Targets: Recognition, Regulation and an Emerging Reciprocal Relationship. *Nat. Rev. Genet.* **2012**, *13*, 271–282. [CrossRef] [PubMed]
13. Kalla, R.; Ventham, N.T.; Kennedy, N.A.; Quintana, J.F.; Nimmo, E.R.; Buck, A.H.; Satsangi, J. MicroRNAs: New Players in IBD. *Gut* **2015**, *64*, 504–513. [CrossRef]
14. Zealy, R.W.; Wrenn, S.P.; Davila, S.; Min, K.-W.; Yoon, J.-H. MicroRNA-Binding Proteins: Specificity and Function. *WIREs RNA* **2017**, *8*, e1414. [CrossRef] [PubMed]
15. Lewis, B.P.; Shih, I.-H.; Jones-Rhoades, M.W.; Bartel, D.P.; Burge, C.B. Prediction of Mammalian MicroRNA Targets. *Cell* **2003**, *115*, 787–798. [CrossRef] [PubMed]
16. Bartel, D.P. MicroRNA Target Recognition and Regulatory Functions. *Cell* **2009**, *136*, 215–233. [CrossRef]
17. Broughton, J.P.; Lovci, M.T.; Huang, J.L.; Yeo, G.W.; Pasquinelli, A.E. Pairing Beyond the Seed Supports MicroRNA Targeting Specificity. *Mol. Cell* **2016**, *64*, 320–333. [CrossRef] [PubMed]
18. Agarwal, V.; Bell, G.W.; Nam, J.-W.; Bartel, D.P. Predicting Effective MicroRNA Target Sites in Mammalian MRNAs. *eLife* **2015**, *4*, e05005. [CrossRef]
19. Kudla, G.; Granneman, S.; Hahn, D.; Beggs, J.D.; Tollervey, D. Cross-Linking, Ligation, and Sequencing of Hybrids Reveals RNA–RNA Interactions in Yeast. *Proc. Natl. Acad. Sci. USA* **2011**, *108*, 10010–10015. [CrossRef]
20. Helwak, A.; Kudla, G.; Dudnakova, T.; Tollervey, D. Mapping the Human MiRNA Interactome by CLASH Reveals Frequent Noncanonical Binding. *Cell* **2013**, *153*, 654–665. [CrossRef]
21. John, B.; Enright, A.J.; Aravin, A.; Tuschl, T.; Sander, C.; Marks, D.S. Human MicroRNA Targets. *PLoS Biol.* **2004**, *2*, e363. [CrossRef] [PubMed]
22. Enright, A.J.; John, B.; Gaul, U.; Tuschl, T.; Sander, C.; Marks, D.S. MicroRNA Targets in Drosophila. *Genome Biol.* **2004**, *5*, R1. [CrossRef] [PubMed]
23. Kertesz, M.; Iovino, N.; Unnerstall, U.; Gaul, U.; Segal, E. The Role of Site Accessibility in MicroRNA Target Recognition. *Nat. Genet.* **2007**, *39*, 1278–1284. [CrossRef] [PubMed]
24. Baek, D.; Villén, J.; Shin, C.; Camargo, F.D.; Gygi, S.P.; Bartel, D.P. The Impact of MicroRNAs on Protein Output. *Nature* **2008**, *455*, 64–71. [CrossRef] [PubMed]
25. Selbach, M.; Schwanhäusser, B.; Thierfelder, N.; Fang, Z.; Khanin, R.; Rajewsky, N. Widespread Changes in Protein Synthesis Induced by MicroRNAs. *Nature* **2008**, *455*, 58–63. [CrossRef] [PubMed]
26. Alexiou, P.; Maragkakis, M.; Papadopoulos, G.L.; Reczko, M.; Hatzigeorgiou, A.G. Lost in Translation: An Assessment and Perspective for Computational MicroRNA Target Identification. *Bioinformatics* **2009**, *25*, 3049–3055. [CrossRef] [PubMed]
27. Ule, J.; Jensen, K.B.; Ruggiu, M.; Mele, A.; Ule, A.; Darnell, R.B. CLIP Identifies Nova-Regulated RNA Networks in the Brain. *Science* **2003**, *302*, 1212–1215. [CrossRef]
28. Karagkouni, D.; Paraskevopoulou, M.D.; Chatzopoulos, S.; Vlachos, I.S.; Tastsoglou, S.; Kanellos, I.; Papadimitriou, D.; Kavakiotis, I.; Maniou, S.; Skoufos, G.; et al. DIANA-TarBase v8: A Decade-Long Collection of Experimentally Supported MiRNA–Gene Interactions. *Nucleic Acids Res.* **2018**, *46*, D239–D245. [CrossRef]
29. Helwak, A.; Tollervey, D. Mapping the MiRNA Interactome by Cross-Linking Ligation and Sequencing of Hybrids (CLASH). *Nat. Protoc.* **2014**, *9*, 711–728. [CrossRef]
30. Moore, M.J.; Scheel, T.K.H.; Luna, J.M.; Park, C.Y.; Fak, J.J.; Nishiuchi, E.; Rice, C.M.; Darnell, R.B. MiRNA–Target Chimeras Reveal MiRNA 3′-End Pairing as a Major Determinant of Argonaute Target Specificity. *Nat. Commun.* **2015**, *6*, 8864. [CrossRef]
31. Riolo, G.; Cantara, S.; Marzocchi, C.; Ricci, C. MiRNA Targets: From Prediction Tools to Experimental Validation. *Methods Protoc.* **2020**, *4*, 1. [CrossRef] [PubMed]
32. Peterson, S.M.; Thompson, J.A.; Ufkin, M.L.; Sathyanarayana, P.; Liaw, L.; Congdon, C.B. Common Features of MicroRNA Target Prediction Tools. *Front. Genet.* **2014**, *5*, 23. [CrossRef] [PubMed]
33. Ekimler, S.; Sahin, K. Computational Methods for MicroRNA Target Prediction. *Genes* **2014**, *5*, 671–683. [CrossRef]
34. Shaker, F.; Nikravesh, A.; Arezumand, R.; Aghaee-Bakhtiari, S.H. Web-based tools for miRNA studies analysis. *Comput. Biol. Med.* **2020**, *127*, 104060. [CrossRef]
35. Betel, D.; Koppal, A.; Agius, P.; Sander, C.; Leslie, C. Comprehensive modeling of microRNA targets predicts functional non-conserved and non-canonical sites. *Genome Biol.* **2010**, *11*, R90. [CrossRef] [PubMed]
36. Maragkakis, M.; Reczko, M.; Simossis, V.A.; Alexiou, P.; Papadopoulos, G.L.; Dalamagas, T.; Giannopoulos, G.; Goumas, G.; Koukis, E.; Kourtis, K.; et al. DIANA-microT web server: Elucidating microRNA functions through target prediction. *Nucleic Acids Res.* **2009**, *37*, W273–W276. [CrossRef]
37. Reczko, M.; Maragkakis, M.; Alexiou, P.; Grosse, I.; Hatzigeorgiou, A.G. Functional microRNA targets in protein coding sequences. *Bioinformatics* **2012**, *28*, 771–776. [CrossRef]
38. Paraskevopoulou, M.D.; Georgakilas, G.; Kostoulas, N.; Vlachos, I.S.; Vergoulis, T.; Reczko, M.; Filippidis, C.; Dalamagas, T.; Hatzigeorgiou, A.G. DIANA-microT web server v5.0: Service integration into miRNA functional analysis workflows. *Nucleic Acids Res.* **2013**, *41*, W169–W173. [CrossRef]

39. Wang, X.; El Naqa, I.M. Prediction of both conserved and nonconserved microRNA targets in animals. *Bioinformatics* **2008**, *24*, 325–332. [CrossRef]
40. Bandyopadhyay, S.; Mitra, R. TargetMiner: microRNA target prediction with systematic identification of tissue-specific negative examples. *Bioinformatics* **2009**, *25*, 2625–2631. [CrossRef]
41. Liu, H.; Yue, D.; Chen, Y.; Gao, S.-J.; Huang, Y. Improving performance of mammalian microRNA target prediction. *BMC Bioinform.* **2010**, *11*, 476. [CrossRef] [PubMed]
42. Eraslan, G.; Avsec, Ž.; Gagneur, J.; Theis, F.J. Deep Learning: New Computational Modelling Techniques for Genomics. *Nat. Rev. Genet.* **2019**, *20*, 389–403. [CrossRef] [PubMed]
43. LeCun, Y.; Bengio, Y.; Hinton, G. Deep Learning. *Nature* **2015**, *521*, 436–444. [CrossRef] [PubMed]
44. Min, S.; Lee, B.; Yoon, S. Deep Learning in Bioinformatics. *Brief. Bioinform.* **2017**, *18*, 851–869. [CrossRef]
45. He, K.; Zhang, X.; Ren, S.; Sun, J. Deep Residual Learning for Image Recognition. In Proceedings of the 2016 IEEE Conference on Computer Vision and Pattern Recognition (CVPR), Las Vegas, NV, USA, 27–30 June 2016; pp. 770–778. [CrossRef]
46. Travis, A.J.; Moody, J.; Helwak, A.; Tollervey, D.; Kudla, G. Hyb: A Bioinformatics Pipeline for the Analysis of CLASH (Crosslinking, Ligation and Sequencing of Hybrids) Data. *Methods* **2014**, *65*, 263–273. [CrossRef] [PubMed]
47. Manakov, S.A.; Shishkin, A.A.; Yee, B.A.; Shen, K.A.; Cox, D.C.; Park, S.S.; Foster, H.M.; Chapman, K.B.; Yeo, G.W.; Nostrand, E.L.V. Scalable and Deep Profiling of MRNA Targets for Individual MicroRNAs with Chimeric ECLIP. *bioRxiv* **2022**. [CrossRef]
48. Database Resources of the National Center for Biotechnology Information. *Nucleic Acids Res.* **2017**, *45*, D12–D17. [CrossRef]
49. Cunningham, F.; Allen, J.E.; Allen, J.; Alvarez-Jarreta, J.; Amode, M.R.; Armean, I.M.; Austine-Orimoloye, O.; Azov, A.G.; Barnes, I.; Bennett, R.; et al. Ensembl 2022. *Nucleic Acids Res.* **2022**, *50*, D988–D995. [CrossRef]
50. Haeussler, M.; Zweig, A.S.; Tyner, C.; Speir, M.L.; Rosenbloom, K.R.; Raney, B.J.; Lee, C.M.; Lee, B.T.; Hinrichs, A.S.; Gonzalez, J.N.; et al. The UCSC Genome Browser Database: 2019 Update. *Nucleic Acids Res.* **2019**, *47*, D853–D858. [CrossRef]
51. Ji, Y.; Zhou, Z.; Liu, H.; Davuluri, R.V. DNABERT: Pre-Trained Bidirectional Encoder Representations from Transformers Model for DNA-Language in Genome. *Bioinformatics* **2021**, *37*, 2112–2120. [CrossRef]
52. Georgakilas, G.K.; Grioni, A.; Liakos, K.G.; Chalupova, E.; Plessas, F.C.; Alexiou, P. Multi-Branch Convolutional Neural Network for Identification of Small Non-Coding RNA Genomic Loci. *Sci. Rep.* **2020**, *10*, 9486. [CrossRef] [PubMed]
53. Guo, H.; Viktor, H.L. Learning from Imbalanced Data Sets with Boosting and Data Generation: The DataBoost-IM Approach. *SIGKDD Explor. Newsl.* **2004**, *6*, 30–39. [CrossRef]
54. Smith, M.R.; Martinez, T.; Giraud-Carrier, C. An Instance Level Analysis of Data Complexity. *Mach Learn* **2014**, *95*, 225–256. [CrossRef]
55. Krüger, J.; Rehmsmeier, M. RNAhybrid: microRNA target prediction easy, fast and flexible. *Nucleic Acids Res.* **2006**, *34*, W451–W454. [CrossRef]
56. Bernhart, S.H.; Tafer, H.; Mückstein, U.; Flamm, C.; Stadler, P.F.; Hofacker, I.L. Partition Function and Base Pairing Probabilities of RNA Heterodimers. *Algorithms Mol. Biol.* **2006**, *1*, 3. [CrossRef]
57. Lorenz, R.; Bernhart, S.H.; Höner zu Siederdissen, C.; Tafer, H.; Flamm, C.; Stadler, P.F.; Hofacker, I.L. ViennaRNA Package 2.0. *Algorithms Mol. Biol.* **2011**, *6*, 26. [CrossRef]
58. Saito, T.; Rehmsmeier, M. The Precision-Recall Plot Is More Informative than the ROC Plot When Evaluating Binary Classifiers on Imbalanced Datasets. *PLoS ONE* **2015**, *10*, e0118432. [CrossRef]
59. Miranda, K.C.; Huynh, T.; Tay, Y.; Ang, Y.-S.; Tam, W.-L.; Thomson, A.M.; Lim, B.; Rigoutsos, I. A Pattern-Based Method for the Identification of MicroRNA Binding Sites and Their Corresponding Heteroduplexes. *Cell* **2006**, *126*, 1203–1217. [CrossRef]

genes

MDPI

Review

Pluripotency-Associated microRNAs in Early Vertebrate Embryos and Stem Cells

Pouneh Maraghechi [1,†], Maria Teresa Salinas Aponte [1,†], András Ecker [1], Bence Lázár [1,2], Roland Tóth [1], Nikolett Tokodyné Szabadi [1] and Elen Gócza [1,*]

1 Department of Animal Biotechnology, Institute of Genetics and Biotechnology, Hungarian University of Agriculture and Life Sciences; Agrobiotechnology and Precision Breeding for Food Security National Laboratory, Szent-Györgyi Albert str. 4, 2100 Gödöllő, Hungary
2 National Centre for Biodiversity and Gene Conservation, Institute for Farm Animal Gene Conservation (NBGK-HGI), Isaszegi str. 200, 2100 Gödöllő, Hungary
* Correspondence: elen.gocza@uni-mate.com
† These authors contributed equally to this work.

Abstract: MicroRNAs (miRNAs), small non-coding RNA molecules, regulate a wide range of critical biological processes, such as proliferation, cell cycle progression, differentiation, survival, and apoptosis, in many cell types. The regulatory functions of miRNAs in embryogenesis and stem cell properties have been extensively investigated since the early years of miRNA discovery. In this review, we will compare and discuss the impact of stem-cell-specific miRNA clusters on the maintenance and regulation of early embryonic development, pluripotency, and self-renewal of embryonic stem cells, particularly in vertebrates.

Keywords: microRNA clusters; embryonic stem cells; embryonic development; cell cycle regulators

1. Introduction

microRNAs (miRNAs) are endogenous, 22–25 nucleotide long non-coding RNAs that post-transcriptionally regulate gene expression by binding to the 3′ untranslated region (UTR) of their target mRNAs and inhibiting their translation or stability. The central regulatory role of miRNAs in multiple biological processes, including cell cycle regulation, apoptosis, aging, cell fate decisions, and different signaling pathways, has been widely described. Developmental studies in vertebrates demonstrated that miRNAs are crucial molecules for embryogenesis (reviewed by [1,2]). Indeed, miRNAs play important roles in stem cell properties by regulating self-renewal and differentiation (reviewed by [3,4]). In this review, we will focus on the impact of stem-cell-specific miRNA clusters on the maintenance and regulation of early embryogenesis and embryonic stem cell (ESC) pluripotency in vertebrates, particularly in humans, mice, rabbits, and chickens.

2. Overview of microRNA Biogenesis

miRNA genes are transcribed by RNA Polymerase II/III into long primary transcripts (pri-miRNA), which can also be polycistronic [5,6]. A significant percentage of miRNAs are intragenic, residing within their host gene's intronic or/and exonic regions [7]. They are transcribed from introns, and occasionally from exons of protein-coding genes, sharing the host gene's expression regulators. Conversely, intergenic miRNAs are located between genes and transcribed independently from their own promoters [6,8]. However, some intergenic miRNAs can also be co-transcribed with their neighboring genes [9]. Many miRNAs are localized as a cluster on the same chromosome in a close proximity of maximum 10 kb [10,11]. They are co-transcribed from the same genomic loci as a single polycistronic transcript [12,13]. They also share the same seed sequence, therefore sharing common target mRNAs and biological functions [10,14]. The biogenesis of miRNA can follow canonical or non-canonical pathways.

In the canonical pathway, pri-miRNAs are mostly transcribed by RNA Polymerase II [5]. Transcribed pri-miRNA forms a hairpin structure. This structure has two domains that are recognized by the Drosha microprocessor complex (composed of Drosha and Dgcr8). Drosha cleaves the stem of the hairpin and produces a shorter hairpin, termed "precursor miRNA" (pre-miRNA) [15]. In the presence of the RanGTP cofactor, Exportin-5 translocates the pre-miRNA from the nucleus into the cytoplasm where Dicer further cleaves it and generates a short, double-stranded miRNA precursor [16]. One of the strands of the generated miRNA duplex is selected as a guide miRNA by Argonaute based on the identities of the 5′ nucleotide of each strand and the relative thermodynamic stability of the two ends of the duplex, while the other strand (miRNA*) is degraded. The guide miRNA strand is then loaded into Argonaute 2 (AGO2), building the RISC complex which is composed of the RNAse III-type nuclease Dicer and the PAZ/PIWI protein (AGO2) [17–19]. Once miRISC is built, it can bind to its target mRNA's 3′-UTR through a complementary sequence. A perfect base pairing results in target degradation, while an imperfect base pairing results in translational silencing of the target mRNA. The miRNA structure contains a well-conserved sequence of six nucleotides called the "seed" that represents the most important feature of target recognition [20,21].

Non-canonical miRNAs are also derived from RNA Polymerase II; however, few miRNAs can be transcribed from RNA Polymerase III. The non-canonical miRNA biogenesis occurs through a Drosha/Dgcr8 and/or Dicer-independent manner, when miRNAs originate from introns (mirtrons) [22,23] or other small RNAs, such as snoRNAs, shRNAs, tRNAs, tRNase Z, and endo-siRNAs [24–28]. miRNAs derived from endo-shRNAs and tRNA precursors are transcribed by RNA Polymerase III [24,27].

In the Drosha/ Dgcr8-independent pathway (also called the mirtron pathway), introns are primarily processed in the nucleus by the spliceosome and a debranching enzyme that produces short hairpins called mirtrons. These hairpins are then exported to the cytoplasm by Exportin-5 and cleaved by Dicer [22,23]. Even though the majority of introns in vertebrates are elongated from several kilobases up to megabases in length, there are short introns of 50–150 nt in length which may form hairpin structures [29,30]. Hence, mitrons constitute only a small fraction of mammalian miRNAs. There are other subclasses of miRNAs with a potential pre-miRNA-like hairpin which are derived by an alternative Drosha/ Dgcr8-independent pathway. The processing of snoRNAs, shRNAs, tRNAs, tRNase Z, and endo-siRNA precursors to miRNAs are also independent of Drosha/ Dgcr8 [24–28].

In the Dicer-independent pathway, the produced pri-miRNAs are processed by Drosha into pre-miRNAs. However, these pre-miRNAs are not suitable substrates for Dicer; therefore, they are loaded into Ago2 instead, and Ago2-dependent cleavage produces an intermediate 3′ end. The final maturation occurs as the 3′-5′ trimming of the 5′ strand is completed by a cellular nuclease [31,32]. In the Dicer-independent pathway, the tRNase Z-dependent pathway precursor tRNA is cleaved by tRNase Z at the 3′ end [33].

3. Role of the miRNAs in Embryonic Development

Many miRNAs function during early embryonic development. The first known miRNA, lin-4, plays a crucial role in the developmental timing of stage-specific cell lineages in *C. elegans* [34]. Shortly after the discovery of lin-4, let-7 was recognized as an essential miRNA to control the pattern of temporal development of larval cell fates during the adult stage, which acts as a temporal switch between larval and adult fates. Loss of the let-7 gene activity leads to abnormalities in the pattern of larval development as a result of the reiteration of larval cell fates during the adult stage, while overexpression of the let-7 gene causes the early presence of adult cell fate during larval stages [35]. Both lin-4 and let-7 miRNAs are highly conserved; miR-125a and miR-125b are the mammalian orthologues of lin-4 [36,37].

The impact of miRNAs in embryonic development has been demonstrated in different organisms either by dissecting the miRNA pathway or by functional and expression pattern analyses of miRNAs. In mice, different miRNAs are present during all stages of embryonic development. Deletion of Dicer, Dgcr8, or Ago2 induced early post-implantation lethality [38–41]. Dicer 1 mutant mouse embryos exhibited a 50% lethality rate at the 7.5-day embryonic stage; surviving embryos exhibited morphological abnormalities and suppressed Oct4 and brachyury expression, indicating an embryonic developmental arrest and deficient pluripotency [38]. Homozygous deletion of the first and second exons of the Dicer gene resulted in severe hypomorphic embryos with retarded phenotype and mid-gestation death, suggesting an essential role for Dicer in normal mouse embryonic development [42]. Moreover, Dgcr8 knockout mouse embryos died early in development and displayed extreme morphological defects [41].

miRNA profiling uncovered developmental stage-specific miRNA expression. Zygotes inherit both maternal and paternal miRNAs [43,44]; however, the maternal miRNA pool is largely depleted between the one- and two-cell stages. Subsequently, miRNA expression is increased upon the activation of zygotic transcription [43]. Remarkably, certain stem-cell-specific miRNAs, such as miR-290 cluster members, are the first de novo expressed miRNAs in early embryogenesis [43,45,46]. Similarly, in rabbits, the expression of ocu-miR-290 cluster members begins at early embryonic stages [47]. Furthermore, ocu-miR-512-5p, which is evolutionarily related to miR-290 and its human homolog miR-371, is also expressed in early stages of rabbit embryonic development [47]. miR-29b is another important development regulator. It is highly expressed in the two-cell-stage mouse embryos and downregulated in the four-cell stage. Inhibition of miR-29b led to early developmental blockade by downregulating Dnmt3a/b and disrupting DNA methylation [48]. The placenta-specific C19MC miRNAs play significant roles in the fate of trophoblasts. For example, miR-519d regulates human trophoblast migration [49]. Additionally, a high expression of ocu-miR-512-5p, another C19MC family member, was reported in rabbit trophoblasts and hypoblasts, in line with its importance in pre-implantation [47]. Likewise, hsa-miR-512-3p miRNAs are among the most highly expressed miRNAs in human blastocysts [50]. miR-376c (member of the hsa-miR-379/miR-656 cluster) promotes cell proliferation and invasion of human trophoblast cells [51], whereas miR-155 inhibits their cell proliferation and migration [52]. The miR-302 cluster is a highly expressed stem-cell-specific miRNA cluster in rabbit pre-implantation embryos [47]. Similar to rabbits, miR-302 cluster members are highly expressed in the human blastocysts [50]. miR-21 is expressed in both human and mouse oocytes and blastocysts [53,54]. It is associated with oocyte maturation, blastocyst formation, and pre-implantation embryonic development [54]. The let-7 family also plays a crucial role in mammalian embryonic development. The expression of let-7 in mouse embryos regulates inner cell mass cell fate determination and reinforces blastocyst formation [55].

4. Discovery of miRNA Clusters in Stem Cells

To date, many studies have demonstrated the regulatory function of miRNAs (both canonical and non-canonical) in the self-renewal and differentiation of ESCs and iPSCs. Many canonical miRNAs, such as miR-302 and miR-290 clusters, reportedly promote ESC properties. miR-320 and miR-702 are the two first-discovered non-canonical miRNAs expressed in mouse ESCs [56]. Interestingly, miR-133a regulates the transcription of the DNMT3B gene in human ESCs through a non-canonical pathway when it is translocated to the nucleus, while it regulates skeletal and cardiac muscle function via canonical target mRNA repression in the cytoplasm [57].

DICER deletion has induced a proliferation defect, hindered teratoma and chimera formation, and constrained the differentiation of mESCs [58,59] and hESCs [60]. Similarly, Dgcr8-knockout mESCs proliferate slowly and exhibit either delayed or reduced differentiation potential, accompanied by defects in embryonic anatomy and teratoma formation [41,61]. While Dicer- and Dgcr8-mutant ESCs contributed to an overall analysis of miRNA functional aspects in ESC properties, cloning and deep sequencing of non-coding small RNAs from stem cells led to the discovery and profiling of stem-cell-specific miRNAs with potential functions in the self-renewal and differentiation of ESCs. ESC-associated transcription factors, including OCT4, NANOG, SOX2, and REX1, regulate ES-specific miRNAs through binding to their promoter region [62,63]. As a prominent example, the miR-290/295 cluster was identified as a stem-cell-specific miRNA cluster in mice by cloning and deep sequencing [64]. Subsequently, the human homolog clusters miR-302/367 and miR-371/373 were also identified by cDNA cloning in human ESCs [65]. Additionally, the deep sequencing of small RNAs showed that the rabbit ES-like cells highly express the ocu-miR-302/367 cluster, which potentially maintains pluripotency by negatively modulating Lefty [47].

4.1. miRNA Clusters Regulate Pluripotency

miR-302/367, miR-290/295, and their human homolog, the hsa-miR-371/373 cluster, are highly expressed in vertebrate ESCs. They modulate self-renewal and pluripotency through regulating cell cycle progression. These clusters are downstream targets of ESC-specific transcription factors, including OCT4, SOX2, and NANOG, which are essential for the self-renewal and maintenance of pluripotency in ESCs [47,63,66–68]. In both mouse and human ESCs, OCT4 and SOX2 bind to conserved promoter regions of miR-302 and activate its transcription. Furthermore, miR-302 directly suppresses differentiation marker genes, resulting in the high-level expression of pluripotent-related transcription factors, such as OCT4 [69]. miR-195 and miR-372a are also highly expressed in hESCs and maintain the proliferative capacity and self-renewal of hESCs through negative cell cycle modulators [70].

4.2. miRNA Clusters Are Involved in Stem Cell Differentiation

miRNAs are also involved in the initiation of stem cell differentiation by suppressing pluripotency-associated pathways. In human ESCs, post-transcriptional repression of *OCT4*, *SOX2*, and *KLF4* mediated by miR-145 leads to the suppression of self-renewal, and consequently the induction of differentiation [71]. Likewise, miR-134, miR-296, and miR-470 promote the differentiation of mouse ESCs by suppressing *Nanog*, *Oct4*, and *Sox2* [72,73]. miR-1305 induces the differentiation of human ESCs through the repression of *POLR3G*, which is an activator of the OCT4/NANOG pathway [74,75]. Other differentiation associated-miRNAs in mice are miR-34a, miR-100, and miR-137, which activate differentiation marker genes via the regulation of epigenetic mediators through targeting *Sirt1*, *Smarca5*, and *Jarid1b* mRNAs, respectively [76]. Furthermore, miR-27a and miR-24 silence self-renewal and promote differentiation by downregulating pluripotent-specific transcription factors and signal transducers of mESC self-renewal networks. The downregulation of Oct4 and Foxo1 and further suppression of signaling pathways through direct targeting of the LIF receptor *gp130* and the two important signal transducers *Smad3* and *Smad4* result in decreased expression of c-Myc, hence creating a mutual negative feedback loop to upregulate the expression of the miRNAs to maintain the differentiated state [77]. The let-7 miRNA family members also regulate the transition from self-renewing stem cells to differentiated cell types in both human and mouse ESCs. Contrary to the above-mentioned pluripotent-specific miRNAs, let-7 miRNAs suppress the self-renewal of ESCs through cell cycle progression [78–80].

It is important to note that miRNAs might have context-dependent functions and can vary across different organisms and cell types. Numerous miRNAs contribute to the maintenance of early embryonic development and embryonic stem cell properties. In the following section, we highlight the well-studied stem-cell-specific miRNA clusters and compare their expression profiles and functions in humans, mice, rabbits, and chickens.

5. Major miRNA Clusters of ESCs

5.1. miR-302/367 Cluster

5.1.1. Characteristics of Cluster

The polycistronic human miR-302/367 cluster is located in the intron of the LARP7 gene at the 4q25 region and consists of ten miRNAs: miR-302a-3p, miR-302a-5p, miR-302b-3p, miR-302b-5p, miR-302c-3p, miR-302c-5p, miR-302d-3p, miR-302d-5p, miR-367-3p, and miR-367-5p [64,65,81]. All 3′ mature miRNAs share the same seed sequence of 5′-AAGTGC-3′, except for miR-367 which harbors a slightly different seed sequence; however, they all have common mRNA targets [68,82]. The cluster is well conserved among the vertebrates (Figure 1). As it is shown in Figure 1, several species exhibit a common cluster structure, comprising four miR-302s (a–b) and one miR-367. Interestingly, the Aves class expresses an extra miRNA within the miR-302/367 cluster, such as gga-miR-1811 in chickens, which shares a highly similar seed sequence with miR-367, suggesting that it might be generated by the tandem duplication of miR-367 [81].

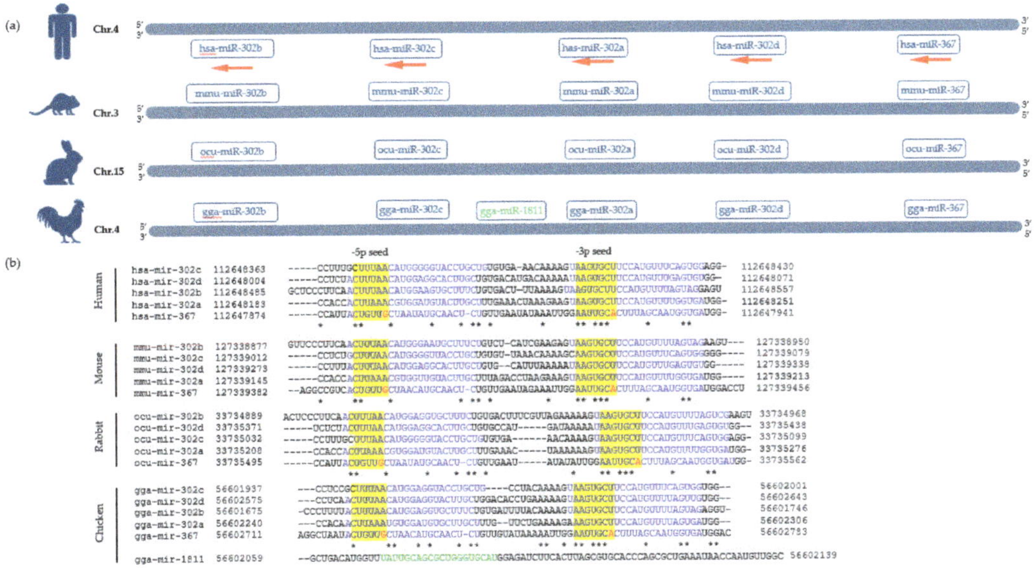

Figure 1. Genome organization and multiple sequence alignment of miR-302/367 precursor coding regions in humans, mice, rabbits, and chickens. (**a**) Humans, mice, rabbits, and chikens exhibit a common cluster structure, comprising four miR-302s and one miR-367. Interestingly, chikens express an extra miRNA, gga-miR-1811, within the miR-302/367 cluster. (**b**) Multiple sequence alignment of miR-302/367 precursor coding regions in humans, mice, rabbits, and chickens represent high similarity of this cluster. The seed sequences are highlighted in yellow; mismatches are shown in red. Mature -5p and -3p miRNA sequences are shown in blue. Mature sequence of gga-miR-1811 is shown in green. The asterisks indicate sequence homology.

The miR-302/367 cluster's promoter itself is a highly conserved region among vertebrates [47,83]. The cluster's expression is regulated by ESC-specific transcription factors (OCT3/4, SOX2, KLF4, MYC, and NANOG) in mice and humans [68,84].

5.1.2. miR-302/367 Expression in Mice, Humans, and Rabbits

In addition to cell cycle regulation, miR-302 cluster members modulate the self-renewal-related genes by repressing epithelial–mesenchymal transition and apoptotic pathways [85]. The miR-302/367 cluster is predominantly expressed in human ESCs, mouse epiblast stem cells (EpiSCs), and rabbit ES-like cells, whereas its expression in mouse ESCs is slightly lower [47,65,86]. In addition, it is highly expressed in human iPSCs and downregulated upon differentiation [87]. Overexpression of the miR-302/367 cluster in both mouse and human somatic cells results in cellular reprogramming the iPSC state and preserves the stemness properties of ESCs [88,89].

In rabbits, miR-302 expression begins at 3.5 dpc of pre-implanted embryos, whereas mouse embryos express miR-302 starting from day 6.5 dpc. The miR-302 cluster reaches the highest expression at the 6 dpc stage in rabbit embryos, and at 7.5 dpc in mouse embryos; its expression is rapidly downregulated by 8.5 dpc in mice [47,68].

5.1.3. Gga-miR-302b-3p/5p Expression in Chicken Embryos and PGCs

Most chicken stem-cell-specific miRNAs are conserved with other vertebrate model species [90]. Gga-miR-302a is highly expressed in the chicken blastoderm and regulates the undifferentiated state of blastodermal cells and primordial germ cells (PGCs) by silencing *Sox11*, a somatic transcription factor [91]. It has also been reported that gga-miR-302b regulates glucose phosphate isomerase (GPI) expression; therefore, it modulates chicken PGC proliferation [92]. A recent study based on an inhibition assay demonstrated that gga-miR-302b (both -3p and -5p miRNAs) promotes the proliferation of PGCs and decreases apoptosis. In addition, the dual inhibition of -3p and -5p miR-302b caused an excessive increase in the number of apoptotic cells [93].

5.1.4. Targets of miR-302/367 Cluster

To achieve the functional characterization of miRNAs, it is crucial to identify their target genes. Cell cycle regulators were the first-identified miR-302/367 targets both in human and mouse ESCs. In human ESCs, miR-302 modulates G1/S transition by directly targeting components of the cell cycle pathway, such as *CYCLIN D1* and *D2*, cyclin-dependent kinase 2 (*CDK2*), *CDK4*, *RB*, *E2F1*, *P130*, and *CDK6* [68]. In addition to repressing CYCLIN D1/D2 and other cell cycle components, it promotes S phase entry through an alternative pathway [94]. Likewise, in mouse ESCs, miR-302/367 suppresses the cyclin-dependent kinase inhibitor 1, also known as *p21*, thus inducing the G1 to S phase transition and cell proliferation [61]. It is also suggested that miR-302a modulates cell cycle progression in mouse ESCs through the negative regulation of *Lats2*, a tumor suppressor gene [95]. Sodium butyrate-mediated upregulation of the miR-302/367 cluster preserves the expression of key cell cycle regulators and supports self-renewal of human ESCs [96]. On the other hand, it also downregulates BNIP3L/Nix, thus inhibiting spontaneous apoptosis in hESCs [97]. *LEFTY1* and *LEFTY2* (Nodal inhibitors) are other prominent targets of miR-302 members. Post-transcriptional downregulation of *LEFTY* by miR-302 inhibits the expression of TGFβ/Activin/ NODAL family proteins and retains the pluripotency of hESCs through activating SMAD2/3 and inducing NANOG expression [98–100]. Similarly, *Lefty* has been identified as a direct target of miR-302 in rabbit ES-like cells by the transient inhibition of ocu-miR-302a [47]. *SMAD7* is a direct target of miR-367 in human pancreatic cancer cells and it promotes the invasion and metastasis of pancreatic cancer cells through the TGF-β signaling pathway [101]. miR-302/367 also promotes BMP signaling through direct targeting of BMP inhibitors, including *TOB2*, *DAZAP2*, and *SLAIN1*, thus maintaining pluripotency by repressing neural differentiation [102,103].

In addition to cell cycle and signaling pathway regulation, the miR-302 cluster modulates multiple key epigenetic regulators in somatic cells, including the lysine-specific histone demethylase enzymes *AOF1* and *AOF2* (also called *KDM1B* and *KDM1A*, respectively) and the methyl-CpG binding proteins *MECP1* and *MECP2*. Post-transcriptional suppression of these epigenetic regulators by miR-302 members induces global DNA demethylation and promotes reprogramming and iPS cell establishment [104]. Furthermore, *MBD2* (methyl-DNA binding domain protein 2) is a direct target of miR-302 to accomplish complete reprogramming of the iPS cell [105]. The main targets of miR-302/367 clusters are schematically presented in Figure 2.

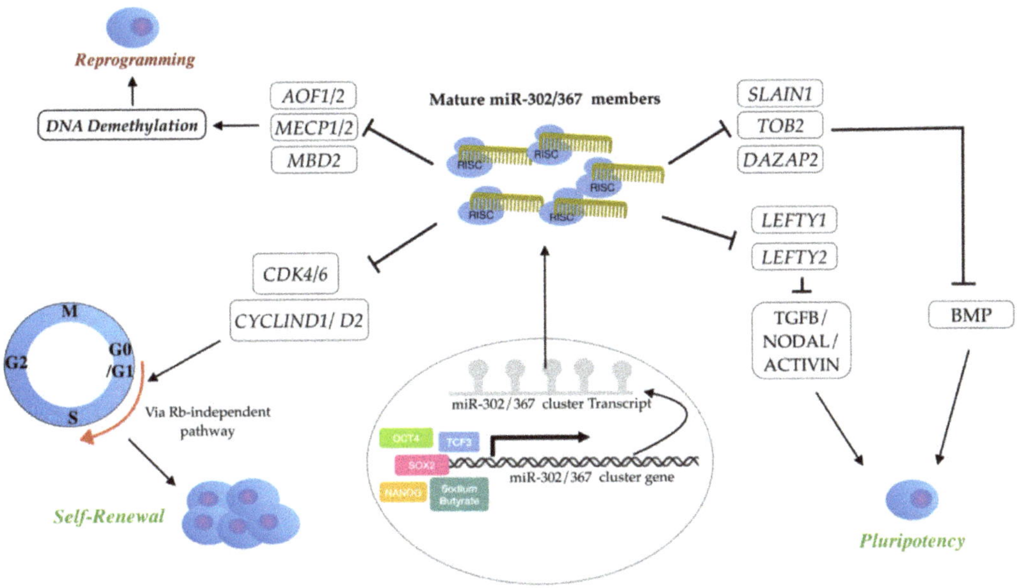

Figure 2. Main targets of miR-302/367 cluster. The miR-302-367 cluster plays a central role in maintaining ESC pluripotency and self-renewal by regulating different signaling pathways. The cytoplasmic mature miR-302/367 gene products positively regulate self-renewal by modulating G1/S transition through direct targeting of the cell cycle regulators, such as *CYCLIN D1/D2* and *CDK4/6*. Additionally, they promote S phase entry via an alternative Rb-independent pathway. These mature miRNA products also maintain the pluripotency of ESCs by targeting Nodal/Activin inhibitors, such as *LEFTY*. They also promote BMP signaling through direct targeting of BMP inhibitors (*TOB2*, *DAZAP2*, and *SLAIN1*), thus maintain pluripotency by repressing neural differentiation. Moreover, miR-302/367 members promote reprogramming and iPS cell establishment through induction of global DNA demethylation by post-transcriptional suppression of epigenetic regulators such as *AOF1/2*, *MECP1/2*, and *MBD2*.

5.2. C19miRNA Cluster

5.2.1. Characteristics of C19M Cluster

C19MC (chromosome 19 miRNA cluster), one of the longest miRNA gene clusters in the human genome, is located on chromosome 19 and extends for about 100 kb. The cluster contains 46 highly homologous pre-miRNA genes, producing 56 mature miRNAs. Despite the high similarity of miRNAs in the cluster, they harbor 16 distinct seed sequences [106]. The cluster itself exhibits a unique genomic structure: most miRNA genes are flanked by ~400–700 bp Alu repeated sequences and short exons with highly repetitive DNA elements [107,108]. The Alu element reportedly mediates gene duplication events in this region, resulting in the expansive miRNA cluster and allowing for its high expression from the multiple copies [107,109]. Reciprocally, free Alu transcripts are post-transcriptionally suppressed by expressed miRNAs from the cluster; thus, the genome self-destruction caused by high rates of duplicative Alu transposition can be prevented [107]. In addition, hsa-miR-371/373 is located in close proximity to C19MC at about 20 kbp downstream of the cluster.

It has been suggested that C19MC miRNAs are primate-specific and have no orthologous regions in the mouse genome; however, we have previously identified three homologs in rabbits, including ocu-miR-512, ocu-miR-498, and ocu-miR-520e [47]. All three mature rabbit miRNAs presented high sequence similarity to their human homologs. They are located 5′ to the rabbit miR-290/295 cluster on the reverse strand of the pseudo-chromosome chrUn0226 in the same order as in humans (Figure 3a,b). Ocu-miR-520e shares the common consensus seed sequence of "AAGTGCT" with its human homologs, but ocu-miR-512 harbors a slightly different seed sequence.

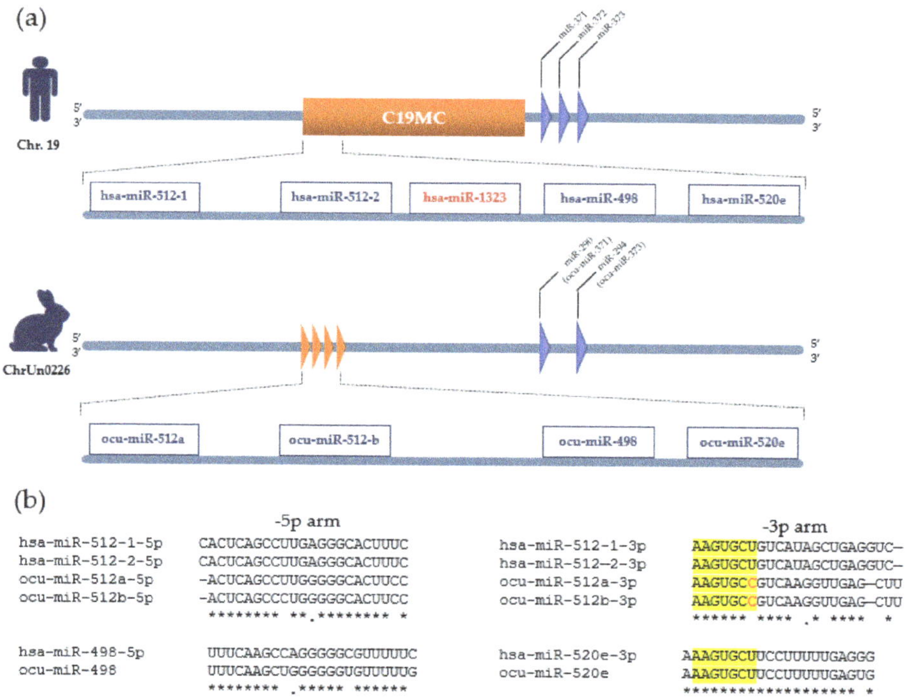

Figure 3. Genome organization and multiple sequence alignment of human C19MC cluster and three rabbit homologs. (**a**) Three rabbit homologues pre-miRNAs of human C19MC including ocu-miR-512, ocu-miR-498, and ocu-miR-520e are located on contig ChrUn0226 in same order as human C19MC on Chr.19. (**b**) Multiple sequence alignment represented high sequence similarity of rabbit mature miRNAs with their human homologues. Ocu-miR-520e shares the common consensus seed sequence of "AAGTGCT" with its human homologs, but ocu-miR-512 harbors a slightly different seed sequence. The seed sequences are highlighted in yellow. Mismatches in seed sequences are shown in red. The asterisks indicate sequence homology.

5.2.2. Expression of Human C19MC Cluster and Its Rabbit Homologs

The C19MC cluster miRNAs are trophoblast-specific and are highly expressed in the placenta [108]. In humans, C19MC is exclusively expressed from the paternal chromosome driven by an upstream promoter region. Mono-allelic expression of the C19MC cluster was demonstrated to be regulated through the DNA demethylation of the CpG-rich promoter region. Transcription of the C19MC cluster results in a primary transcript harboring the entire C19MC gene cluster, which is further processed into precursor miRNA and subsequently generates mature miRNAs via the Dgcr8-Drosha complex [110]. A recent study demonstrated that the pri-miRNA maturation of C19CM is tissue-specific and enhancer-mediated. A strong association of Drosha with the promoter/enhancer regions in human results is necessary for the efficient pri-miRNA maturation and to achieve a high expression level in placenta cells compared to the stem cells [111].

C19MC miRNAs are present in primary human-trophoblast-derived vesicles, particularly in the villous trophoblasts (VTs) [49,112]. They have also been detected in human extravillous trophoblasts (EVTs), where they may regulate migration [49]. Among

C19MC-related miRNAs, ocu-miR-512 showed a high expression level in trophoblasts and hypoblasts of rabbit pre-implanted embryos [47].

In addition to the placenta, C19MC miRNAs are expressed in human ESCs and contribute to stemness [113,114]. Almost all C19MC miRNAs are active in naïve human ESCs, display high expression levels, and are epigenetically silenced in primed hESCs [115]. Unlike hESCs, rabbit ES-like cells do not exhibit a significant expression level of C19MC, which may reflect the primed state of rabbit ES-like cells [47]. The cluster is also expressed in human iPSCs and enhances reprogramming by suppressing epithelial-to-mesenchymal transition [116,117].

Furthermore, C19MC miRNAs are selectively expressed in various cancer types, such as breast cancer, hepatocellular carcinoma, embryonal brain tumors, infantile hemangioma, testicular germ cell tumors, parathyroid tumors, and thyroid adenomas. The overexpression of cancer-derived, circulating C19MC miRNAs, which commonly correspond to tumor size and proliferative state, makes them attractive potential biomarkers for diagnosis and treatment response monitoring [118–125].

5.2.3. Targets of C19MC Cluster

Computational miRNA target prediction tools identified 4734 target genes for the C19MC cluster [114,115]. Among them, genes involved in chromatin structure modifications [114], the p53 pathway (including *CCNG2, CDKN1A, PMAIP1, TP53INP1,* and *ZMAT3*), and the extracellular matrix (ECM) [115] were enriched in hESCs.

miR-524 enhances somatic cell reprogramming by targeting *TP53INP1* (Tumor Protein P53 Inducible Nuclear Protein 1), promotes cell proliferation, and inhibits apoptosis [116]. Moreover, miR-524 downregulates *ZEB2* and *SMAD4* (epithelial–mesenchymal-transition-related genes) and promotes mesenchymal-to-epithelial transition, which is required for the initiation of reprogramming [126,127].

5.3. Mmu-miR-290/295, hsa-miR-371/373, and ocu-miR-290/295 Clusters

5.3.1. Cluster Characteristics

The mmu-miR-290/295, hsa-miR-371/373, and ocu-miR-290/295 clusters are highly conserved in humans, mice, rabbits, chimpanzees, rats, dogs, and cows [128,129]; however, some differences are present in their structure, number of members, and genomic location.

The mmu-miR-290/295 cluster, which is the most abundant miRNA cluster in mouse ESCs [130], is coded by a 2.2 kb region located on chromosome 7 (Figure 4a). The single spliced primary transcript is processed into 14 mature miRNAs: miR-290-5p, miR-290-3p, miR-291a-5p, miR-291a-3p, miR-291b-5p, miR-291b-3p, miR-292-5p, miR-292-3p, miR-293-5p, miR-293-3p, miR-294-5p, miR-294-3p, miR-29-5p, and miR-295-3p [64,128]. All pre-miRNAs of the cluster share the same "AAGTGC" seed sequence, except for the miR-293 pre-miRNA (Figure 4b). The cluster's transcription is regulated by a 332 nt intragenic enhancer region within the cluster [131]. Additionally, pluripotency-associated transcription factors, such as Oct4, Sox2, Snai, c-Myc, and Nanog, directly control the cluster expression by binding an upstream promoter element [66,132].

miR-371/373, a human ortholog of the mmu-miR-290/295 cluster, is located in close proximity to C19MC on chromosome 19 (Figure 4a) [64,125]. The hsa-miR-371/373 cluster is transcribed into four pre-miRNAs, pre-miR-371a, pre-miR-371b, pre-miR-372, and pre-miR-373, and further processed into six mature mRNAs: miR-371a-3p, miR-371a-5p, miR-371b-3p, miR-371b-5p, miR-372, and miR-373 [65,133,134]. The ESC-specific seed sequence "AAGTGC" can be found in all of the cluster's miRNAs (Figure 4b).

Rabbit homologs of the mmu-miR-290/295 cluster are located on the reverse strand of a short pseudo-chromosome (chrUn0226) and are comprised of three mature miRNAs: ocu-miR290-5p, ocu-miR-292-3p, and ocu-miR-294-3p (Figure 4a). The ocu-miR-294-3p miRNA has also been mapped on chromosome 2 of the rabbit genome. Like hsa-miR-371/373, the ocu-miR290/295 cluster is located 3′ to the rabbit C19MC (Figure 4a). Interestingly, based on putative secondary structure prediction, ocu-miR-290-5p and ocu-miR-292-3p form a single pre-miRNA. Only ocu-miR-294 contains the stem-cell-specific seed sequence "AAGTGCT" (Figure 4b) [47].

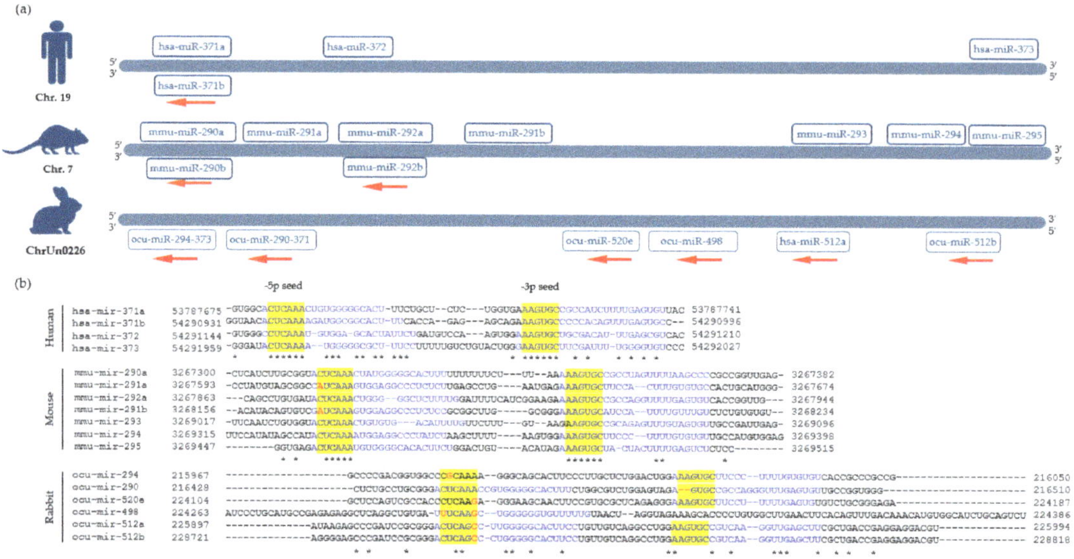

Figure 4. Genome organization and multiple sequence alignment of mmu-miR-290/295, ocu-miR-290/295, and hsa-miR-371/373 precursors in mice, humans, and rabbits. (**a**) miR-371/373, a human ortholog of the mmu-miR-290/295 cluster, is located in close proximity to C19MC on chromosome 19 and comprises of four pre-miRNAs. Rabbit homologs of the mmu-miR-290/295 cluster are located on the reverse strand of a short pseudo-chromosome (chrUn0226) and are comprised of three mature miRNAs. Similar to hsa-miR-371/373, the ocu-miR290/295 cluster is located 3′ to the rabbit C19MC. (**b**) All pre-miRNAs of mmu-miR-290/295 and hsa-miR-371/373 share ESC-specific seed sequence of "AAGTGC" except for mmu- miR-293 pre-miRNA. From rabbit cluster only ocu-miR-294 contains the same seed sequence. The seed sequences are highlighted in yellow. Mismatches in seed sequences are shown in red. Mature -5p and -3p miRNA sequences are shown in blue. The asterisks indicate sequence homology.

5.3.2. Evolution of mmu-miR-290/295, hsa-miR-371/373, and ocu-miR-290/295 Clusters

It is speculated that clustered miRNAs evolved via gene duplication, genomic rearrangements, and specific gene repressions. Evolving novel clusters might have comprised highly homologous miRNAs or could be joined with other miRNA families. In addition, miRNA family members can either be located in a cluster family or distributed randomly throughout the genome [135–137].

A cluster sequence and repeat analysis showed a close evolutionary relationship between human, mouse, and rabbit miRNA clusters. The miR-290/295 cluster has likely evolved from miR-290 or miR-291a and appears to be the evolutionary precursor. The individual pre-miRNA hairpin sequences of mmu-miR-290/295 and hsa-miR-371/373 clusters are homologous to each other. It appears that mouse miR-290/295 is the possible origin of the human miR-371/373 cluster, while the hsa-miR-371/373 cluster itself gave rise to another miRNA cluster, the hsa-miR-512 cluster, located in close proximity to its ancestor on human chromosome 19 [67]. Moreover, the enthalpy levels of pre-miR-372 and pre-miR-373 are similar to mmu-miR-291a, mmu-miR-291b, and mmu-miR-294 [67]. It has also been demonstrated that hsa-miR-373 is a member of the miR-520/373 family, composed of the hsa-miR-302/367, hsa-miR-371/373, and hsa-miR-520 clusters that share identical seed sequences [138–140].

Furthermore, according to the seed sequence and target interaction analysis, it seems that mmu-miR-290/295 and hsa-miR-371/373 share the same seed repertoire, given that in mice the cluster transcribes seven pre-miRNAs and in humans the cluster transcribes just three pre-miRNAs but the pre-miR-371 carries all the seeds to form the orthologs pre-miR-290, pre-miR-292, and pre-miR-293 in mice [141].

As mentioned above, the rabbit miR-290/295 cluster resides in only three miRNA members, representing a low evolutionary conservation with its homolog cluster in mice. Moreover, ocu-miR-290-5p and ocu-miR-292-3p form a single pre-miRNA, in contrast to the mouse homolog cluster which encodes three pre-miRNAs (pre-miR-290, pre-miR-291a, and pre-miR-292) in the same genomic region [47]. In addition, in proximity to the ocu-miR-290/295 cluster on the contig chromosome, there are three miRNAs of the C19MC cluster of ocu-miR-512, ocu-miR-520e, and ocu-miR-498, similar to the human and mouse cluster distribution. Hence, it is possible that these rabbit miRNAs located on the contig chromosome might recapitulate the evolution of human homologs and derive from the same ancestor by a duplication event [47].

5.3.3. The Expression of mmu-miR-290/295, hsa-miR-371/373, and ocu-miR-290/295 Clusters

Mmu-miR-290/295 miRNAs are the first de novo expressed miRNAs in mouse embryos, starting at the two- to four-cell stage and exhibiting high expression in blastocysts [43,45,46]. The cluster is abundantly expressed in undifferentiated mESCs and maintains the pluripotent state of mESCs [66,130]. The overexpression of mmu-miR290/295 miRNAs under serum starvation retains the stem cell properties of mESCs and regulates the cell cycle by preventing a G1 stop and extending the survival of the stem cell population [67]. In addition, the simultaneous introduction of mmu-miR-290/295 along with Oct4, Sox2, and Klf4 (OSK) enhances the reprogramming of mouse embryonic fibroblasts into mouse iPSCs [142].

In contrast to mmu-miR-290/295, hsa-miR-371/373 expression in human ESCs is considerably lower and comprises only 1% of the total miRNA pool in hESCs [65,143]. Hsa-miR-372 and hsa-miR-373 display a high level of expression in human testicular germ cell tumors, indicating their oncogenic potential [133]. The introduction of hsa-miR-372 along with hsa-miR-302b enhanced the reprogramming efficiency of human fibroblasts [144].

The ocu-miR-290/295 miRNAs show low expression levels in rabbit ES-like cells, whereas they are highly expressed during preimplantation embryonic development, implying their potential role during early embryonic development [47].

Taken all together, the expression patterns of mmu- and ocu-miR-290/295 and hsa-miR-371/373 in ESCs and in early embryos suggest a prominent function in early embryonic development.

5.3.4. miR-290/295 Cluster Targets

The mmu-miR-290/295 cluster is a well-described cell cycle regulator miRNA cluster. A short G1 phase and lack of a G1/S checkpoint are characteristics of the ESC cell cycle. The mouse miR-290/295 cluster regulates G1/S transition through the downregulation of cell cycle inhibitors such as *p21*, *Lats2*, *Wee1*, and *Fbxl5*, thus maintaining the pluripotent state of mouse ESCs [56,61,67,145]. Additionally, rapid G1/S transition promoted by mmu-miR-294 miRNA is Retinoblastoma gene (Rb)-independent. The accumulation of cells in G1 is regulated by direct targeting of the Rb family under nutrient starvation [94]. On the other hand, the mmu-miR-290/295 cluster reportedly represses S/G2 transition through *p53* and *Cyclin D1* [67,146]. Therefore, the mmu-miR-290/295 cluster contributes to pluripotency by accelerating G1 and extending the S phase in mESC.

Furthermore, mmu-miR-290/295 modulates DNA methylation in the pre-implantation embryos by the targeting of retinoblastoma-like 2 (*Rbl2*) and the subsequent inhibition of *Dnmt3b* [62,147]. On the one hand, pluripotent-associated transcription factors such as *Oct4*, *Sox2*, and *Nanog* either directly or indirectly regulate the transcription of DNMT3B.

On the other hand, the core promoter region of the mmu-miR-290/295 cluster is a direct target for pluripotent-associated transcription factors. Altogether, pluripotent-associated transcription factors, the mmu-miR-290/295 cluster, *Rbl2*, and *Dnmt3b* build a regulatory network that regulates DNA methylation in early embryonic development and in ESCs [148]. Moreover, mmu-miR-290/295 directly targets *Dkk-1*, which is a Wnt pathway inhibitor, resulting in the upregulation of c-Myc, a downstream target of the Wnt signaling pathway [149,150]. Thus, mmu-miR-290/295 miRNAs partially favor pluripotency upon differentiation.

The mmu-miR-290/295 cluster together with miR-302/376 promote transition from naïve to primed pluripotency by enhancing the activity of the MEK pathway through the direct repression of *Akt1* [151]. In addition, mmu-miR-290/295 miRNAs are required for the incorporation of two core components of polycomb repressive complexes 2 (*Prc2*), *Ezh2*, and *Suz12* into promoters of bivalent differentiation genes in order to maintain the pluripotent state of mouse ESCs [152,153].

The mmu-miR-290/295 cluster also enhances the reprogramming of somatic cells and the quality of mouse iPSCs [89,142]. Mmu-miR-291 post-transcriptionally suppresses the methyltransferase *Ash1l* expression and consequently downregulates *HOX* genes, which in turn promote reprogramming through polycomb-mediated gene silencing [153].

In summary, the mmu-miR-290/295 cluster modulates both stemness and differentiation characteristics of mouse ESCs and embryos by regulating the cell cycle, de novo DNA methylation, apoptosis, and the transcription of pluripotent-associated transcription factors (Figure 5). To date, no target gene has been predicted for the rabbit miR-290/295 cluster. Certainly, target recognition and the investigation of functional characteristics are indispensable to discovering and describing the impact of the ocu-miR-290/295 cluster in early embryonic development and stem cell biology.

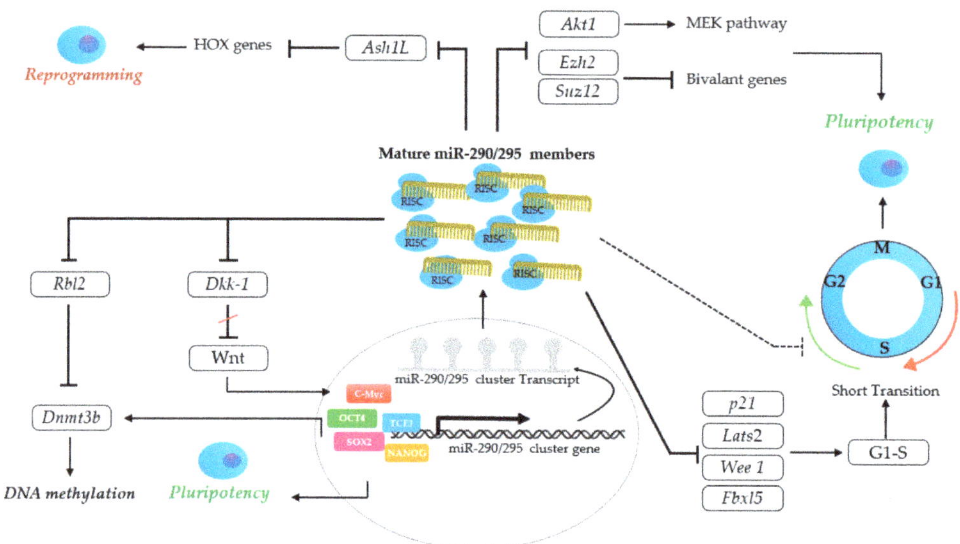

Figure 5. Main targets of miR-290/294 cluster. The miR-290/295 cluster modulates both stemness and differentiation characteristics of mouse ESCs and embryos by regulating cell cycle, de novo DNA methylation, apoptosis, and transcription of pluripotent-associated transcription factors. The cytoplasmic mature miR-290/295 members maintain pluripotency of mESC by accelerating G1 and extending S phase through direct targeting of cell cycle inhibitors such as *p21*, *Lats2*, *Wee1*, and *Fbxl5*, and also repressing S/G2 transition. These mature miRNAs also incorporate *Prc2*, *Ezh2* and *Suz12* into promoters of bivalent differentiation genes. The miR-290/295 cluster together with miR-302/376

promote transition from naïve to primed pluripotency by enhancing the activity of MEK pathway through direct repression of *Akt1*. These mature miRNAs partially favor pluripotency upon differentiation by direct targeting of *Dkk-1*, a Wnt pathway inhibitor, resulting in upregulation of c-Myc. The mature miR-290/295 gene products also modulate DNA methylation in the pre-implantation embryos by targeting of *Rbl2* and subsequent inhibition of *Dnmt3b*. The miR-290/295 cluster also enhances the reprogramming of somatic cells and the quality of mouse iPSCs through post-transcriptional suppression of *Ash1l* expression and consequently downregulation of *HOX* genes, which in turn promote reprogramming through polycomb-mediated gene silencing.

5.3.5. Targets of hsa-miR-371/373 Cluster

To date, very little known is about how the hsa-miR-371-373 cluster can integrate into the pluripotency regulation of stem cells. The miR-372 miRNA together with miR-195 can partially rescue DICER knockdown phenotypes in hESCS but this is not sufficient to influence the cell cycle kinetics in wild-type hESCs. However, miR-372 represses *P21* (a CDK inhibitor); thus, it indirectly controls the G1/S transition and consequently regulates hESC division [70].

Due to its high level of expression and its potential biomarker capacity in different human tumor cells, its mechanism of action has been further dissected in relation to its oncogenic function. The miR-371/373 cluster activates WNT/β-catenin signaling through the repression of *DKK1*, which in turn induces cell growth and the invasive activity of tumor cells in humans [154]. However, it is not determined if the miR-371-373 cluster preserves the pluripotency of hESCs by activation of the WNT/β-catenin signaling pathway. In testicular germ cell tumor cells, the suppression of *LATS2* by miR-371/373 miRNAs prevents p53-mediated CDK inhibition and promotes tumorigenic growth [133]. Moreover, miR-373 miRNA represses *CD44*, thus leading to tumor migration and invasion in human breast cancer [138].

It also has been reported that miR-371/373 regulates glycolysis through the repression of *MBD2I*, resulting in promoting the expression of MYC and thus increasing human fibroblast reprogramming [155].

6. Conclusions

miRNAs are well-recognized, major post-transcriptional regulators of most cellular events including proliferation, cell cycle progression, differentiation, survival, and apoptosis in many cell types. Extensive research on the regulatory function of miRNAs indicates that miRNAs play crucial roles in stem cell properties. Pluripotent stem cells have enormous potential in the field of regenerative medicine, disease modeling, and new drug discoveries. In this review, we aimed to compare and discuss the regulatory functions of stem-cell-specific miRNA clusters and their action on the maintenance and regulation of ESC biology in vertebrates. We also discussed the expression profile of stem-cell-specific miRNAs in early embryogenesis, which may play an important role in the establishment and maintenance of pluripotent progenitor cells. Furthermore, we compared the stem-cell-specific human miRNAs with our other vertebrate model species.

Several miRNA clusters have been identified to regulate the stemness and pluripotency of stem cells. The most important miRNA clusters which are highly expressed in ESCs of vertebrates are miR-302/367 and miR-290/295 and its human homologue miR-371/373. They regulate the self-renewal and pluripotency of ESCs through the modulation of cell cycle progression. These miRNA clusters share a highly conserved seed sequence, "AAGUGCU" (referred to as stem-cell-specific miRNA) (Figures 2 and 4), and induce the pluripotency of ESCs by direct targeting of the 3′ UTR of key ESC transcription factors, such as OCT4, SOX2, and NANOG. They also play a predominant role in vertebrate early development. Likewise, C19MC miRNAs are highly expressed in human ESCs, contribute to the stemness state of hESCs, and prevent their differentiation. Some members of C19MC miRNAs and their rabbit homologs also share the same stem-cell-specific "AAGTGCT" seed sequence (Figure 3).

The profiles of ESC-specific miRNA clusters and their relative functions in vertebrates which have been discussed through this review are summarized in Table 1.

Table 1. Summary of ESC-specific miRNAs and their relative function.

Cell Type	miRNAs	Expression Level	Function	Signaling Pathways	Targets
Mouse ESCs	miR-302/367 cluster	Low expression	Regulation of pluripotency, self-renewal, and reprogramming	Cdkn1a [61] TGF-β signaling pathway BMP signaling pathway	NA Lats2 [95]
	miR-290/295 cluster	High expression	Regulation of cell cycle progression and naïve pluripotency	ATM/ATR chek2-p53 pathway Wnt signaling pathway MEK Pathway activation	*Cyclin D1, p21, Lats2* [67,146] *Dkk-1* [150] *Akt1* [151]
			Early phases of differentiation	DNA methylation *HOX* gene inactivation	*Rbl2* [62,147] *Ash1l* [153]
Human ESCs	miR-302/367 cluster	High expression	Regulation of pluripotency, self-renewal, and reprogramming	Cyclin D1/D2 and Cdk2 [68] Nodal–Activin pathway	*LEFTY* [98–100] *BNIP3L/Nix* [97]
	miR-371/373 cluster	High expression	Cell cycle regulation Reprogramming	G1/S transition P53 pathway	*P21* [70] *MBD2* [155]
	C19MC	High expression in naïve hESCs and iPCs	Pluripotency maintenance	Chromatin structure modification pathway [114]	NA
			Prevent differentiation	ECM- related pathway [115]	NA
			Derivation and maintenance of human trophoblast stem cells	P53 pathway	*CCNG2 DKN1A PMAIP1 ZMAT3 TP53INP1* [115]
Rabbit ES-like cells	miR-302/367 cluster	High expression	Regulation of pluripotency and self-renewal	Nodal–Activin pathway	*Lefty* [47]
	miR-290/295 cluster	Low expression	Early embryonic development	NA	NA
	C19MC	Low expression	Early embryonic development	NA	NA
Chicken PGC line	miR-302b miR-302a	High expression	Cell proliferation Prevent differentiation	Glycolysis metabolism Somatic gene silencing	*Gpi3* [92] *Sox11* [91]

NA: not addressed.

This review gives outstanding insights into the cell cycle regulation of stem cells by miRNAs. A deep understanding of regulatory molecular networks in which miRNAs are interacting will greatly enhance our knowledge of miRNAs' contribution to stem cell biology and therefore open promising avenues toward stem cell therapy.

Author Contributions: Conceptualization: E.G., P.M. and M.T.S.A.; writing—original draft preparation: P.M., M.T.S.A. and E.G.; writing—review and editing: B.L., A.E. and R.T.; visualization: M.T.S.A. and N.T.S.; supervision: E.G.; funding acquisition: E.G. All authors have read and agreed to the published version of the manuscript.

Funding: This study was supported by National Research, Development and Innovation Office, grants RRF-2.3.1-21-2022-00007 and 2019-2.1.11-TÉT-2019-00036.

Institutional Review Board Statement: All applied methods in NBGK-HGI were approved by the Directorate of Food Safety and Animal Health of the Government Office of Pest County (License number: PE/EA197-4/2016) and by the Institutional Ethical Review Board.

Informed Consent Statement: Not applicable.

Data Availability Statement: The data presented in this study are available on request from the corresponding author.

Conflicts of Interest: The authors declare no conflict of interest.

References

1. Lee, C.T.; Risom, T.; Strauss, W.M. MicroRNAs in mammalian development. *Birth Defects Res. C Embryo Today* **2006**, *78*, 129–139. [CrossRef] [PubMed]
2. Rosa, A.; Brivanlou, A.H. MicroRNAs in early vertebrate development. *Cell Cycle* **2009**, *8*, 3513–3520. [CrossRef] [PubMed]
3. Bartel, D.P. MicroRNAs: Genomics, biogenesis, mechanism, and function. *Cell* **2004**, *116*, 281–297. [CrossRef] [PubMed]
4. Bartel, D.P. MicroRNAs: Target recognition and regulatory functions. *Cell* **2009**, *136*, 215–233. [CrossRef]
5. Lee, Y.; Kim, M.; Han, J.; Yeom, K.H.; Lee, S.; Baek, S.H.; Kim, V.N. MicroRNA genes are transcribed by RNA polymerase II. *EMBO J.* **2004**, *23*, 4051–4060. [CrossRef]
6. Ha, M.; Kim, V.N. Regulation of microRNA biogenesis. *Nat. Rev. Mol. Cell Biol.* **2014**, *15*, 509–524. [CrossRef]
7. Franca, G.S.; Vibranovski, M.D.; Galante, P.A. Host gene constraints and genomic context impact the expression and evolution of human microRNAs. *Nat. Commun.* **2016**, *7*, 11438. [CrossRef]
8. Kim, Y.K.; Kim, V.N. Processing of intronic microRNAs. *EMBO J.* **2007**, *26*, 775–783. [CrossRef]
9. Georgakilas, G.; Vlachos, I.S.; Paraskevopoulou, M.D.; Yang, P.; Zhang, Y.; Economides, A.N.; Hatzigeorgiou, A.G. microTSS: Accurate microRNA transcription start site identification reveals a significant number of divergent pri-miRNAs. *Nat. Commun.* **2014**, *5*, 5700. [CrossRef]
10. Guo, L.; Zhao, Y.; Zhang, H.; Yang, S.; Chen, F. Integrated evolutionary analysis of human miRNA gene clusters and families implicates evolutionary relationships. *Gene* **2014**, *534*, 24–32. [CrossRef]
11. Pal, A.S.; Kasinski, A.L. Animal Models to Study MicroRNA Function. *Adv. Cancer Res.* **2017**, *135*, 53–118. [CrossRef]
12. Baskerville, S.; Bartel, D.P. Microarray profiling of microRNAs reveals frequent coexpression with neighboring miRNAs and host genes. *RNA* **2005**, *11*, 241–247. [CrossRef]
13. Lagos-Quintana, M.; Rauhut, R.; Lendeckel, W.; Tuschl, T. Identification of novel genes coding for small expressed RNAs. *Science* **2001**, *294*, 853–858. [CrossRef]
14. Seitz, H.; Royo, H.; Bortolin, M.L.; Lin, S.P.; Ferguson-Smith, A.C.; Cavaille, J. A large imprinted microRNA gene cluster at the mouse Dlk1-Gtl2 domain. *Genome Res.* **2004**, *14*, 1741–1748. [CrossRef]
15. Yeom, K.H.; Lee, Y.; Han, J.; Suh, M.R.; Kim, V.N. Characterization of DGCR8/Pasha, the essential cofactor for Drosha in primary miRNA processing. *Nucleic Acids Res.* **2006**, *34*, 4622–4629. [CrossRef]
16. Yi, R.; Qin, Y.; Macara, I.G.; Cullen, B.R. Exportin-5 mediates the nuclear export of pre-microRNAs and short hairpin RNAs. *Genes. Dev.* **2003**, *17*, 3011–3016. [CrossRef]
17. Hutvagner, G.; Zamore, P.D. A microRNA in a multiple-turnover RNAi enzyme complex. *Science* **2002**, *297*, 2056–2060. [CrossRef]
18. Chendrimada, T.P.; Gregory, R.I.; Kumaraswamy, E.; Norman, J.; Cooch, N.; Nishikura, K.; Shiekhattar, R. TRBP recruits the Dicer complex to Ago2 for microRNA processing and gene silencing. *Nature* **2005**, *436*, 740–744. [CrossRef]
19. Medley, J.C.; Panzade, G.; Zinovyeva, A.Y. microRNA strand selection: Unwinding the rules. *Wiley Interdiscip. Rev. RNA* **2021**, *12*, e1627. [CrossRef]
20. Brennecke, J.; Stark, A.; Russell, R.B.; Cohen, S.M. Principles of microRNA-target recognition. *PLoS Biol.* **2005**, *3*, e85. [CrossRef]
21. Davis-Dusenbery, B.N.; Hata, A. Mechanisms of control of microRNA biogenesis. *J. Biochem.* **2010**, *148*, 381–392. [CrossRef] [PubMed]
22. Okamura, K.; Hagen, J.W.; Duan, H.; Tyler, D.M.; Lai, E.C. The mirtron pathway generates microRNA-class regulatory RNAs in Drosophila. *Cell* **2007**, *130*, 89–100. [CrossRef] [PubMed]
23. Ruby, J.G.; Jan, C.H.; Bartel, D.P. Intronic microRNA precursors that bypass Drosha processing. *Nature* **2007**, *448*, 83–86. [CrossRef]
24. Babiarz, J.E.; Ruby, J.G.; Wang, Y.; Bartel, D.P.; Blelloch, R. Mouse ES cells express endogenous shRNAs, siRNAs, and other Microprocessor-independent, Dicer-dependent small RNAs. *Genes. Dev.* **2008**, *22*, 2773–2785. [CrossRef] [PubMed]
25. Ender, C.; Krek, A.; Friedlander, M.R.; Beitzinger, M.; Weinmann, L.; Chen, W.; Pfeffer, S.; Rajewsky, N.; Meister, G. A human snoRNA with microRNA-like functions. *Mol. Cell* **2008**, *32*, 519–528. [CrossRef]
26. Saraiya, A.A.; Wang, C.C. snoRNA, a novel precursor of microRNA in Giardia lamblia. *PLoS Pathog.* **2008**, *4*, e1000224. [CrossRef]
27. Bogerd, H.P.; Karnowski, H.W.; Cai, X.; Shin, J.; Pohlers, M.; Cullen, B.R. A mammalian herpesvirus uses noncanonical expression and processing mechanisms to generate viral MicroRNAs. *Mol. Cell* **2010**, *37*, 135–142. [CrossRef]
28. Miyoshi, K.; Miyoshi, T.; Hartig, J.V.; Siomi, H.; Siomi, M.C. Molecular mechanisms that funnel RNA precursors into endogenous small-interfering RNA and microRNA biogenesis pathways in Drosophila. *RNA* **2010**, *16*, 506–515. [CrossRef]
29. Yu, J.; Yang, Z.; Kibukawa, M.; Paddock, M.; Passey, D.A.; Wong, G.K. Minimal introns are not "junk". *Genome Res.* **2002**, *12*, 1185–1189. [CrossRef]
30. Berezikov, E.; Chung, W.J.; Willis, J.; Cuppen, E.; Lai, E.C. Mammalian mirtron genes. *Mol. Cell* **2007**, *28*, 328–336. [CrossRef]
31. Yang, J.S.; Maurin, T.; Robine, N.; Rasmussen, K.D.; Jeffrey, K.L.; Chandwani, R.; Papapetrou, E.P.; Sadelain, M.; O'Carroll, D.; Lai, E.C. Conserved vertebrate mir-451 provides a platform for Dicer-independent, Ago2-mediated microRNA biogenesis. *Proc. Natl. Acad. Sci. USA* **2010**, *107*, 15163–15168. [CrossRef]
32. Cheloufi, S.; Dos Santos, C.O.; Chong, M.M.; Hannon, G.J. A dicer-independent miRNA biogenesis pathway that requires Ago catalysis. *Nature* **2010**, *465*, 584–589. [CrossRef]
33. Weiner, A.M. E Pluribus Unum: 3′ end formation of polyadenylated mRNAs, histone mRNAs, and U snRNAs. *Mol. Cell* **2005**, *20*, 168–170. [CrossRef]

34. Wightman, B.; Ha, I.; Ruvkun, G. Posttranscriptional regulation of the heterochronic gene lin-14 by lin-4 mediates temporal pattern formation in C. elegans. *Cell* **1993**, *75*, 855–862. [CrossRef]
35. Reinhart, B.J.; Slack, F.J.; Basson, M.; Pasquinelli, A.E.; Bettinger, J.C.; Rougvie, A.E.; Horvitz, H.R.; Ruvkun, G. The 21-nucleotide let-7 RNA regulates developmental timing in Caenorhabditis elegans. *Nature* **2000**, *403*, 901–906. [CrossRef]
36. Pasquinelli, A.E.; Reinhart, B.J.; Slack, F.; Martindale, M.Q.; Kuroda, M.I.; Maller, B.; Hayward, D.C.; Ball, E.E.; Degnan, B.; Muller, P.; et al. Conservation of the sequence and temporal expression of let-7 heterochronic regulatory RNA. *Nature* **2000**, *408*, 86–89. [CrossRef]
37. Lagos-Quintana, M.; Rauhut, R.; Yalcin, A.; Meyer, J.; Lendeckel, W.; Tuschl, T. Identification of tissue-specific microRNAs from mouse. *Curr. Biol.* **2002**, *12*, 735–739. [CrossRef]
38. Bernstein, E.; Kim, S.Y.; Carmell, M.A.; Murchison, E.P.; Alcorn, H.; Li, M.Z.; Mills, A.A.; Elledge, S.J.; Anderson, K.V.; Hannon, G.J. Dicer is essential for mouse development. *Nat. Genet.* **2003**, *35*, 215–217. [CrossRef]
39. Liu, J.; Carmell, M.A.; Rivas, F.V.; Marsden, C.G.; Thomson, J.M.; Song, J.J.; Hammond, S.M.; Joshua-Tor, L.; Hannon, G.J. Argonaute2 is the catalytic engine of mammalian RNAi. *Science* **2004**, *305*, 1437–1441. [CrossRef]
40. Murchison, E.P.; Stein, P.; Xuan, Z.; Pan, H.; Zhang, M.Q.; Schultz, R.M.; Hannon, G.J. Critical roles for Dicer in the female germline. *Genes. Dev.* **2007**, *21*, 682–693. [CrossRef]
41. Wang, Y.; Medvid, R.; Melton, C.; Jaenisch, R.; Blelloch, R. DGCR8 is essential for microRNA biogenesis and silencing of embryonic stem cell self-renewal. *Nat. Genet.* **2007**, *39*, 380–385. [CrossRef] [PubMed]
42. Yang, W.J.; Yang, D.D.; Na, S.; Sandusky, G.E.; Zhang, Q.; Zhao, G. Dicer is required for embryonic angiogenesis during mouse development. *J. Biol. Chem.* **2005**, *280*, 9330–9335. [CrossRef] [PubMed]
43. Tang, F.; Kaneda, M.; O'Carroll, D.; Hajkova, P.; Barton, S.C.; Sun, Y.A.; Lee, C.; Tarakhovsky, A.; Lao, K.; Surani, M.A. Maternal microRNAs are essential for mouse zygotic development. *Genes. Dev.* **2007**, *21*, 644–648. [CrossRef] [PubMed]
44. Liu, W.M.; Pang, R.T.; Chiu, P.C.; Wong, B.P.; Lao, K.; Lee, K.F.; Yeung, W.S. Sperm-borne microRNA-34c is required for the first cleavage division in mouse. *Proc. Natl. Acad. Sci. USA* **2012**, *109*, 490–494. [CrossRef] [PubMed]
45. Zeng, F.; Schultz, R.M. RNA transcript profiling during zygotic gene activation in the preimplantation mouse embryo. *Dev. Biol.* **2005**, *283*, 40–57. [CrossRef]
46. Svoboda, P.; Flemr, M. The role of miRNAs and endogenous siRNAs in maternal-to-zygotic reprogramming and the establishment of pluripotency. *EMBO Rep.* **2010**, *11*, 590–597. [CrossRef]
47. Maraghechi, P.; Hiripi, L.; Toth, G.; Bontovics, B.; Bosze, Z.; Gocza, E. Discovery of pluripotency-associated microRNAs in rabbit preimplantation embryos and embryonic stem-like cells. *Reproduction* **2013**, *145*, 421–437. [CrossRef]
48. Zhang, J.; Wang, Y.; Liu, X.; Jiang, S.; Zhao, C.; Shen, R.; Guo, X.; Ling, X.; Liu, C. Expression and potential role of microRNA-29b in mouse early embryo development. *Cell Physiol. Biochem.* **2015**, *35*, 1178–1187. [CrossRef]
49. Xie, L.; Mouillet, J.F.; Chu, T.; Parks, W.T.; Sadovsky, E.; Knofler, M.; Sadovsky, Y. C19MC microRNAs regulate the migration of human trophoblasts. *Endocrinology* **2014**, *155*, 4975–4985. [CrossRef]
50. Rosenbluth, E.M.; Shelton, D.N.; Sparks, A.E.; Devor, E.; Christenson, L.; Van Voorhis, B.J. MicroRNA expression in the human blastocyst. *Fertil. Steril.* **2013**, *99*, 855–861.e853. [CrossRef]
51. Fu, G.; Ye, G.; Nadeem, L.; Ji, L.; Manchanda, T.; Wang, Y.; Zhao, Y.; Qiao, J.; Wang, Y.L.; Lye, S.; et al. MicroRNA-376c impairs transforming growth factor-beta and nodal signaling to promote trophoblast cell proliferation and invasion. *Hypertension* **2013**, *61*, 864–872. [CrossRef]
52. Dai, Y.; Qiu, Z.; Diao, Z.; Shen, L.; Xue, P.; Sun, H.; Hu, Y. MicroRNA-155 inhibits proliferation and migration of human extravillous trophoblast derived HTR-8/SVneo cells via down-regulating cyclin D1. *Placenta* **2012**, *33*, 824–829. [CrossRef]
53. Assou, S.; Al-edani, T.; Haouzi, D.; Philippe, N.; Lecellier, C.H.; Piquemal, D.; Commes, T.; Ait-Ahmed, O.; Dechaud, H.; Hamamah, S. MicroRNAs: New candidates for the regulation of the human cumulus-oocyte complex. *Hum. Reprod.* **2013**, *28*, 3038–3049. [CrossRef]
54. Dehghan, Z.; Mohammadi-Yeganeh, S.; Rezaee, D.; Salehi, M. MicroRNA-21 is involved in oocyte maturation, blastocyst formation, and pre-implantation embryo development. *Dev. Biol.* **2021**, *480*, 69–77. [CrossRef]
55. Liu, W.; Chen, J.; Yang, C.; Lee, K.F.; Lee, Y.L.; Chiu, P.C.; Zhang, Y.; Duan, Y.G.; Liu, K.; Yeung, W.S. Expression of microRNA let-7 in cleavage embryos modulates cell fate determination and formation of mouse blastocystsdagger. *Biol. Reprod.* **2022**, *107*, 1452–1463. [CrossRef]
56. Kim, V.N. Cell cycle micromanagement in embryonic stem cells. *Nat. Genet.* **2008**, *40*, 1391–1392. [CrossRef]
57. Di Mauro, V.; Crasto, S.; Colombo, F.S.; Di Pasquale, E.; Catalucci, D. Wnt signalling mediates miR-133a nuclear re-localization for the transcriptional control of Dnmt3b in cardiac cells. *Sci. Rep.* **2019**, *9*, 9320. [CrossRef]
58. Kanellopoulou, C.; Muljo, S.A.; Kung, A.L.; Ganesan, S.; Drapkin, R.; Jenuwein, T.; Livingston, D.M.; Rajewsky, K. Dicer-deficient mouse embryonic stem cells are defective in differentiation and centromeric silencing. *Genes. Dev.* **2005**, *19*, 489–501. [CrossRef]
59. Murchison, E.P.; Partridge, J.F.; Tam, O.H.; Cheloufi, S.; Hannon, G.J. Characterization of Dicer-deficient murine embryonic stem cells. *Proc. Natl. Acad. Sci. USA* **2005**, *102*, 12135–12140. [CrossRef]
60. Teijeiro, V.; Yang, D.; Majumdar, S.; Gonzalez, F.; Rickert, R.W.; Xu, C.; Koche, R.; Verma, N.; Lai, E.C.; Huangfu, D. DICER1 Is Essential for Self-Renewal of Human Embryonic Stem Cells. *Stem Cell Rep.* **2018**, *11*, 616–625. [CrossRef]
61. Wang, Y.; Baskerville, S.; Shenoy, A.; Babiarz, J.E.; Baehner, L.; Blelloch, R. Embryonic stem cell-specific microRNAs regulate the G1-S transition and promote rapid proliferation. *Nat. Genet.* **2008**, *40*, 1478–1483. [CrossRef] [PubMed]

62. Sinkkonen, L.; Hugenschmidt, T.; Berninger, P.; Gaidatzis, D.; Mohn, F.; Artus-Revel, C.G.; Zavolan, M.; Svoboda, P.; Filipowicz, W. MicroRNAs control de novo DNA methylation through regulation of transcriptional repressors in mouse embryonic stem cells. *Nat. Struct. Mol. Biol.* **2008**, *15*, 259–267. [CrossRef] [PubMed]
63. Barroso-delJesus, A.; Romero-Lopez, C.; Lucena-Aguilar, G.; Melen, G.J.; Sanchez, L.; Ligero, G.; Berzal-Herranz, A.; Menendez, P. Embryonic stem cell-specific miR302-367 cluster: Human gene structure and functional characterization of its core promoter. *Mol. Cell Biol.* **2008**, *28*, 6609–6619. [CrossRef] [PubMed]
64. Houbaviy, H.B.; Murray, M.F.; Sharp, P.A. Embryonic stem cell-specific MicroRNAs. *Dev. Cell* **2003**, *5*, 351–358. [CrossRef]
65. Suh, M.R.; Lee, Y.; Kim, J.Y.; Kim, S.K.; Moon, S.H.; Lee, J.Y.; Cha, K.Y.; Chung, H.M.; Yoon, H.S.; Moon, S.Y.; et al. Human embryonic stem cells express a unique set of microRNAs. *Dev. Biol.* **2004**, *270*, 488–498. [CrossRef]
66. Marson, A.; Levine, S.S.; Cole, M.F.; Frampton, G.M.; Brambrink, T.; Johnstone, S.; Guenther, M.G.; Johnston, W.K.; Wernig, M.; Newman, J.; et al. Connecting microRNA genes to the core transcriptional regulatory circuitry of embryonic stem cells. *Cell* **2008**, *134*, 521–533. [CrossRef]
67. Lichner, Z.; Pall, E.; Kerekes, A.; Pallinger, E.; Maraghechi, P.; Bosze, Z.; Gocza, E. The miR-290-295 cluster promotes pluripotency maintenance by regulating cell cycle phase distribution in mouse embryonic stem cells. *Differentiation* **2011**, *81*, 11–24. [CrossRef]
68. Card, D.A.; Hebbar, P.B.; Li, L.; Trotter, K.W.; Komatsu, Y.; Mishina, Y.; Archer, T.K. Oct4/Sox2-regulated miR-302 targets cyclin D1 in human embryonic stem cells. *Mol. Cell Biol.* **2008**, *28*, 6426–6438. [CrossRef]
69. Li, H.L.; Wei, J.F.; Fan, L.Y.; Wang, S.H.; Zhu, L.; Li, T.P.; Lin, G.; Sun, Y.; Sun, Z.J.; Ding, J.; et al. miR-302 regulates pluripotency, teratoma formation and differentiation in stem cells via an AKT1/OCT4-dependent manner. *Cell Death Dis.* **2016**, *7*, e2078. [CrossRef]
70. Qi, J.; Yu, J.Y.; Shcherbata, H.R.; Mathieu, J.; Wang, A.J.; Seal, S.; Zhou, W.; Stadler, B.M.; Bourgin, D.; Wang, L.; et al. microRNAs regulate human embryonic stem cell division. *Cell Cycle* **2009**, *8*, 3729–3741. [CrossRef]
71. Xu, N.; Papagiannakopoulos, T.; Pan, G.; Thomson, J.A.; Kosik, K.S. MicroRNA-145 regulates OCT4, SOX2, and KLF4 and represses pluripotency in human embryonic stem cells. *Cell* **2009**, *137*, 647–658. [CrossRef]
72. Tay, Y.; Zhang, J.; Thomson, A.M.; Lim, B.; Rigoutsos, I. MicroRNAs to Nanog, Oct4 and Sox2 coding regions modulate embryonic stem cell differentiation. *Nature* **2008**, *455*, 1124–1128. [CrossRef]
73. Tay, Y.M.; Tam, W.L.; Ang, Y.S.; Gaughwin, P.M.; Yang, H.; Wang, W.; Liu, R.; George, J.; Ng, H.H.; Perera, R.J.; et al. MicroRNA-134 modulates the differentiation of mouse embryonic stem cells, where it causes post-transcriptional attenuation of Nanog and LRH1. *Stem Cells* **2008**, *26*, 17–29. [CrossRef]
74. Jin, S.; Collin, J.; Zhu, L.; Montaner, D.; Armstrong, L.; Neganova, I.; Lako, M. A Novel Role for miR-1305 in Regulation of Pluripotency-Differentiation Balance, Cell Cycle, and Apoptosis in Human Pluripotent Stem Cells. *Stem Cells* **2016**, *34*, 2306–2317. [CrossRef]
75. Wong, R.C.; Pollan, S.; Fong, H.; Ibrahim, A.; Smith, E.L.; Ho, M.; Laslett, A.L.; Donovan, P.J. A novel role for an RNA polymerase III subunit POLR3G in regulating pluripotency in human embryonic stem cells. *Stem Cells* **2011**, *29*, 1517–1527. [CrossRef]
76. Tarantino, C.; Paolella, G.; Cozzuto, L.; Minopoli, G.; Pastore, L.; Parisi, S.; Russo, T. miRNA 34a, 100, and 137 modulate differentiation of mouse embryonic stem cells. *FASEB J.* **2010**, *24*, 3255–3263. [CrossRef]
77. Ma, Y.; Yao, N.; Liu, G.; Dong, L.; Liu, Y.; Zhang, M.; Wang, F.; Wang, B.; Wei, X.; Dong, H.; et al. Functional screen reveals essential roles of miR-27a/24 in differentiation of embryonic stem cells. *EMBO J.* **2015**, *34*, 361–378. [CrossRef]
78. Shim, J.; Nam, J.W. The expression and functional roles of microRNAs in stem cell differentiation. *BMB Rep.* **2016**, *49*, 3–10. [CrossRef]
79. Cimadamore, F.; Amador-Arjona, A.; Chen, C.; Huang, C.T.; Terskikh, A.V. SOX2-LIN28/let-7 pathway regulates proliferation and neurogenesis in neural precursors. *Proc. Natl. Acad. Sci. USA* **2013**, *110*, E3017–E3026. [CrossRef]
80. Melton, C.; Judson, R.L.; Blelloch, R. Opposing microRNA families regulate self-renewal in mouse embryonic stem cells. *Nature* **2010**, *463*, 621–626. [CrossRef]
81. Gao, Z.; Zhu, X.; Dou, Y. The miR-302/367 cluster: A comprehensive update on its evolution and functions. *Open Biol.* **2015**, *5*, 150138. [CrossRef] [PubMed]
82. Kaid, C.; Silva, P.B.; Cortez, B.A.; Rodini, C.O.; Semedo-Kuriki, P.; Okamoto, O.K. miR-367 promotes proliferation and stem-like traits in medulloblastoma cells. *Cancer Sci.* **2015**, *106*, 1188–1195. [CrossRef] [PubMed]
83. Barroso-del Jesus, A.; Lucena-Aguilar, G.; Menendez, P. The miR-302-367 cluster as a potential stemness regulator in ESCs. *Cell Cycle* **2009**, *8*, 394–398. [CrossRef] [PubMed]
84. Rahimi, K.; Fuchtbauer, A.C.; Fathi, F.; Mowla, S.J.; Fuchtbauer, E.M. Expression of the miR-302/367 microRNA cluster is regulated by a conserved long non-coding host-gene. *Sci. Rep.* **2021**, *11*, 11115. [CrossRef]
85. Guo, W.T.; Wang, X.W.; Yan, Y.L.; Li, Y.P.; Yin, X.; Zhang, Q.; Melton, C.; Shenoy, A.; Reyes, N.A.; Oakes, S.A.; et al. Suppression of epithelial-mesenchymal transition and apoptotic pathways by miR-294/302 family synergistically blocks let-7-induced silencing of self-renewal in embryonic stem cells. *Cell Death Differ.* **2015**, *22*, 1158–1169. [CrossRef]
86. Jouneau, A.; Ciaudo, C.; Sismeiro, O.; Brochard, V.; Jouneau, L.; Vandormael-Pournin, S.; Coppee, J.Y.; Zhou, Q.; Heard, E.; Antoniewski, C.; et al. Naive and primed murine pluripotent stem cells have distinct miRNA expression profiles. *RNA* **2012**, *18*, 253–264. [CrossRef]
87. Laurent, L.C. MicroRNAs in embryonic stem cells and early embryonic development. *J. Cell Mol. Med.* **2008**, *12*, 2181–2188. [CrossRef]

88. Liao, B.; Bao, X.; Liu, L.; Feng, S.; Zovoilis, A.; Liu, W.; Xue, Y.; Cai, J.; Guo, X.; Qin, B.; et al. MicroRNA cluster 302-367 enhances somatic cell reprogramming by accelerating a mesenchymal-to-epithelial transition. *J. Biol. Chem.* **2011**, *286*, 17359–17364. [CrossRef]
89. Anokye-Danso, F.; Trivedi, C.M.; Juhr, D.; Gupta, M.; Cui, Z.; Tian, Y.; Zhang, Y.; Yang, W.; Gruber, P.J.; Epstein, J.A.; et al. Highly efficient miRNA-mediated reprogramming of mouse and human somatic cells to pluripotency. *Cell Stem Cell* **2011**, *8*, 376–388. [CrossRef]
90. Shao, P.; Zhou, H.; Xiao, Z.D.; He, J.H.; Huang, M.B.; Chen, Y.Q.; Qu, L.H. Identification of novel chicken microRNAs and analysis of their genomic organization. *Gene* **2008**, *418*, 34–40. [CrossRef]
91. Lee, S.I.; Lee, B.R.; Hwang, Y.S.; Lee, H.C.; Rengaraj, D.; Song, G.; Park, T.S.; Han, J.Y. MicroRNA-mediated posttranscriptional regulation is required for maintaining undifferentiated properties of blastoderm and primordial germ cells in chickens. *Proc. Natl. Acad. Sci. USA* **2011**, *108*, 10426–10431. [CrossRef]
92. Rengaraj, D.; Park, T.S.; Lee, S.I.; Lee, B.R.; Han, B.K.; Song, G.; Han, J.Y. Regulation of glucose phosphate isomerase by the 3′UTR-specific miRNAs miR-302b and miR-17-5p in chicken primordial germ cells. *Biol. Reprod.* **2013**, *89*, 33. [CrossRef]
93. Lazar, B.; Szabadi, N.T.; Anand, M.; Toth, R.; Ecker, A.; Urban, M.; Aponte, M.T.S.; Stepanova, G.; Hegyi, Z.; Homolya, L.; et al. Effect of miR-302b MicroRNA Inhibition on Chicken Primordial Germ Cell Proliferation and Apoptosis Rate. *Genes* **2021**, *13*, 82. [CrossRef]
94. Wang, Y.; Melton, C.; Li, Y.P.; Shenoy, A.; Zhang, X.X.; Subramanyam, D.; Blelloch, R. miR-294/miR-302 promotes proliferation, suppresses G1-S restriction point, and inhibits ESC differentiation through separable mechanisms. *Cell Rep.* **2013**, *4*, 99–109. [CrossRef]
95. Liang, Y.; Li, Y.; Li, Z.; Liu, Z.; Zhang, Z.; Chang, S.; Wu, J. Mechanism of folate deficiency-induced apoptosis in mouse embryonic stem cells: Cell cycle arrest/apoptosis in G1/G0 mediated by microRNA-302a and tumor suppressor gene Lats2. *Int. J. Biochem. Cell Biol.* **2012**, *44*, 1750–1760. [CrossRef]
96. Ware, C.B.; Wang, L.; Mecham, B.H.; Shen, L.; Nelson, A.M.; Bar, M.; Lamba, D.A.; Dauphin, D.S.; Buckingham, B.; Askari, B.; et al. Histone deacetylase inhibition elicits an evolutionarily conserved self-renewal program in embryonic stem cells. *Cell Stem Cell* **2009**, *4*, 359–369. [CrossRef]
97. Zhang, Z.; Hong, Y.; Xiang, D.; Zhu, P.; Wu, E.; Li, W.; Mosenson, J.; Wu, W.S. MicroRNA-302/367 cluster governs hESC self-renewal by dually regulating cell cycle and apoptosis pathways. *Stem Cell Rep.* **2015**, *4*, 645–657. [CrossRef]
98. James, D.; Levine, A.J.; Besser, D.; Hemmati-Brivanlou, A. TGFbeta/activin/nodal signaling is necessary for the maintenance of pluripotency in human embryonic stem cells. *Development* **2005**, *132*, 1273–1282. [CrossRef]
99. Vallier, L.; Reynolds, D.; Pedersen, R.A. Nodal inhibits differentiation of human embryonic stem cells along the neuroectodermal default pathway. *Dev. Biol.* **2004**, *275*, 403–421. [CrossRef]
100. Ramos-Mejia, V.; Melen, G.J.; Sanchez, L.; Gutierrez-Aranda, I.; Ligero, G.; Cortes, J.L.; Real, P.J.; Bueno, C.; Menendez, P. Nodal/Activin signaling predicts human pluripotent stem cell lines prone to differentiate toward the hematopoietic lineage. *Mol. Ther.* **2010**, *18*, 2173–2181. [CrossRef]
101. Zhu, Z.; Xu, Y.; Zhao, J.; Liu, Q.; Feng, W.; Fan, J.; Wang, P. miR-367 promotes epithelial-to-mesenchymal transition and invasion of pancreatic ductal adenocarcinoma cells by targeting the Smad7-TGF-beta signalling pathway. *Br. J. Cancer* **2015**, *112*, 1367–1375. [CrossRef] [PubMed]
102. Lipchina, I.; Elkabetz, Y.; Hafner, M.; Sheridan, R.; Mihailovic, A.; Tuschl, T.; Sander, C.; Studer, L.; Betel, D. Genome-wide identification of microRNA targets in human ES cells reveals a role for miR-302 in modulating BMP response. *Genes. Dev.* **2011**, *25*, 2173–2186. [CrossRef] [PubMed]
103. Lipchina, I.; Studer, L.; Betel, D. The expanding role of miR-302-367 in pluripotency and reprogramming. *Cell Cycle* **2012**, *11*, 1517–1523. [CrossRef] [PubMed]
104. Lin, S.L.; Chang, D.C.; Lin, C.H.; Ying, S.Y.; Leu, D.; Wu, D.T. Regulation of somatic cell reprogramming through inducible mir-302 expression. *Nucleic Acids Res.* **2011**, *39*, 1054–1065. [CrossRef] [PubMed]
105. Lee, M.R.; Prasain, N.; Chae, H.D.; Kim, Y.J.; Mantel, C.; Yoder, M.C.; Broxmeyer, H.E. Epigenetic regulation of NANOG by miR-302 cluster-MBD2 completes induced pluripotent stem cell reprogramming. *Stem Cells* **2013**, *31*, 666–681. [CrossRef]
106. Bentwich, I.; Avniel, A.; Karov, Y.; Aharonov, R.; Gilad, S.; Barad, O.; Barzilai, A.; Einat, P.; Einav, U.; Meiri, E.; et al. Identification of hundreds of conserved and nonconserved human microRNAs. *Nat. Genet.* **2005**, *37*, 766–770. [CrossRef]
107. Lehnert, S.; Van Loo, P.; Thilakarathne, P.J.; Marynen, P.; Verbeke, G.; Schuit, F.C. Evidence for co-evolution between human microRNAs and Alu-repeats. *PLoS ONE* **2009**, *4*, e4456. [CrossRef]
108. Bortolin-Cavaille, M.L.; Dance, M.; Weber, M.; Cavaille, J. C19MC microRNAs are processed from introns of large Pol-II, non-protein-coding transcripts. *Nucleic Acids Res.* **2009**, *37*, 3464–3473. [CrossRef]
109. Zhang, R.; Wang, Y.Q.; Su, B. Molecular evolution of a primate-specific microRNA family. *Mol. Biol. Evol.* **2008**, *25*, 1493–1502. [CrossRef]
110. Noguer-Dance, M.; Abu-Amero, S.; Al-Khtib, M.; Lefevre, A.; Coullin, P.; Moore, G.E.; Cavaille, J. The primate-specific microRNA gene cluster (C19MC) is imprinted in the placenta. *Hum. Mol. Genet.* **2010**, *19*, 3566–3582. [CrossRef]
111. Fothi, A.; Biro, O.; Erdei, Z.; Apati, A.; Orban, T.I. Tissue-specific and transcription-dependent mechanisms regulate primary microRNA processing efficiency of the human chromosome 19 MicroRNA cluster. *RNA Biol.* **2021**, *18*, 1170–1180. [CrossRef] [PubMed]

112. Donker, R.B.; Mouillet, J.F.; Chu, T.; Hubel, C.A.; Stolz, D.B.; Morelli, A.E.; Sadovsky, Y. The expression profile of C19MC microRNAs in primary human trophoblast cells and exosomes. *Mol. Hum. Reprod.* **2012**, *18*, 417–424. [CrossRef] [PubMed]
113. Stadler, B.; Ivanovska, I.; Mehta, K.; Song, S.; Nelson, A.; Tan, Y.; Mathieu, J.; Darby, C.; Blau, C.A.; Ware, C.; et al. Characterization of microRNAs involved in embryonic stem cell states. *Stem Cells Dev.* **2010**, *19*, 935–950. [CrossRef] [PubMed]
114. Ren, J.; Jin, P.; Wang, E.; Marincola, F.M.; Stroncek, D.F. MicroRNA and gene expression patterns in the differentiation of human embryonic stem cells. *J. Transl. Med.* **2009**, *7*, 20. [CrossRef] [PubMed]
115. Kobayashi, N.; Okae, H.; Hiura, H.; Kubota, N.; Kobayashi, E.H.; Shibata, S.; Oike, A.; Hori, T.; Kikutake, C.; Hamada, H.; et al. The microRNA cluster C19MC confers differentiation potential into trophoblast lineages upon human pluripotent stem cells. *Nat. Commun.* **2022**, *13*, 3071. [CrossRef]
116. Nguyen, P.N.N.; Choo, K.B.; Huang, C.J.; Sugii, S.; Cheong, S.K.; Kamarul, T. miR-524-5p of the primate-specific C19MC miRNA cluster targets TP53IPN1- and EMT-associated genes to regulate cellular reprogramming. *Stem Cell Res. Ther.* **2017**, *8*, 214. [CrossRef]
117. Mong, E.F.; Yang, Y.; Akat, K.M.; Canfield, J.; VanWye, J.; Lockhart, J.; Tsibris, J.C.M.; Schatz, F.; Lockwood, C.J.; Tuschl, T.; et al. Chromosome 19 microRNA cluster enhances cell reprogramming by inhibiting epithelial-to-mesenchymal transition. *Sci. Rep.* **2020**, *10*, 3029. [CrossRef]
118. Jinesh, G.G.; Flores, E.R.; Brohl, A.S. Chromosome 19 miRNA cluster and CEBPB expression specifically mark and potentially drive triple negative breast cancers. *PLoS ONE* **2018**, *13*, e0206008. [CrossRef]
119. Fornari, F.; Milazzo, M.; Chieco, P.; Negrini, M.; Marasco, E.; Capranico, G.; Mantovani, V.; Marinello, J.; Sabbioni, S.; Callegari, E.; et al. In hepatocellular carcinoma miR-519d is up-regulated by p53 and DNA hypomethylation and targets CDKN1A/p21, PTEN, AKT3 and TIMP2. *J. Pathol.* **2012**, *227*, 275–285. [CrossRef]
120. Augello, C.; Vaira, V.; Caruso, L.; Destro, A.; Maggioni, M.; Park, Y.N.; Montorsi, M.; Santambrogio, R.; Roncalli, M.; Bosari, S. MicroRNA profiling of hepatocarcinogenesis identifies C19MC cluster as a novel prognostic biomarker in hepatocellular carcinoma. *Liver Int.* **2012**, *32*, 772–782. [CrossRef]
121. Kleinman, C.L.; Gerges, N.; Papillon-Cavanagh, S.; Sin-Chan, P.; Pramatarova, A.; Quang, D.A.; Adoue, V.; Busche, S.; Caron, M.; Djambazian, H.; et al. Fusion of TTYH1 with the C19MC microRNA cluster drives expression of a brain-specific DNMT3B isoform in the embryonal brain tumor ETMR. *Nat. Genet.* **2014**, *46*, 39–44. [CrossRef]
122. Strub, G.M.; Kirsh, A.L.; Whipple, M.E.; Kuo, W.P.; Keller, R.B.; Kapur, R.P.; Majesky, M.W.; Perkins, J.A. Endothelial and circulating C19MC microRNAs are biomarkers of infantile hemangioma. *JCI Insight* **2016**, *1*, e88856. [CrossRef]
123. Flor, I.; Spiekermann, M.; Loning, T.; Dieckmann, K.P.; Belge, G.; Bullerdiek, J. Expression of microRNAs of C19MC in Different Histological Types of Testicular Germ Cell Tumour. *Cancer Genom. Proteom.* **2016**, *13*, 281–289.
124. Vaira, V.; Elli, F.; Forno, I.; Guarnieri, V.; Verdelli, C.; Ferrero, S.; Scillitani, A.; Vicentini, L.; Cetani, F.; Mantovani, G.; et al. The microRNA cluster C19MC is deregulated in parathyroid tumours. *J. Mol. Endocrinol.* **2012**, *49*, 115–124. [CrossRef]
125. Rippe, V.; Dittberner, L.; Lorenz, V.N.; Drieschner, N.; Nimzyk, R.; Sendt, W.; Junker, K.; Belge, G.; Bullerdiek, J. The two stem cell microRNA gene clusters C19MC and miR-371-3 are activated by specific chromosomal rearrangements in a subgroup of thyroid adenomas. *PLoS ONE* **2010**, *5*, e9485. [CrossRef]
126. Buganim, Y.; Faddah, D.A.; Jaenisch, R. Mechanisms and models of somatic cell reprogramming. *Nat. Rev. Genet.* **2013**, *14*, 427–439. [CrossRef]
127. David, L.; Polo, J.M. Phases of reprogramming. *Stem Cell Res.* **2014**, *12*, 754–761. [CrossRef]
128. Houbaviy, H.B.; Dennis, L.; Jaenisch, R.; Sharp, P.A. Characterization of a highly variable eutherian microRNA gene. *RNA* **2005**, *11*, 1245–1257. [CrossRef]
129. Eddy, S.R.; Mitchison, G.; Durbin, R. Maximum discrimination hidden Markov models of sequence consensus. *J. Comput. Biol.* **1995**, *2*, 9–23. [CrossRef]
130. Calabrese, J.M.; Seila, A.C.; Yeo, G.W.; Sharp, P.A. RNA sequence analysis defines Dicer's role in mouse embryonic stem cells. *Proc. Natl. Acad. Sci. USA* **2007**, *104*, 18097–18102. [CrossRef]
131. Tata, P.R.; Tata, N.R.; Kuhl, M.; Sirbu, I.O. Identification of a novel epigenetic regulatory region within the pluripotency associated microRNA cluster, EEmiRC. *Nucleic Acids Res.* **2011**, *39*, 3574–3581. [CrossRef] [PubMed]
132. Chen, X.; Xu, H.; Yuan, P.; Fang, F.; Huss, M.; Vega, V.B.; Wong, E.; Orlov, Y.L.; Zhang, W.; Jiang, J.; et al. Integration of external signaling pathways with the core transcriptional network in embryonic stem cells. *Cell* **2008**, *133*, 1106–1117. [CrossRef] [PubMed]
133. Voorhoeve, P.M.; le Sage, C.; Schrier, M.; Gillis, A.J.; Stoop, H.; Nagel, R.; Liu, Y.P.; van Duijse, J.; Drost, J.; Griekspoor, A.; et al. A genetic screen implicates miRNA-372 and miRNA-373 as oncogenes in testicular germ cell tumors. *Cell* **2006**, *124*, 1169–1181. [CrossRef] [PubMed]
134. Pan, B.; He, B.; Xu, X.; Liu, X.; Xu, T.; Xu, M.; Chen, X.; Zeng, K.; Lin, K.; Hu, X.; et al. MicroRNA-371-3 cluster as biomarkers for the diagnosis and prognosis of cancers. *Cancer Manag. Res.* **2019**, *11*, 5437–5457. [CrossRef]
135. Chen, K.; Rajewsky, N. The evolution of gene regulation by transcription factors and microRNAs. *Nat. Rev. Genet.* **2007**, *8*, 93–103. [CrossRef]
136. Chen, L.; Heikkinen, L.; Emily Knott, K.; Liang, Y.; Wong, G. Evolutionary conservation and function of the human embryonic stem cell specific miR-302/367 cluster. *Comp. Biochem. Physiol. Part. D Genom. Proteom.* **2015**, *16*, 83–98. [CrossRef]
137. Mohammed, J.; Siepel, A.; Lai, E.C. Diverse modes of evolutionary emergence and flux of conserved microRNA clusters. *RNA* **2014**, *20*, 1850–1863. [CrossRef]

138. Huang, Q.; Gumireddy, K.; Schrier, M.; le Sage, C.; Nagel, R.; Nair, S.; Egan, D.A.; Li, A.; Huang, G.; Klein-Szanto, A.J.; et al. The microRNAs miR-373 and miR-520c promote tumour invasion and metastasis. *Nat. Cell Biol.* **2008**, *10*, 202–210. [CrossRef]
139. Liu, P.; Wilson, M.J. miR-520c and miR-373 upregulate MMP9 expression by targeting mTOR and SIRT1, and activate the Ras/Raf/MEK/Erk signaling pathway and NF-kappaB factor in human fibrosarcoma cells. *J. Cell Physiol.* **2012**, *227*, 867–876. [CrossRef]
140. Keklikoglou, I.; Koerner, C.; Schmidt, C.; Zhang, J.D.; Heckmann, D.; Shavinskaya, A.; Allgayer, H.; Guckel, B.; Fehm, T.; Schneeweiss, A.; et al. MicroRNA-520/373 family functions as a tumor suppressor in estrogen receptor negative breast cancer by targeting NF-kappaB and TGF-beta signaling pathways. *Oncogene* **2012**, *31*, 4150–4163. [CrossRef]
141. Wu, S.; Aksoy, M.; Shi, J.; Houbaviy, H.B. Evolution of the miR-290-295/miR-371-373 cluster family seed repertoire. *PLoS ONE* **2014**, *9*, e108519. [CrossRef]
142. Judson, R.L.; Babiarz, J.E.; Venere, M.; Blelloch, R. Embryonic stem cell-specific microRNAs promote induced pluripotency. *Nat. Biotechnol.* **2009**, *27*, 459–461. [CrossRef] [PubMed]
143. Morin, R.D.; O'Connor, M.D.; Griffith, M.; Kuchenbauer, F.; Delaney, A.; Prabhu, A.L.; Zhao, Y.; McDonald, H.; Zeng, T.; Hirst, M.; et al. Application of massively parallel sequencing to microRNA profiling and discovery in human embryonic stem cells. *Genome Res.* **2008**, *18*, 610–621. [CrossRef]
144. Subramanyam, D.; Lamouille, S.; Judson, R.L.; Liu, J.Y.; Bucay, N.; Derynck, R.; Blelloch, R. Multiple targets of miR-302 and miR-372 promote reprogramming of human fibroblasts to induced pluripotent stem cells. *Nat. Biotechnol.* **2011**, *29*, 443–448. [CrossRef]
145. Dalton, S. Exposing hidden dimensions of embryonic stem cell cycle control. *Cell Stem Cell* **2009**, *4*, 9–10. [CrossRef]
146. Gonzales, K.A.; Liang, H.; Lim, Y.S.; Chan, Y.S.; Yeo, J.C.; Tan, C.P.; Gao, B.; Le, B.; Tan, Z.Y.; Low, K.Y.; et al. Deterministic Restriction on Pluripotent State Dissolution by Cell-Cycle Pathways. *Cell* **2015**, *162*, 564–579. [CrossRef]
147. Benetti, R.; Gonzalo, S.; Jaco, I.; Munoz, P.; Gonzalez, S.; Schoeftner, S.; Murchison, E.; Andl, T.; Chen, T.; Klatt, P.; et al. A mammalian microRNA cluster controls DNA methylation and telomere recombination via Rbl2-dependent regulation of DNA methyltransferases. *Nat. Struct. Mol. Biol.* **2008**, *15*, 268–279. [CrossRef]
148. Yuan, K.; Ai, W.B.; Wan, L.Y.; Tan, X.; Wu, J.F. The miR-290-295 cluster as multi-faceted players in mouse embryonic stem cells. *Cell Biosci.* **2017**, *7*, 38. [CrossRef]
149. Niehrs, C. Function and biological roles of the Dickkopf family of Wnt modulators. *Oncogene* **2006**, *25*, 7469–7481. [CrossRef]
150. Zovoilis, A.; Smorag, L.; Pantazi, A.; Engel, W. Members of the miR-290 cluster modulate in vitro differentiation of mouse embryonic stem cells. *Differentiation* **2009**, *78*, 69–78. [CrossRef]
151. Gu, K.L.; Zhang, Q.; Yan, Y.; Li, T.T.; Duan, F.F.; Hao, J.; Wang, X.W.; Shi, M.; Wu, D.R.; Guo, W.T.; et al. Pluripotency-associated miR-290/302 family of microRNAs promote the dismantling of naive pluripotency. *Cell Res.* **2016**, *26*, 350–366. [CrossRef] [PubMed]
152. Graham, B.; Marcais, A.; Dharmalingam, G.; Carroll, T.; Kanellopoulou, C.; Graumann, J.; Nesterova, T.B.; Bermange, A.; Brazauskas, P.; Xella, B.; et al. MicroRNAs of the miR-290-295 Family Maintain Bivalency in Mouse Embryonic Stem Cells. *Stem Cell Rep.* **2016**, *6*, 635–642. [CrossRef] [PubMed]
153. Kanellopoulou, C.; Gilpatrick, T.; Kilaru, G.; Burr, P.; Nguyen, C.K.; Morawski, A.; Lenardo, M.J.; Muljo, S.A. Reprogramming of Polycomb-Mediated Gene Silencing in Embryonic Stem Cells by the miR-290 Family and the Methyltransferase Ash1l. *Stem Cell Rep.* **2015**, *5*, 971–978. [CrossRef]
154. Zhou, A.D.; Diao, L.T.; Xu, H.; Xiao, Z.D.; Li, J.H.; Zhou, H.; Qu, L.H. beta-Catenin/LEF1 transactivates the microRNA-371-373 cluster that modulates the Wnt/beta-catenin-signaling pathway. *Oncogene* **2012**, *31*, 2968–2978. [CrossRef]
155. Cao, Y.; Guo, W.T.; Tian, S.; He, X.; Wang, X.W.; Liu, X.; Gu, K.L.; Ma, X.; Huang, D.; Hu, L.; et al. miR-290/371-Mbd2-Myc circuit regulates glycolytic metabolism to promote pluripotency. *EMBO J.* **2015**, *34*, 609–623. [CrossRef]

genes

MDPI

Review

Essential Role of the 14q32 Encoded miRNAs in Endocrine Tumors

Lilla Krokker [1], Attila Patócs [1,2,3] and Henriett Butz [1,2,3,*]

[1] Department of Laboratory Medicine, Semmelweis University, H-1089 Budapest, Hungary; krkkr.lilla@gmail.com (L.K.); patocs.attila@med.semmelweis-univ.hu (A.P.)
[2] Hereditary Cancers Research Group, Hungarian Academy of Sciences-Semmelweis University, H-1089 Budapest, Hungary
[3] Department of Molecular Genetics, National Institute of Oncology, H-1122 Budapest, Hungary
* Correspondence: butz.henriett@med.semmelweis-univ.hu

Abstract: Background: The 14q32 cluster is among the largest polycistronic miRNA clusters. miRNAs encoded here have been implicated in tumorigenesis of multiple organs including endocrine glands. Methods: Critical review of miRNA studies performed in endocrine tumors have been performed. The potential relevance of 14q32 miRNAs through investigating their targets, and integrating the knowledge provided by literature data and bioinformatics predictions have been indicated. Results: Pituitary adenoma, papillary thyroid cancer and a particular subset of pheochromocytoma and adrenocortical cancer are characterized by the downregulation of miRNAs encoded by the 14q32 cluster. Pancreas neuroendocrine tumors, most of the adrenocortical cancer and medullary thyroid cancer are particularly distinct, as 14q32 miRNAs were overexpressed. In pheochromocytoma and growth-hormone producing pituitary adenoma, however, both increased and decreased expression of 14q32 miRNAs cluster members were observed. In the background of this phenomenon methodological, technical and biological factors are hypothesized and discussed. The functions of 14q32 miRNAs were also revealed by bioinformatics and literature data mining. Conclusions: 14q32 miRNAs have a significant role in the tumorigenesis of endocrine organs. Regarding their stable expression in the circulation of healthy individuals, further investigation of 14q32 miRNAs could provide a potential for use as biomarkers (diagnostic or prognostic) in endocrine neoplasms.

Keywords: miRNA; 14q32; miRNA cluster; DLK1-MEG3 locus; endocrine tumor; pituitary adenoma; adrenocortical cancer; neuroendocrine tumor; pheochromocytoma; thyroid cancer

1. Introduction

MicroRNAs (miRNAs) are single-stranded, small (~17–22 nucleotide long), protein non-coding RNA molecules that regulate gene expression post-transcriptionally by RNA interference. According to the canonical miRNA biogenesis, the mature miRNA is generated from a hairpin RNA precursor molecule produced by RNA polymerase II or III [1].

After biogenesis, the mature miRNA incorporates into a protein complex called miRISC (miRNA-induced silencing complex) [2]. In the miRISC complex miRNAs lead to translational repression, mRNA destabilization or mRNA cleavage through miRNA-mRNA interaction via base complementarity. MiRNAs target mRNAs mainly at 3′ untranslated regions but even the coding sequence or 5′UTR have been described to be miRNA target regions [2]. Recently, it has been discovered that in some particular cases miRNAs can even enhance gene expression [2].

Approximately 30–50% of all protein-coding genes are thought to be controlled by miRNAs [3]. As one miRNA targets several transcripts, and one mRNA is regulated by numerous miRNAs, the net physiological outcome is the result of a miRNA-target network. The role of miRNAs is primarily considered to set the gene expression to an optimal level as an adaptive process called "fine tuning" [4].

MiRNAs have been shown to be involved in the control of many physiological and pathophysiological processes, such as proliferation, differentiation, metabolism and apoptosis through the modulation of target gene expression. Altered miRNA expression has been identified in several endocrine diseases including neoplasms [5,6]. Depending on their target molecules, miRNAs are considered as oncomiRs or tumor suppressor miRNAs, and therefore they are often considered potentially useful biomarkers. MiRNAs are highly tissue-specific, and they may be unique identifiers of certain tumor types, even having different effects in different cell/tissue types.

The 14q32 miRNA cluster is among the largest polycistronic clusters comprising almost a hundred small non-coding RNAs, including a significant number of miRNAs [7]. MiRNAs located in this region cover over 5% of the known human miRNA genes [8,9].

The 14q32 region is called the *DLK1-DIO3* domain or *DLK1-MEG3* cluster. Indeed, this cluster contains protein-coding (Delta-like 1 (*DLK1*), Deiodinase Iodothyronine Type III (DIO3) and Retrotransposon-like Gene 1 (*RTL1*)) and nonprotein-coding genes (ncRNAs, such as Maternally Expressed Gene 3 (*MEG3*), Maternally Expressed Gene 8 (*MEG8*) and *RTL1* antisense (*RTL1-AS*)), small nucleolar RNAs (snoRNAs) and miRNAs. This approximately 300 kilobase miRNA region can be divided into two parts: cluster A and cluster B. Cluster A includes *MEG3* and *RTL1* genes, while cluster B can be found 5′ from *MEG9* and *DIO3* genes (Figure 1A).

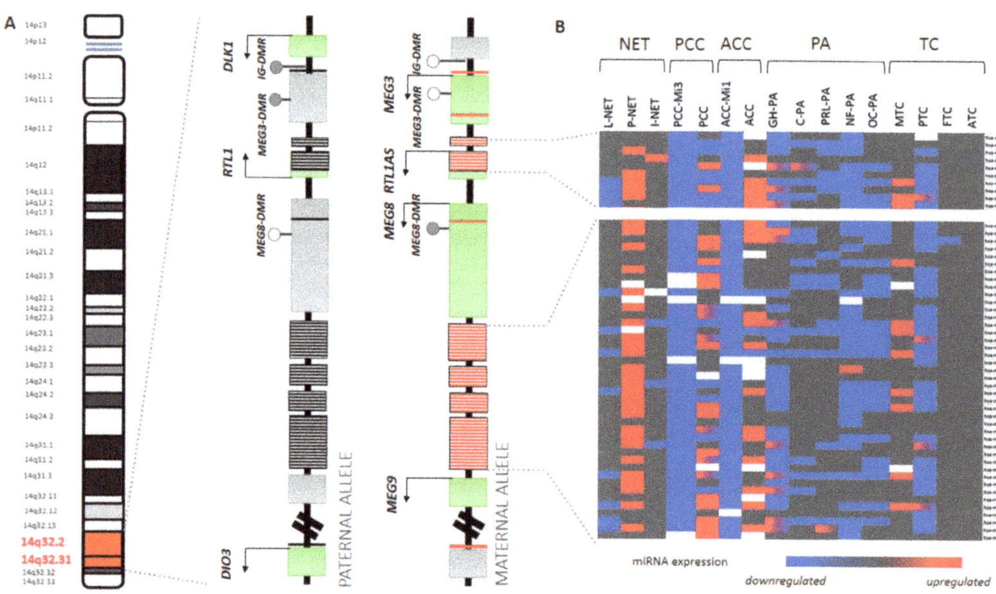

Figure 1. The imprinted 14q32 region. (**A**) On the paternal and maternal alleles middle: exons colored by grey indicate imprinted genes, exons colored by green indicate transcriptionally active state. DMR status is presented by grey (methylated) and white (unmethylated) circles. Black stripes represent imprinted miRNA genes, red stripes represent transcriptionally active miRNA genes. (**B**) On the heatmap (right): Representative expressional change of 14q32 miRNAs in endocrine tumors (downregulation and overexpression are presented compared to physiological, monoallelic expression. Heatmap was constructed using the data of 59 studies (see details in the methods section)). MiRNAs indicated on the heatmap are listed in Table 1 indicating the exact chromosomal localization. Colors indicate expression (blue color was used for low and *red* for high expression, grey represents monoallelic expression, *white*: no data available).

As miRNA expression can be influenced by several factors, Goossens and colleagues investigated the effect of the most common confounding factors: sex and age on the miRNA expression profile of this region [10]. The finding that 14q32 miRNA expression did not differ between men and women, and that no correlation with age was observed, highlighted the importance of this miRNA cluster in cell biology [10]. This was further supported by others who described that the 14q32 maternally imprinted locus was a major source of longitudinally stable circulating miRNAs as measured by small RNA sequencing of healthy individuals [11]. In addition, the serum level of 14q32 miRNAs was not significantly affected by common confounders such as age, body mass index (BMI) or time of centrifugation, nor alternative methods of expression data normalization [11].

14q32 miRNAs are frequently described deregulated in human diseases and cancers [12–14]. In line with the finding that miRNAs are highly tissue-specific [7,15], 14q32 miRNAs are considered both oncogenic and tumor suppressing depending on cell type [14,16].

The expression of 14q32 genes is regulated by genomic imprinting. The differential expression of alleles inherited from a mother or father at the genomic imprinted loci is realized by different methylation. The regulatory loci of the methylated nucleotides are called differentially methylated regions (DMRs). Differential methylation patterns at DMRs provide monoallelic expression from either maternal or paternal allele. Generally, imprinted genes have a key role in regulating growth and other physiological functions during embryonic development. Germline deletions and uniparental disomy of this locus in humans associate with developmental abnormalities and dysmorphism, suggesting that the 14q32 locus might have significant importance in development [17–24]. As several maternally imprinted genes limit growth during development, they usually possess a tumor-suppressor role in human cancer [25]. In line with this, the *DLK1-MEG3* cluster was frequently affected by allelic loss or epigenetic changes in various cancers [26–29]. In the 14q32 region, the paternally expressed protein coding genes are *DLK1*, *DIO3* and *RTL1* [30]. From the maternal allele *MEG3*, *MEG8* and *RTL1-AS* long noncoding RNAs are expressed [9,31]. This imprinted gene expression of this locus is under the control of three DMRs [20,32]: a DMR located 11 kb upstream of *MEG3* (also called intergenic differentially methylated region, IG DMR), a DMR 1.3 kb upstream of the *MEG3* transcription start site (*MEG3*-DMR) and a DMR found in the *DLK1* promoter (*DLK1*-DMR) [33] (Figure 1A).

The *MEG3*-IG DMR, which is methylated on the paternal allele and unmethylated on the maternal allele, functions as the primary imprinting control region (ICR) for the entire locus during development [34], whereas *MEG3*-DMR serves as the principal regulator in adult tissues [20]. This imprinted methylation pattern provides the reciprocal expression of *DLK1* and *MEG3*.

Accordingly, 14q32 miRNAs are also involved in the imprinting regulation [7,35] and altered 14q32 miRNA expression have been described in several diseases including malignancies [36–41]. Several studies have shown downregulation of miRNAs from the 14q32 region in different types of cancer, such as ovarian, breast, prostate, bladder, osteosarcoma, and gastrointestinal stromal, with significant correlations to poor prognosis and aggressiveness [42–48]. The tumor suppressor role has been recognized in several of the downregulated 14q32 miRNAs through targeting key oncogenes in glioblastoma, neuroblastoma, metastatic lung cancer, hepatic cancer and rhabdomyosarcoma [25,49–51]. In contrast, miRNAs from the 14q32 region may act as oncogenes as well [52–54], suggesting that these miRNAs may have different biological roles depending on the tissue of origin and genetic background.

14q32 miRNAs also influence prognosis of various cancers. Oshima et al. presented that the expression of 14q32 miRNAs was a favorable prognostic factor in patients with metastatic cancer [41]. Based on studies of Lussier et al. 2011 and 2012, the term oligomiR was introduced [55,56]. There were miRNAs differentially expressed between patients with limited numbers and slow progression of metastases (oligometastases) compared to patients with widely disseminated or rapidly progressive metastatic disease [55,56]. Interestingly, miRNAs encoded in the 14q32 were significantly enriched among oligomiRs [22]. Additionally, 14q32 miRNAs suppressed lung and liver metastases and correlated with improved clinical outcomes. In osteosarcoma 14q32 miRNAs also had prognostic significance, as an inverse correlation was described between aggressive tumor behavior (increased metastatic potential and accelerated time to death) and the residual expression of this miRNA locus [47].

Despite of the relevance of 14q32 miRNAs in other malignancies, regarding endocrine tumors, there has no comprehensive review published about the role of 14q32 miRNA cluster. In this work, the authors aimed to collect high-throughput miRNA studies performed in endocrine tumor samples, to extract the role and potential relevance of 14q32 miRNAs through investigating miRNA targets and to integrate the knowledge provided by literature data and bioinformatics predictions.

2. Materials and Methods

Literature mining was performed using Pubmed database using the following keywords: "14q32" or "miRNA" or "DLK1-MEG3" and combined with either of "endocrine tumors", "neuroendocrine tumor", "pituitary adenoma", "adrenocortical tumor", "pheochromocytoma" or "thyroid cancer". Publication focused on 14q32 miRNAs regarding endocrine tumors, and high-throughput miRNA profiling studies of endocrine tumors were selected to construct an expression heatmap (Figure 1B, Table 1). Downregulated/over- expressed miRNAs were included when at least one study reported it significantly without conflict (conflict was considered when another study reported the opposite change). When conflicting information was observed between studies, gradient colour was used. Unfortunately, as in many studies, raw data were not available and only the significant lists were reported; only "downregulated" and "overexpressed" characteristics could be considered irrespectively of fold change. The heatmap itself was generated in Microsoft Excel (Microsoft Office Professional Plus 2013).

Table 1. MiRNA studies used for heatmap generation.

Tumor Type		Study	miRNA Profiling Platform
NET	Lung	Yoshimoto et al., 2018 [57]	microarray
		Mairinger et al., 2014 [58]	TaqMan array
		Deng et al., 2014 [59]	microarray
		Rapa et al., 2015 [60]	PCR array
		Wong et al., 2020 [61]	NGS
	Pancreas	Zimmermann et al., 2018 [62]	TaqMan array
		Roldo et al., 2006 [63]	microarray
		Jiang et al., 2015 [64]	PCR array
		Zhou et al., 2016 [65]	microarray (GSE43796) reanalysis
		Lee et al., 2015 [66]	Nanostring nCounter
	small intestinal	Yoshimoto et al., 2018 [57]	microarray
		Arvidsson et al., 2018 [67]	microarray
		Li et al., 2013 [68]	microarray
		Miller et al., 2016	Nanostring nCounter
PPGL		Castro-Vega et al., 2015 [69]	NGS
		Tömböl et al., 2010 [70]	TaqMan array
		de Cubas et al., 2013 [71]	microarray
		Meyer-Rochow et al., 2010 [72]	microarray
		Calsina et al., 2019 [73]	individual qPCR
ACC		Tömböl et al., 2009 [74]	TaqMan array
		Chabre et al., 2012 [54]	microarray
		Özata et al., 2011 [75]	microarray
		Assié et al., 2014 [76]	NGS

Table 1. *Cont.*

Tumor Type		Study	miRNA Profiling Platform
Pituitary	GH	Mao et al., 2010 [77]	microarray
		D'Angelo et al., 2012 [78]	microarray
		Bottoni et al., 2007 [79]	microarray
		Butz et al., 2009 [80]	TaqMan array
		Cheunsuchon et al., 2011 [81]	individual qPCR
		He et al., 2019 [82]	NGS
	ACTH	Gentilin et al., 2013 [83]	individual TaqMan assay
		Amaral et al., 2009 [84]	individual TaqMan assay
		Stilling et al., 2010 [85]	microarray
		Cheunsuchon et al., 2011 [81]	individual qPCR
	PRL	He et al., 2019 [82]	NGS
		Müssnich et al., 2015 [86]	microarray
		Chen et al., 2012 [87]	NGS
		Cheunsuchon et al., 2011 [81]	individual qPCR
	NFPA	He et al., 2019 [82]	NGS
		Darvasi et al., 2019 [88]	NGS, TaqMan array and microarray
		Butz et al., 2011 [89]	TaqMan array
		Liang et al., 2015 [90]	individual qPCR
		Cheunsuchon et al., 2011 [81]	individual qPCR
		Müssnich et al., 2015 [86]	microarray
		Bottoni et al., 2007 [79]	microarray
	OC	Krokker et al., 2019 [91]	NGS
Thyroid	MTC	Lassalle et al., 2016 [92]	microarray
		Hudson et al., 2013 [93]	Taqman array
		Nikiforova et al., 2008 [94]	Taqman array
	PTC	Geraldo et al., 2017 [95]	NGS (obtained from The Cancer Genome Atlas dataset)
		Rosignolo et al., 2017 [96]	Taqman array
		Tetzlaff et al., 2007 [97]	microarray
		Linwah et al., 2011 [98]	microarray
		Jacques et al., 2013 [99]	microarray
		Lassalle et al., 2011 [100]	microarray
		Mancikova et al., 2015 [101]	NGS
		Peng et al., 2014 [102]	microarray
		Riesco-Eizaguirre et al., 2015 [103]	NGS
		Saiselet et al., 2015 [104]	NGS
		Swierniak et al., 2013 [105]	NGS
	FTC	Nikiforova et al., 2008 [94]	TaqMan array
		Rossing et al., 2012 [106]	microarray
		Dettmer et al., 2013 [107]	Taqman array
		Jacques et al., 2013 [99]	microarray
		Lassalle et al., 2011 [100]	microarray
		Mancikova et al., 2015 [101]	NGS
		Wojtas et al., 2014 [108]	microarray
	ATC	Hébrant et al., 2014 [109]	microarray
		Visone et al., 2007 [110]	microarray
		Boufraqech et al., 2015 [111]	microarray
		Braun et al., 2010 [112]	microarray

ACC: adrenocortical carcinoma; ACTH: corticotroph adenoma; ATC: anaplastic thyroid cancer; FTC: follicular thyroid carcinoma; GH: growth hormone; MTC: medullary thyroid carcinoma; NET: neuroendocrine tumor; NFPA: nonfunctional pituitary adenoma; OC: oncocytoma; PPGL: pheochromocytoma-paraganglioma; PRL: prolactin; PTC: papillary thyroid carcinoma.

Also, miRbase MIMAT IDs of dominant mature 14q32 miRNAs were used for bioinformatics analysis (http://www.mirbase.org/, access date: 25 February 2021). Pathway analysis for 14q32 miRNA targets by gene set enrichment analysis for KEGG pathways was performed using DIANA TOOLS miRPath v.3 following target prediction by microT-CDS algorithm (http://snf-515788.vm.okeanos.grnet.gr/, access date: 25 February 2021). Gene ontology was assessed using miRDB Target Ontology Analysis module by selecting enrichment for Biological Process and Molecular Function (http://mirdb.org/ontology.html, access date: 25 February 2021).

3. miRNAs in Endocrine Tumors

3.1. Neuroendocrine Tumors (NET)

Neuroendocrine tumors (NETs) are a group of heterogeneous neoplasms arising from neuroendocrine cells throughout the body (most commonly from the gastrointestinal system or lungs). Although gastro-entero-pancreatic NETs (GEP-NETs) represent less than 1% of all digestive system cancers it consists 7–21% of all neuroendocrine neoplasms [113]. Lung NETs originate from pulmonary neuroendocrine cells accounting for approximately 25% of primary lung neoplasms and they are classified into typical carcinoids (TCs, well differentiated, low-grade), atypical carcinoids (ACs, well-differentiated, intermediate-grade), large cell neuroendocrine carcinomas (LCNECs, poorly differentiated, high-grade) and small cell lung cancer (SCLCs, poorly differentiated, high-grade) subtypes [114].

In GEP-NETs tissue, however, several miRNA studies have been published [115,116], scarce information was available regarding 14q32 miRNA profiles. In the study by Jiang et al., 29 overexpressed miRNAs derived from 14q32 were identified in insulinomas vs. normal islets, and several showed high abundance in insulinoma cells [64,117]. Unfortunately, this finding was not reported by others [65]. MiRNAs of this region were associated with prognosis, since miR-485–3p encoded at 14q32 region was significantly elevated in the metastatic tumors compared to the primary pancreatic NETs (pNET) (Table 2) [66]. Investigating small intestine (siNET) and colorectal NET, studies did not identify differentially expressed miRNAs encoded at 14q32 [118–121]. However, similar to pNET, miRNAs were linked to progression as 14q32 encoded miR-494 was significantly overexpressed in metastases compared to primary siNET [68]. Interestingly, by using miRNA expression profiling, Yoshimoto et al. described a similar pattern of miRNAs of carcinoids of the lung and gastrointestinal NETs, which was different from adenocarcinomas, small cell lung cancers and normal mucosal cells, suggesting a common origin of systemic carcinoids/NETs [57]. Also, miR-494 was downregulated in carcinoid vs. adenocarcinoma/normal tissue group [57]. Regarding lung NET types, Rapa et al. detected several 14q32 miRNAs (miR-409-3p, miR-409-5p, miR-376a, miR-376b, miR-381) upregulated in typical compared to atypical lung carcinoids [60]. Measured on 46 lung carcinoid tumors, a more extensive list of miRNAs expressed from 14q32 cluster detected as downregulated compared to their adjacent normal tissue pair samples by Deng et al. (miR-487b, miR-410, miR-369, miR-376a, miR-432, miR-409, miR-494, miR-136, miR-370, miR-127 and miR-154) [59].

Table 2. Summary of the most important dysregulated, 14q32 encoded miRNAs in different endocrine neoplasms.

NET	
miR-485-3p	increased in the metastatic tumors compared to the primary pNET
miR-494	overexpressed in metastases compared to primary siNET downregulated in carcinoid vs. adenocarcinoma/normal lung tissue
miR-376a, miR-376b, miR-381, miR-409-3p, miR-409-5p,	upregulated in typical compared to atypical lung carcinoids
miR-127, miR-136, miR-154, miR-369, miR-370, miR-376a, miR-410, miR-432, miR-409, miR-487b, miR-494	downregulated in lung carcinoid compared to adjacent normal tissue
miR-409-3p, miR-409-5p, miR-411, miR-431-5p, miR-485 and miR-539	downregulated in metastatic carcinoids compared to non-metastatic lung NET
miR-127, miR-136, miR-154, miR-485, miR-770-5p	negative correlation with tumor biology of lung NET

Table 2. *Cont.*

PPGL	
miR-493-5p	commonly downregulated in all PCC molecular subtypes (based on germline mutation)
miR-127-3p, miR-136, miR-154-3p/5p, miR-323a-3p, miR-337-5p/-3p, miR-369-5p, miR-370, miR-376a-5p, miR-376c, miR-377, miR-382, miR-409-5p, miR-410, miR-485-3p és 5p, miR-487a, miR-495, miR-539, miR-543, miR-758, miR-889	downregulation in MAX-related PPGLs and a subset of sporadic PCC
miR-154-3p, hsa-miR-369-5p, hsa-miR-485-5p, hsa-miR-487a, hsa-miR-495, hsa-miR-543, hsa-miR-656, hsa-miR-889	overexpression in TMEM127-related PPGL cases
miR-541	overexpressed in VHL-related PCC vs. sporadic PCC, decreased expression in recurrent tumors compared to primary tumors
miR-154, miR-337-3p	upregulated in a subset of metastatic PCC compared to non-metastatic cases
miR-409-3p, miR-369-3p	downregulation in a subset of metastatic PCC compared to benign PCC
miR-431	upregulated in malignant tumors compared to benign
Adrenocortical Tumors	
miR-370	overexpressed in APA compared to non-APA adrenal tumors
miR-299	downregulated in KCNJ5 mutant APA vs. non-KCNJ5 mutant samples
14q32 miRNA cluster	whole miRNA cluster downregulation in Mi1 subset of ACC
miR-136, miR-127-3p, miR-487b, miR-376c and miR-432	overexpressed in ACC compared to normal adrenal cortex
miR-376a, miR-376b	overexpression in ACC vs. ACA
miR-376a	downregulated in ACC vs. NF adenoma, CPA and normal adrenal cortex
miR-299-5p, miR-485-5p	overexpressed in ACC vs. NF adenoma, CPA and normal adrenal cortex
miR-370, miR-376a, miR-376b, miR-376c, miR-377, miR-379, miR-382, miR-411, miR-487a, miR-494, miR-495	downregulated in non-aggressive ACC as compared to aggressive ones
miRNA-665	overexpressed in ACC as compared to benign adrenocortical tumors
miR-431	implicated in adjuvant therapy response in ACC
PitNET	
miR-127-3p, miR-154, miR-329, miR-337, miR-369-5p, miR-376c, miR-432, miR-433	downregulated in PRL adenoma vs. normal
miR-410	overexpressed in PRL adenoma vs. normal
miR-411-3p	overexpressed in GH adenoma vs. normal
miR-381, miR-654-3p	downregulated in GH adenoma vs. normal
miR-127, miR-134, miR-136, miR-154, miR-323a, miR-337, miR-369, miR-370, miR-376a-1, miR-376a-2, miR-376b, miR-376c, miR-379, miR-380, miR-381, miR-382, miR-409, miR-410, miR-411, miR-431, miR-432, miR-433, miR-487b, miR-493, miR-494, miR-495, miR-539, miR-543, miR-544a, miR-654, miR-656, miR-770-5p, miR-889	downregulated in NF adenoma vs. normal
miR-1185-1-3p	upregulated in NF adenoma vs. normal
miR-127-3p, miR-136, miR-154, miR-299-5p, miR-323-5p, miR-329, miR-369-3p, miR-369-5p, miR-376c, miR-377, miR-411-3p, miR-431-3p, miR-433, miR-493	downregulated in corticitroph adenoma vs. normal
miR-431, miR-493	overexpressed in corticotroph carcinoma vs. adenoma
miR-127, miR-136, miR-154, miR-299, miR-323a, miR-323b, miR-329-1, miR-329-2, miR-369, miR-370, miR-376a-1, miR-376a-2, miR-376b, miR-376c, miR-379, miR-381, miR-382, miR-409, miR-411, miR-431, miR-485, miR-487b, miR-494, miR-539, miR-654, miR-889	downregulated in oncocytoma vs. normal
Thyroid Carcinoma	
miR-9, miR-127, miR-136, miR-154, miR-323, miR-376a,c, miR-370, miR-487b	upregulated in MTC vs. normal
miR-299	downregulated in FTC

Table 2. *Cont.*

miR-134, miR-136, miR-409, miR-654	overexpressed in PTC
miR-134, miR-300, miR-379, miR-382, miR-494-3p, miR-494-5p, miR-495	downregulated in PTC
miR-654-3p	inverse correlation with PTC progression

ACA: adrenocortical adenoma; ACC: adrenocortical carcinoma; APA: aldosterone producing adenoma; CPA: cortisol producing adenoma; FTC: follicular thyroid carcinoma; GH: growth hormone; KCNJ5: potassium inwardly rectifying channel subfamily J member 5; MAX: MYC associated factor X; MTC: medullary thyroid carcinoma; NET: neuroendocrine tumor; NF: nonfunctional; PCC: pheochromocytoma; pNET: pancreatic neuroendocrine tumor; PPGL: pheochromocytoma-paraganglioma; PRL: prolactin; PTC: papillary thyroid carcinoma; siNET: small intestinal neuroendocrine tumor; TMEM127: transmembrane protein 127; VHL: von Hippel-Lindau tumor suppressor.

Several studies evaluated the role of miRNAs as prognostic markers [116]. Among others, miR-409-3p, miR-409-5p, miR-431-5p, miR-411, miR-485 and miR-539 encoded at 14q32 were significantly downregulated in metastatic carcinoids compared to non-metastatic lung NETs, while miR-409-3p, miR-409-5p and miR-431-5p were found downregulated in cases with vascular invasion [60,116].

Also among 14q32 miRNAs, the expression of miR-127, miR-136, miR-154, miR-485, miR-770-5p showed negative correlation with tumor biology of lung NET, and miR-377* was identified, showing a significant impact on survival time [58].

A recent study reported that among the most abundant miRNAs in lung NET types, miR-127 encoded at 14q32 showed high expression in typical carcinoids tumors [61]. Besides, no 14q32 miRNA was identified as discriminatory miRNAs characteristic to typical carcinoid, atypical carcinoid, small cell lung cancer or large cell neuroendocrine carcinomas in the study of Wong et al. [61].

3.2. *Pheochromocytoma-Paraganglioma (PPGL)*

The rare pheochromocytomas (PCC) and paragangliomas (PGL) (together PPGL, incidence 1–8:1,000,000) arise from the same type of neural crest tissue of the sympathetic and parasympathetic paraganglia [122]. While tumors of the adrenal medulla are called PCCs, neoplasms developing from the head and neck, thoracic, abdominal or pelvic regions paraganglia are referred as PGLs. These tumors are usually benign and the 10-year overall survival is around ~96%, but 10% of PCC and even 40% of PGL occur as metastatic disease resulting in a 5-year survival below 50% [122]. Interestingly, PPGL has an extremely high rate of genetic susceptibility, when a germline mutation leads to autosomal dominant genetic syndromes (multiple endocrine neoplasia type 2A and 2B caused by *RET* mutations, von Hippel Lindau syndrome due to *VHL* mutations, neurofibromatosis type 1 with *NF1* mutations or hereditary PG syndrome caused by mutations of succinate dehydrogenase (SDH) genes, PPGL genes including *KIF1b*, *PHD2*, *TMEM127*, *MAX*, *FH*, *MDH2*, *GOT2* and *SLC25A11* [123]. Unfortunately, there are neither clear histopathological signs of malignant behavior or efficient therapy for malignant PPGL. Therefore, miRNAs have been good candidates being potential biomarkers.

Expectedly, miRNA profile in different genetic subtypes was also distinct and based on a high-throughput miRNA profiling [70,72,124] several 14q32 miRNAs were dysregulated in PPGL [71]. 14q32 encoded miR-493* was commonly downregulated in all molecular subtypes based on germline mutation [71]. The 14q32 miRNA profile (20 miRNAs) showed significant downregulation in *MAX*-related PPGLs and a subset of sporadic PC samples as well [71]. In *TMEM127*-related cases overexpression of eight 14q32 miRNAs were detected [71]. MiR-541 was found significantly overexpressed in *VHL*-related PCCs vs. sporadic counterparts, and these miRNAs had a lower level of expression in recurrent tumors compared to primary PCC [70]. As hypermethylation of *DLK-MEG3* locus was reported in approximately 10% of PCC samples [69], the pathogenic role of downregulated miRNAs located here was also proposed [125]. Indeed, in a comprehensive multi-omic approach, miRNA profiling by next generation sequencing (NGS) revealed 7 homogeneous subgroups of PCC. PCC samples of the Mi1,2 and Mi4-7 clusters exhibited higher 14q32 miRNA expression compared to Mi3 [69], while Mi3 subgroup was characterized by a strong silencing of the imprinted *DLK1-MEG3* cluster. In this study 15 of 17 tumors belonging to cluster Mi3 displayed loss-of-heterozygosity (LOH) at the 14q32 locus harboring *DLK1-MEG3*. The authors hypothesized that the loss of the maternal unmethylated allele might explain the repression of this imprinted miRNA cluster that was also supported by the methylation analysis of *MEG3* promoter [69]. Interestingly, PCC samples belonging to this Mi3 cluster were not associated with any germline mutation (they were all sporadic tumors) and they belonged to a distinct mRNA expression cluster (C2B) [69]. In line with these results, another large-scale study found the upregulation of miR-154, miR-337-3p in a subset of metastatic PCC compared to non-metastatic cases [73]. The downregulation of miR-409-3p, miR-369-3p was also identified in a different subset of metastatic tumors compared to benign ones [73]. In another study comparing benign and malignant cases, miR-431 was detected as upregulated in malignant tumors [72].

3.3. Adrenocortical Tumors

In aldosterone producing adenomas (APA), no differentially expressed miRNAs encoded at 14q32 were detected [126–128]. Interestingly, another study identified miR-410 and miR-433 as Wnt/β-catenin signaling regulatory miRNAs with significantly different expression between APA and peritumoral adrenal tissues using microarray [129]. A study investigating APA compared to non-APA adrenal tumors (adrenocortical adenoma (ACA), subclinical hypercortisolism (SH), non-functioning adrenal adenoma (NF)) identified miR-370 as overexpressed in aldosterone producing tumors. Also, similar to pheochromocytoma, in APA miRNA signature was also reflected in germline mutation carrier status [130], and miR-299 from locus 14q32 was found downregulated in *KCNJ5* mutant APA vs. non-*KCNJ5* mutant samples [130].

Regarding adrenocortical carcinoma (ACC), a combined genomic approach classified tumor samples into 3 clusters (Mi1-3) based on miRNA expression pattern [76]. ACC samples in cluster Mi1 showed the largest miRNA expression differences relative to normal adrenal samples. Samples in this Mi cluster were characterized by the downregulation of 38 miRNAs expressed from 14q32 locus and by the upregulation of miRNAs belonging to the Xq27.3 miRNA cluster [76]. By using SNP array and DNA methylation analysis, this study identified LOH of chromosome arm 14q in all Mi1 ACC tumors associated with *MEG3* promoter methylation. The authors suggested that the loss of the maternal unmethylated allele resulted in silencing of the 14q32 miRNA cluster in Mi1 ACC tumors, suggesting that this region had a key role in ACC pathogenesis [76]. Özata and his colleagues, however, found 5 miRNAs from 14q32 (miR-136, miR-127-3p, miR-487b, miR-376c and miR-432) overexpressed in ACC compared to normal adrenal cortex, but no 14q32 miRNAs were identified in association with survival [75]. Additionally, miR-376a and miR-376b overexpression were also described in ACC vs. ACA samples [54]. Interestingly, while miR-376a was detected as downregulated miRNA, miR-299-5p and miR-485-5p were found overexpressed in ACC vs. hormonally nonfunctioning adenoma, cortisol-producing adenoma and normal adrenal cortex [74].

Regarding ACC behavior miR-370, miR-376a, miR-376b, miR-376c, miR-377, miR-379, miR-382, miR-411, miR-487a, miR-494, and miR-495 encoded at 14q32 miRNA cluster were downregulated in non-aggressive ACC as compared to aggressive ones [54]. In another study, miRNA-665 was overexpressed in ACC as compared to benign adrenocortical tumors [131]. MiR-431 was also reported to be underexpressed in patients with ACC with progressive disease undergoing adjuvant therapy (mitotane, chemotherapy, and radiotherapy) compared to therapy responders [132]. Restoration of miR-431 increased cell responses to adjuvant therapy and led to cell cycle arrest at S phase. Authors demonstrated that Zinc Finger E-Box Binding Homeobox 1 (*ZEB1*), a target of miR-431, was implicated in reversal of the epithelial-mesenchymal transition (EMT), leading to increased cell responses to adjuvant therapies in ACC [132].

Interestingly, *DLK1* was found as a marker of adrenal gland tumor, which was in line with findings that 14q32 miRNAs (except for Mi1 subgroup) were upregulated in

ACC suggesting a common transcriptional regulation of the entire locus in ACC (Figure 1) [133].

3.4. Pituitary Neuroendocrine Tumors (PitNET)

Pituitary adenomas are among the most frequent intracranial tumors with a high incidence rate of approximately 10–15% [134]. Although the great majority of them are benign, they represent significant morbidity by mass effect or by hormonal disturbance. Generally, pituitary adenomas are sporadic; only 5% of them occur as part of genetic syndromes such as MEN1, MEN4, Carney complex or McCune-Albright syndrome. Interestingly, miRNAs has been extensively investigated in pituitary tumors, including not only expressional reports, but functional studies [6]. As more than one hundred original publications have reported, findings have been extensively summarized by excellent reviews [6,135,136]. Here, authors aimed to only highlight the role of 14q32 miRNAs in pituitary adenomas using high-throughput studies comparing pituitary adenoma samples to normal pituitary tissues.

Prolactinomas. In a work using next generation sequencing, no 14q32 miRNAs were reported differentially expressed in prolactinomas [82]. However, with a targeted approach, Cheunsuchon et al. found 7 of 18 investigated 14q32 miRNAs in prolactin (PRL)-secreting tumors significantly down-regulated [81]. In line with these results, D'Angelo et al. detected downregulated miR-432 in PRL adenoma tissues and using functional in-vitro assays, high-mobility group AT-hook 2 mRNA (*HMGA2)* proved to be a miR-432 target [78]. On the contrary, Chen et al. detected miR-432 and miR-493 upregulation compared to normal anterior pituitary gland samples; moreover, they reported a significant positive correlation between the expression of the two miRNAs and the serum level of prolactin [87]. Additionally, of the 14q32 miRNA cluster, miR-410 was found to be upregulated in prolactinomas [86], as well as in 6 out of the 12 GH-secreting adenomas. This finding suggested that a reduced miR-410 expression seemed to be restricted to gonadotroph adenomas.

Growth hormone (GH) producing tumors. Numerous underexpressed miRNA located at 14q32 were identified using NGS and PCR array [80,82] (Figure 1). However, this was not entirely supported by other studies [77,78,81]. While some reported overexpression of miR-136 in GH-producing adenomas based on microarray profiling [77], a recent study using NGS and Bottoni et al. reported its downregulation compared to normal pituitary [79,82]. Nevertheless, miR-411-3p was overexpressed and miR-381 with miR-654-3p were downregulated from 14q32 locus [77,82]. MiR-370-3p was detected to be underexpressed in pituitary adenomas compared to normal pituitary and in non-functional pituitary adenomas (NFPA) compared to functional ones [137]. Furthermore, its level showed correlation with GH expression determined by immunohistochemistry [137]. Palumbo et al. identified 17 miRNAs to be differentially expressed in GH-producing pituitary tumors; however, none were encoded at 14q32 [138]. Pituitary tumor-transforming 1 (*PTTG1*) was identified as a target of miR-126 and miR-381 encoded at 14q32 cluster [139]. Also, Liang et al. demonstrated that overexpression of 4 14q32 miRNAs (miR-655, miR-300, miR-381 and miR-329) inhibited proliferation, migration and invasion, but induced apoptosis in GH3 and MMQ rat pituitary cells and regulated the *PTTG1* expression [90]. The authors suggested a negative feedback loop between *PTTG1* targeting miRNAs, *PTTG1* and p53 where p53 transcriptionally activated the expression of the four miRNAs, while *PTTG1* inhibited the transcriptional activity of p53 [90]. Among 14q32 miRNAs the downregulated miR-432 inhibited cell proliferation of GH3 cells and has a negative role on the growth regulation of pituitary adenoma by targeting *HMGA2* [78].

Non-functioning pituitary adenomas (NFPA). Several miRNAs mapped to 14q32 showed significant underexpression compared to normal pituitary in NFPA detected by different platforms [86,88,89]. Besides 32 downregulated miRNAs, *MEG3* and *DLK1* also showed underexpression in NFPA samples [89]. Indeed, another study investigating the silencing of the imprinted *DLK1-MEG3* locus in human NFPAs [81] also reported numerous 14q32 encoded miRNAs expression as lost or significantly diminished compared to normal pituitary. Furthermore, the authors identified these miRNA expression alterations together with increased methylation of *MEG3*-IG DMR [81,140,141]. This was in line with the finding that *MEG3* was not expressed in NFPAs; therefore the authors suggested that the silencing of the *DLK1-MEG3* locus played an important role in human NFPA pathogenesis [81,140,141]. Among 14q32 miRNAs miR-1185-1-3p was identified upregulated, while miR-493 downregulated [82]. Bottoni et al. found that miR-154, miR-127 and miR-134 were downregulated in NFPA and were predictive miRNAs for the histotype [79]. Others identified downregulation of miR-432 encoded at 14q32 in NFPA and gonadotroph adenomas [90]. The functional role of miR-432 was investigated in HP75 human pituitary adenoma cells, and miRNA transfection led to a significant reduction of cell number compared to controls [78]. Regarding gonadotroph adenomas reduced miR-410 expression seemed to be restricted to gonadotroph adenomas compared to other adenoma types [86]. Authors validated cyclin B1 (*CCNB1*) as target of miRNA-410 since its overexpression reduced *CCNB1* at protein and mRNA levels, decreasing cell proliferation.

Corticotroph adenomas. Cheunsuchon et al. investigated 18 members of the 14q32 miRNA cluster, among which several miRNAs identified as significantly downregulated (miR-127-3p, miR-136, miR-154, miR-299-5p, miR-329, miR-369-3p, miR-369-5p, miR-376c, miR-377 and miR-433) and only miR-431 was found overexpressed in tumors compared to normal tissues [81]. Although more than a few downregulated miRNAs were detected in adrenocorticotropin (ACTH)-secreting tumors, their expression levels were considered significantly higher compared to those found in NFPAs [81]. Accordingly, *DLK1* was found downregulated in corticotroph tumors [81]. While Stilling et al. detected 5 other miRNAs significantly downregulated located at 14q32 (miR-323-5p, miR-136*, miR-411*, miR-431*, miR-493) in corticotroph adenomas [85], others failed to detect any differentially expressed miRNAs from 14q32 region [84].

Pituitary carcinomas. In corticotroph carcinomas, miR-323-5p was downregulated in comparison to normal pituitary, and miR-493 was upregulated in carcinoma vs. adenoma [85]. It was suggested that miRNA-493 interacted with galectin-3 (*LGALS3*, lectin, galactoside-binding, soluble, 3) and Runt-related transcription factor 2 (*RUNX2*) genes, [142–145]. These data also showed that galectin-3 had a role in regulating cell proliferation and apoptosis of pituitary cells.

Pituitary oncocytoma. In pituitary oncocytoma, numerous underexpressed miRNAs (40% of all downregulated miRNAs) compared to normal control were mapped to 14q32 region [91].

3.5. Thyroid Carcinoma

Thyroid cancer is the most frequent malignant endocrine tumor. The majority of them (~95%) arise from follicular cells and classified as papillary (PTC, 75–80%), follicular (FTC, 10–15%) or anaplastic thyroid cancer (ATC, 0.2–2%) [146]. Tumors developing from calcitonin secreting parafollicular C cells are a distinct entity, and called medullary thyroid cancer (MTC) representing ~5–10% of all thyroid cancers [146]. This subtype commonly occur sporadically; however, approximately 10–25% of them are hereditary and appear as part of MEN2 syndrome, caused by germline mutations of the *RET* proto-oncogene [146]. Most of the well differentiated thyroid cancer (including PTC, FTC) has excellent prognosis; however, patients with ATC have 6–12 months median survival [147].

Nikiforova et al. detected markedly different profiles of miRNA expression between MTC and all other thyroid tumors that derives from follicular cells, reflecting tissue-specific characteristics of miRNAs [94]. Among these, several 14q32 miRNAs were overexpressed in MTC compared to normal and other thyroid cancer types [94]. Expectedly, Lassalle et al. detected numerous miRNAs differentially expressed between sporadic and hereditary MTC cases including miR-136, miR-487b, miR-376a,c, and miR-127 located at 14q32 miRNA cluster [92]. Interestingly, the highly expressed miR-375 was revealed as a novel circulating prognostic marker for MTC patients as well, as MTC patients had significantly higher miR-375 plasma levels than healthy controls and subjects in remission [148]. Additionally, high circulating miR-375 level was associated with significantly reduced overall survival and was a strong prognostic factor of poor prognosis [148].

Numerous miRNAs were described in non-medullary thyroid cancer types, however, with controversial results. Therefore, a comprehensive re-analysis integrating 21 thyroid cancer miRNA studies by Saiselet et al. determined the commonly reported differentially expressed miRNAs in non-medullary thyroid carcinomas compared to normal tissues [149]. Of the investigated studies, in FTC and ATC, no differentially expressed miRNA encoded at 14q32 miRNAs occurred except the downregulated miR-299 in FTC [149,150]. However, 4 overexpressed (miR-134, miR-136, miR-409, miR-654) and several underexpressed (miR-124, miR-134, miR-300, miR-379, miR-382 and miR-494-3p, miR-494-5p and miR-495) 14q32 miRNA were identified in PTC samples. In a more recent study, the global downregulation of miRNAs from the 14q32 region in human PTC was also confirmed [95]. The decreased miR-654-3p levels with long-term PTC progression in Tg-Braf mice was also observed and the level of miR-654-3p inversely correlated with epithelial-mesenchymal transition (EMT) [95]. The in-vitro restoration of miR-654-3p inhibited cell proliferation and migration and induced reprogramming of metastasis-related genes, supporting the tumor suppressor role for this miRNA [95]. Interestingly, in another study analyzing miRNA expression profiles in classical-type PTC, follicular-variant PTC, and tall-cell variant, no 14q32 miRNA was detected compared normal adjacent thyroid tissues [96].

From a clinical point of view miRNAs are suggested as potential biomarkers, as cytology following fine-needle aspiration biopsy (FNAB) are interpreted as indeterminate without definitive diagnosis regarding thyroid tumors in 3–6% to 10–25% [147]. However, miRNAs located at 14q32 did not help in discriminating benign vs. malignant thyroid lesions from FNAB samples [6].

4. Different Expression of 14q32 miRNA Cluster Members

14q32 locus contains more than forty miRNAs, and previously it had been thought that they were generated from one polycistronic transcript containing the whole miRNA cluster under a coordinated regulation with the *MEG3* non-coding RNA located upstream [7,12,151]. Also, hyper-methylation of the 14q32 DMRs was described to associate with decreased 14q32 miRNA expression and vice versa, suggesting that the entire imprinted cluster is regulated jointly [37,44,45,152].

However, in several endocrine tumors, the pattern of 14q32 miRNAs were not so homogenous. Indeed, in other tissues and tumor types, similar findings were described [10,153,154]. Also, in non-tumorous cells not all of the members of 14q32 miRNA cluster were expressed in all tissues and 14q32 miRNAs demonstrated varying level of expression, suggesting other possible regulating mechanisms [155]. Indeed, the expression of protein-coding and non-coding genes encoded at the 14q32 locus was regulated by epigenetic changes, but the exact mechanism behind controlling this process is not entirely known [21,36,46,68,156,157].

However, several mechanisms have been identified in the context of this variable expression of the cluster members, among which *methylation* was the most obvious. Genomic imprinting imbalance could result in the differential modulation of paternally and maternally expressed genes from the 14q32 region that might serve as an explanation, at least in part, for the increased levels of *DIO3* observed in some papillary thyroid cancer samples [95,153,158]. Various DNA methylation patterns of the 14q32 locus were observed in different blood vessel types, which were not associated with miRNA expression [10]. Direct correlation was not possible to be proven between 14q32 estimated methylation fraction of multiple cytosines followed by guanine residues (CpG) in the 3 DMRs located along 14q32 and 14q32 miRNA expression [10]. Moreover, neither DNMT gene expression or DNA methylation did not correlate with primary or mature miRNA expression [10]. In urothelial carcinoma, distinctive epigenetic alterations were again observed at the three regions controlling *DLK1* and *MEG3* expression [154]. The authors suggested that altered nucleosomal positioning could account for the irregular patterning of DNA methylation; namely, that one specific CpG site became significantly hypomethylated in cancer cells, while methylation of flanking sites rather increased [154].

Recent studies have shown that *chromatin remodeling by lncRNA-mediated mechanisms,* may also participate in regulating the expression of the 14q32-encoded miRNAs [46,68].

Additionally, Greife and colleagues demonstrated the loss of active and gain of repressive histone modifications at all regulatory sequences using chromatin immunoprecipitation [154].

Differences in miRNA *splicing, primary transcript processing or pre-miRNA cleavage and maturation* were also reported related to 14q32 miRNAs [159]. Some suggested that the expression of miRNA clustered on 14q32 might be particularly sensitive to changes in the miRNA biogenesis pathway [159–161]. A large proportion of 14q32 encoded miRNAs contained structural features associated with Dicer-independent processing [162], therefore Ago2-dependent pre-miRNA processing [162,163] was particularly important for the biogenesis of miRNA in this cluster. Goossens et al. reinforced that miRNA-specific expression fingerprints implied individual regulation of 14q32 miRNA expression [10].

RNA Binding Proteins (RBPs) were other post-transcriptional regulators of miRNA expression. RBPs bound precursor miRNAs and promoted or inhibited their maturation. For instance, Myocyte Enhancer Factor 2A (*MEF2A*) was such an RBP regulating miR-329 and miR-494 encoded at 14q32 chromosomal region [164]. Cold-inducible RNA-binding protein (*CIRBP*) and hydroxyacyl-CoA dehydrogenase trifunctional multienzyme complex subunit β (*HADHB*) were also RNA binding proteins that regulated 14q32 miRNA expression [165].

The different expressional pattern regarding this miRNA cluster was also observed by Manodoro et al., who attributed it to the presence of the binding sites of *CCCTC-binding factor (CTCF)* which was implicated in transcriptional activation/repression and imprinting [32,157,166]. CTCF exerted its regulatory function by binding to unmethylated DNA in an allele-specific manner [167,168]. Interestingly, it was found that different CTCF binding sites display a different influence on 14q32 miRNA expression depending on the position [157].

Altogether, these data suggest that multiple mechanisms other than genetic mutations or chromosomal loss might be involved in the regulation of 14q32-encoded miRNAs.

5. Functional Impact of 14q32 miRNAs

By analyzing function of individual 14q32 miRNAs, besides several molecular functions and biological processes, TGFβ and Wnt signaling were also identified, which are frequently involved in tumor development (Supplementary Table S1).

However, as 14q32 miRNAs more or less function in cooperation, we performed target prediction and gene set enrichment analysis to investigate the net effect of their co-expressional pattern. Several cancer-related pathways (including TGF-β signaling, Ras signaling, ErbB signaling), pathways involved in invasiveness and metastasis development (e.g., proteoglycans in cancer, adherens junction) or influencing pluripotency and stemness were identified as a potentially functional role (Table 3).

Literature/experimental data also suggested the regulation of axon guidance, actin cytoskeleton, focal adhesion, mammalian target of rapamycin, calcium, mitogen-activated protein kinase and ErbB signaling pathways by 14q32 miRNAs [159]. Liu et al. found that these miRNAs were highly associated with cellular pluripotency [35]. Interestingly, the transcription factor, *MEF2A* regulated the expression of 14q32 miRNAs being a direct target of miR-329 [169]. Uppal and colleagues, using mRNA profiling and bioinformatics, demonstrated that 14q32 miRNAs target genes in PI3K/AKT/mTOR and TGF-β pathways were involved in focal adhesion, cell–extracellular matrix interactions, gap junctions and actin cytoskeleton, resulting in impaired adhesion, invasion and migration, processes that were essential for the development of metastases [22,170,171]. The regulation of PI3K/AKT/mTOR and TGF-β pathways by 14q32 miRNAs was also strengthened by Qian and colleagues in hemopoietic stem cells as well [172].

Table 3. Top 20 significant signaling pathway regulated by 14q32 miRNAs.

KEGG Pathway	*p*-Value	# Genes	# miRNAs
Hippo signaling pathway (hsa04390)	2.635×10^{-7}	103	47
Proteoglycans in cancer (hsa05205)	2.507×10^{-6}	132	48
Pathways in cancer (hsa05200)	3.424×10^{-6}	255	48
Adherens junction (hsa04520)	1.345×10^{-5}	57	41
TGF-β signaling pathway (hsa04350)	1.582×10^{-5}	58	45
Axon guidance (hsa04360)	2.465×10^{-5}	88	45
Rap1 signaling pathway (hsa04015)	3.946×10^{-5}	141	48
Glioma (hsa05214)	4.825×10^{-5}	47	43
Ras signaling pathway (hsa04014)	4.825×10^{-5}	146	49
Circadian rhythm (hsa04710)	6.429×10^{-5}	27	37
Lysine degradation (hsa00310)	9.643×10^{-5}	33	43
Signaling pathways regulating pluripotency of stem cells (hsa04550)	0.0001	96	50
FoxO signaling pathway (hsa04068)	0.0001	92	46
Thyroid hormone signaling pathway (hsa04919)	0.0001	79	46
Ubiquitin mediated proteolysis (hsa04120)	0.0004	93	44
Dorso-ventral axis formation (hsa04320)	0.0006	24	36
Prion diseases (hsa05020)	0.0009	17	26
ErbB signaling pathway (hsa04012)	0.0011	63	45
Renal cell carcinoma (hsa05211)	0.0015	48	41
Pancreatic cancer (hsa05212)	0.0023	48	43

In osteosarcoma, the decrease of 14q32 miRNA levels stabilized c-MYC protooncogene expression and consequently increased the level of oncogenic miR-17-92 miRNA cluster [173]. Cell cycle and epithelial-mesenchymal transition (EMT) were also proved to be influenced by 14q32 miRNAs [38,47,95,174]. Genes involved in metastasis development were also enriched among 14q32 miRNA targets [47]. Cyclin dependent kinase 5 (*CDK5*) and Twist Family BHLH Transcription Factor 1 (*TWIST1*) have been reported to increase metastasis through regulating cell cycle and EMT and they were found to be upregulated in osteosarcoma tumors with low levels of 14q32 miRNAs [47,174,175]. Furthermore, thymidine kinase 1 (*TK1*), that expressed at high levels in proliferating cells and appeared to correlate with high risk in multiple cancer types, was negatively correlated with 14q32 miRNA expression [47,176–178].

In thyroid cancer, the role of 14q32 miRNAs was particularly investigated. Geraldo et al. showed that 14q32 miRNAs contributed to tumor progression and metastasis by targeting key regulators of cell adhesion, migration, proliferation, hypoxic response and wound healing [95]. The reintroduction of miR-654-3p reversed EMT by targeting, hence increasing the expression of cadherin 1 (*CDH1*) and catenin α 1 (*CTNNA1*), and decreasing the expression of Snail Family Transcriptional Repressor 2 (*SNAI2*) [95]. Also, it was demonstrated that genes involved in tumor progression (ECM-remodeling and metastasis) were restored after transfection of a miR-654-3p mimetic [95].

Angiogenesis and neovascularization were also significantly regulated by 14q32 miRNAs. This was confirmed by identifying vascular endothelial growth factor A (*VEGFA*) as a target of miR-494 [169], as well as of miR-127 [179]. Furthermore, miR-495 were demonstrated to target C-C motif chemokine ligand 2 (*CCL2*), through which it affects proliferation and apoptosis of human umbilical vein endothelial cells (HUVECs) [180]. Forkhead box O1 (*FOXO1*) influencing endothelial growth and proliferation [181], wound closure and vascular density was also identified as a target molecule of miR-544 in colorectal cancer [181–183]. The role of 4 14q32 miRNAs was additionally proved in in vivo experiments, as the inhibition of miR-329, miR-494, miR-487b and miR-495 in mice stimulated neovascularization after hind limb ischemia [10,169].

6. Summary and Discussion

Altogether, 14q32 miRNAs have an important role in development and tumorigenesis. The importance of this miRNA cluster regulatory function is represented by the finding that their expression is stable in healthy cells, and even in cell-free serum samples. Their expression is also independent of the most common confounding factors, such as age, sex or BMI. Several studies have shown downregulation of miRNAs from the 14q32 region in different types of cancer; however, 14q32 miRNAs are overexpressed in some cancer types reflecting the tissue-specificity of miRNA function.

Among different endocrine tumors in pituitary adenoma and oncocytoma, papillary thyroid cancer and a particular subset of pheochromocytoma and adrenocortical cancer are characterized by the downregulation of almost all miRNAs encoded by the 14q32 cluster. In the subgroups of ACC and PCC, the silencing of the imprinted 14q32 cluster due to LOH of chromosome arm 14q or 14q32 locus. In other tumor types including NFPA and/or gonadotroph pituitary adenomas increased methylation of the region DMRs could explain the orchestrated downregulation of the coding and non-coding gene expression of the entire region.

Interestingly, pancreas NET, most of the adrenocortical cancer cases and medullary thyroid cancer are particularly distinct, as 14q32 miRNAs are overexpressed in these tumors. The role of these overexpressed miRNAs should be further investigated in relation to tumorigenesis.

In the third group of endocrine tumor types such as pheochromocytoma and growth-hormone producing tumors, and based on the expression pattern of 14q32 miRNAs, however, both increased and decreased expression of 14q32 miRNAs cluster members were observed. In the background of this phenomenon, methodological, technical and biological factors can be hypothesized as well. Different researcher groups applied different study design, different sample numbers, RNA extraction and miRNA quantification methods which all could lead to controversial results. Also, as detailed above, several biological explanations have been revealed in the context of the various expressional pattern of the different members of 14q32 miRNA clusters; e.g., different methylation pattern, chromatin remodeling, histone modifications, alteration of miRNA biogenesis, the effects of RNA binding proteins and transcription factors. These factors await being further investigated in pheochromocytoma and growth-hormone producing pituitary adenomas.

Unfortunately, the function of 14q32 in endocrine tumors is not so broadly investigated compared to other tumor types. However, in different types of pituitary adenoma cell lines, 14q32 miRNAs proved their tumor suppressor role by inducing cell cycle arrest and cell growth inhibition by targeting *PTTG1* and *HMGA2*. In PTC, the significance of another 14q32 miRNA, miR-654-3p, was demonstrated in in vivo experiments as it influenced proliferation, migration, metastasis-related gene expression and EMT.

14q32 miRNAs are also associated with disease prognosis. In endocrine tumors several 14q32 miRNAs were identified as prognostic markers in pancreatic, small intestinal and lung NET. Some of these miRNAs are also linked to patient survival in lung NET. In PCC 14q32 miRNAs indicated metastatic cases compared to non-metastatic cases, assisting the discrimination of benign and malignant tumors. Also, the higher expression of miRNAs encoded at 14q32 were associated with aggressive ACC cases. Finally, while in MTC, 14q32 miRNA miR-375 has not only been reported as a strong prognostic factor of poor prognosis, but its higher level was associated with reduced overall survival, while on the contrary, the decreased 14q32 miRNA in PTC was associated with long-term progression.

Based on data presented in Figure 1, we found some miRNAs unique among endocrine tumor types (miR-337 and miR-758 overexpression in pNET, miR-329 and miR-541 overexpression in PCC/PGL and miR-376c overexpression in MTC). However, there is no full consensus among miRNA profiling studies; i.e., not all studies identified the same miRNAs differentially expressed, and therefore these findings should be further validated. The discrepancy can be due to different study design, difference in sample number (statistical power) and also due to technical factors (e.g., different platforms for high-throughput profiling). Analyzing expressional pattern of 14q32 miRNA cluster instead of individual miRNAs, characteristic/unique expression profile was described in MTC compared to other thyroid carcinoma type [94] or in different types of pituitary adenoma [79,81]. Furthermore, even the same type of endocrine tumor, pheochromocytoma can be grouped by different miRNA pattern according to germline mutational background [71]. As 14q32 miRNAs were found to be dysregulated in several cancer types, globally, they cannot be considered as unique tissue biomarkers. Another level of use can be their application in liquid biopsy samples e.g., in circulation. This is supported by the finding that 14q32 miRNAs were stable in serum and their level was not significantly affected by common confounder factors [11]. Already, the potential use of circulating miRNAs has been suggested in endocrine tumors. In ACC a unique, tissue specific miRNA, miR-483-5p has been suggested as a potential candidate as predictive marker for recurrence [54], and miR-146a-5p and miR-221-3p as serum biomarkers for post-treatment monitoring of PTC patients [96]. Unfortunately, to our best knowledge, no 14q32 mapped miRNA in circulation has been investigated in endocrine tumors. However, in non-endocrine tumor types their expression indicate prognosis and survival [155]. Accordingly, this miRNA cluster was proved to influence EMT process and metastasis development [95,132], hence, they may be used as prognostic biomarker in endocrine tumors as well.

The evolutional role and constraint can be considered another point of view of imprinted miRNAs [184–186]. It was suggested that imprinted noncoding RNAs was under distinctive selective forces when regulating transcripts of the allele inherited from the other parent [184–187]. Accordingly, when an mRNA had sequence complementary to an imprinted miRNA, the complementary miRNA-mRNA sequences pair originated from different alleles. This can be considered a communication between the maternal and paternal alleles, hence the two alleles coordinate their activities [186]. The kinship theory considers genomic imprinting as a mechanism to change gene dosage, because it has a differential effect on the fitness of matrilineal and patrilineal relatives [184–187]. Additionally, Haig and Mainieri suggested that when an imprinted miRNA targets an unimprinted mRNA, their interaction may have different fitness consequences for the loci encoding the miRNA and mRNA [184]. In a recent study, *HMGA2* was proposed as an attractive candidate to be one of the original targets because its 3′ regulatory region contained multiple predicted target sites for 14q32 miRNAs, with some of these target sites evolutionarily older than the 14q32 miRNA cluster [184]. *HMGA2* have special role in this context as it has reported to be overexpressed in several cancer type including endocrine tumors and often regulated by miRNAs [188,189], which also highlights the role of 14q32 miRNAs in tumorigenesis.

7. Conclusions

Similar to other cancer types, 14q32 miRNAs have a significant role in the tumorigenesis of endocrine glands. In different endocrine tumor types this miRNA cluster reflects the general tissue specificity of miRNAs regarding expression pattern, tumor suppressor or oncogene function, and they have a significant impact on prognosis as well. Regarding the stable expression of 14q32 miRNAs in healthy individuals in circulation, the further investigation of this miRNA cluster could provide an option to use them as diagnostic or prognostic biomarkers in endocrine neoplasms.

Supplementary Materials: The following are available online at https://www.mdpi.com/2073-4425/12/5/698/s1, Table S1: 14q32 miRNAs and their function.

Author Contributions: Conceptualization, H.B. and A.P.; literature mining, H.B. and L.K.; bioinformatics analysis, H.B. and L.K.; data curation, H.B.; writing—original draft preparation, H.B.; writing—review and editing, A.P.; visualization, L.K.; supervision, A.P.; funding acquisition, H.B. All authors have read and agreed to the published version of the manuscript.

Funding: The research was supported by a Hungarian Scientific Research Grant of the National Research, Development and Innovation Office (NKFI FK 135065) to Henriett Butz.

Institutional Review Board Statement: Not applicable.

Informed Consent Statement: Not applicable.

Data Availability Statement: All relevant data are included in the manuscript.

Conflicts of Interest: The authors declare no conflict of interest.

References

1. Yates, L.A.; Norbury, C.J.; Gilbert, R.J.C. The Long and Short of MicroRNA. *Cell* **2013**, *153*, 516–519. [CrossRef]
2. Valinezhad Orang, A.; Safaralizadeh, R.; Kazemzadeh-Bavili, M. Mechanisms of MiRNA-Mediated Gene Regulation from Common Downregulation to MRNA-Specific Upregulation. *Int. J. Genom.* **2014**, *2014*, 970607. [CrossRef]
3. Lewis, B.P.; Burge, C.B.; Bartel, D.P. Conserved Seed Pairing, Often Flanked by Adenosines, Indicates That Thousands of Human Genes Are MicroRNA Targets. *Cell* **2005**, *120*, 15–20. [CrossRef] [PubMed]
4. Mattick, J.S.; Makunin, I.V. Small Regulatory RNAs in Mammals. *Hum. Mol. Genet.* **2005**, *14* (Suppl. S1), R121–R132. [CrossRef] [PubMed]
5. Peng, C.; Wang, Y.-L. Editorial: MicroRNAs as New Players in Endocrinology. *Front. Endocrinol.* **2018**, *9*, 459. [CrossRef] [PubMed]
6. Butz, H.; Patócs, A. MicroRNAs in Endocrine Tumors. *EJIFCC* **2019**, *30*, 146–164. [PubMed]
7. Seitz, H. A Large Imprinted MicroRNA Gene Cluster at the Mouse Dlk1-Gtl2 Domain. *Genome Res.* **2004**, *14*, 1741–1748. [CrossRef]
8. Cavaillé, J.; Seitz, H.; Paulsen, M.; Ferguson-Smith, A.C.; Bachellerie, J.-P. Identification of Tandemly-Repeated C/D SnoRNA Genes at the Imprinted Human 14q32 Domain Reminiscent of Those at the Prader-Willi/Angelman Syndrome Region. *Hum. Mol. Genet.* **2002**, *11*, 1527–1538. [CrossRef]
9. Royo, H.; Cavaillé, J. Non-Coding RNAs in Imprinted Gene Clusters. *Biol. Cell* **2008**, *100*, 149–166. [CrossRef]
10. Goossens, E.A.C.; de Vries, M.R.; Simons, K.H.; Putter, H.; Quax, P.H.A.; Nossent, A.Y. MiRMap: Profiling 14q32 MicroRNA Expression and DNA Methylation throughout the Human Vasculature. *Front. Cardiovasc. Med.* **2019**, *6*. [CrossRef]
11. Valbuena, G.N.; Apostolidou, S.; Roberts, R.; Barnes, J.; Alderton, W.; Harper, L.; Jacobs, I.; Menon, U.; Keun, H.C. The 14q32 Maternally Imprinted Locus Is a Major Source of Longitudinally Stable Circulating MicroRNAs as Measured by Small RNA Sequencing. *Sci. Rep.* **2019**, *9*, 15787. [CrossRef]
12. Glazov, E.A.; McWilliam, S.; Barris, W.C.; Dalrymple, B.P. Origin, Evolution, and Biological Role of MiRNA Cluster in DLK-DIO3 Genomic Region in Placental Mammals. *Mol. Biol. Evol.* **2008**, *25*, 939–948. [CrossRef] [PubMed]
13. Benetatos, L.; Voulgaris, E.; Vartholomatos, G. DLK1-MEG3 Imprinted Domain MicroRNAs in Cancer Biology. *Crit. Rev. Eukaryot. Gene Expr.* **2012**, *22*, 1–15. [CrossRef]
14. Benetatos, L.; Vartholomatos, G.; Hatzimichael, E. MEG3 Imprinted Gene Contribution in Tumorigenesis. *Int. J. Cancer* **2011**, *129*, 773–779. [CrossRef] [PubMed]
15. Wienholds, E. MicroRNA Expression in Zebrafish Embryonic Development. *Science* **2005**, *309*, 310–311. [CrossRef] [PubMed]
16. Swarbrick, A.; Woods, S.L.; Shaw, A.; Balakrishnan, A.; Phua, Y.; Nguyen, A.; Chanthery, Y.; Lim, L.; Ashton, L.J.; Judson, R.L.; et al. MiR-380-5p Represses P53 to Control Cellular Survival and Is Associated with Poor Outcome in MYCN Amplified Neuroblastoma. *Nat. Med.* **2010**, *16*, 1134–1140. [CrossRef]
17. Holder, J.L.; Lotze, T.E.; Bacino, C.; Cheung, S.-W. A Child with an Inherited 0.31 Mb Microdeletion of Chromosome 14q32.33: Further Delineation of a Critical Region for the 14q32 Deletion Syndrome. *Am. J. Med. Genet. A* **2012**, *158A*, 1962–1966. [CrossRef]
18. Briggs, T.A.; Lokulo-Sodipe, K.; Chandler, K.E.; Mackay, D.J.G.; Temple, I.K. Temple Syndrome as a Result of Isolated Hypomethylation of the 14q32 Imprinted DLK1/MEG3 Region. *Am. J. Med. Genet. A* **2016**, *170A*, 170–175. [CrossRef]
19. Bens, S.; Kolarova, J.; Gillessen-Kaesbach, G.; Buiting, K.; Beygo, J.; Caliebe, A.; Ammerpohl, O.; Siebert, R. The Differentially Methylated Region of MEG8 Is Hypermethylated in Patients with Temple Syndrome. *Epigenomics* **2015**, *7*, 1089–1097. [CrossRef]
20. Kagami, M.; O'Sullivan, M.J.; Green, A.J.; Watabe, Y.; Arisaka, O.; Masawa, N.; Matsuoka, K.; Fukami, M.; Matsubara, K.; Kato, F.; et al. The IG-DMR and the MEG3-DMR at Human Chromosome 14q32.2: Hierarchical Interaction and Distinct Functional Properties as Imprinting Control Centers. *PLoS Genet.* **2010**, *6*, e1000992. [CrossRef] [PubMed]
21. Kagami, M.; Sekita, Y.; Nishimura, G.; Irie, M.; Kato, F.; Okada, M.; Yamamori, S.; Kishimoto, H.; Nakayama, M.; Tanaka, Y.; et al. Deletions and Epimutations Affecting the Human 14q32.2 Imprinted Region in Individuals with Paternal and Maternal Upd(14)-like Phenotypes. *Nat. Genet.* **2008**, *40*, 237–242. [CrossRef]
22. Uppal, A.; Wightman, S.C.; Mallon, S.; Oshima, G.; Pitroda, S.P.; Zhang, Q.; Huang, X.; Darga, T.E.; Huang, L.; Andrade, J.; et al. 14q32-Encoded MicroRNAs Mediate an Oligometastatic Phenotype. *Oncotarget* **2015**, *6*, 3540–3552. [CrossRef]

23. Rosenfeld, J.A.; Fox, J.E.; Descartes, M.; Brewer, F.; Stroud, T.; Gorski, J.L.; Upton, S.J.; Moeschler, J.B.; Monteleone, B. Neill, N.J.; et al. Clinical Features Associated with Copy Number Variations of the 14q32 Imprinted Gene Cluster. *Am. J Med. Genet. Part A* **2015**, *167*, 345–353. [CrossRef] [PubMed]
24. Ogata, T.; Kagami, M. Kagami–Ogata Syndrome: A Clinically Recognizable Upd(14)Pat and Related Disorder Affecting the Chromosome 14q32.2 Imprinted Region. *J. Hum. Genet.* **2016**, *61*, 87–94. [CrossRef]
25. Sun, M.; Xia, R.; Jin, F.; Xu, T.; Liu, Z.; De, W.; Liu, X. Downregulated Long Noncoding RNA MEG3 Is Associated with Poor Prognosis and Promotes Cell Proliferation in Gastric Cancer. *Tumour Biol.* **2014**, *35*, 1065–1073. [CrossRef] [PubMed]
26. Béna, F.; Gimelli, S.; Migliavacca, E.; Brun-Druc, N.; Buiting, K.; Antonarakis, S.E.; Sharp, A.J. A Recurrent 14q32.2 Microdeletion Mediated by Expanded TGG Repeats. *Hum. Mol. Genet.* **2010**, *19*, 1967–1973. [CrossRef] [PubMed]
27. Kawakami, T.; Chano, T.; Minami, K.; Okabe, H.; Okada, Y.; Okamoto, K. Imprinted DLK1 Is a Putative Tumor Suppressor Gene and Inactivated by Epimutation at the Region Upstream of GTL2 in Human Renal Cell Carcinoma. *Hum. Mol. Genet.* **2006**, *15* 821–830. [CrossRef]
28. Sanlaville, D.; Aubry, M.C.; Dumez, Y.; Nolen, M.C.; Amiel, J.; Pinson, M.P.; Lyonnet, S.; Munnich, A.; Vekemans, M.; Morichon-Delvallez, N. Maternal Uniparental Heterodisomy of Chromosome 14: Chromosomal Mechanism and Clinical Follow Up. *J. Med Genet.* **2000**, *37*, 525–528. [CrossRef]
29. Temple, I.K.; Shrubb, V.; Lever, M.; Bullman, H.; Mackay, D.J.G. Isolated Imprinting Mutation of the DLK1/GTL2 Locus Associated with a Clinical Presentation of Maternal Uniparental Disomy of Chromosome 14. *J. Med. Genet.* **2007**, *44*, 637–640. [CrossRef]
30. Da Rocha, S.T.; Edwards, C.A.; Ito, M.; Ogata, T.; Ferguson-Smith, A.C. Genomic Imprinting at the Mammalian Dlk1-Dio3 Domain. *Trends Genet.* **2008**, *24*, 306–316. [CrossRef] [PubMed]
31. Hagan, J.P.; O'Neill, B.L.; Stewart, C.L.; Kozlov, S.V.; Croce, C.M. At Least Ten Genes Define the Imprinted Dlk1-Dio3 Cluster on Mouse Chromosome 12qF1. *PLoS ONE* **2009**, *4*, e4352. [CrossRef] [PubMed]
32. Wylie, A.A.; Murphy, S.K.; Orton, T.C.; Jirtle, R.L. Novel Imprinted DLK1/GTL2 Domain on Human Chromosome 14 Contains Motifs That Mimic Those Implicated in IGF2/H19 Regulation. *Genome Res.* **2000**, *10*, 1711–1718. [CrossRef] [PubMed]
33. Takada, S.; Paulsen, M.; Tevendale, M.; Tsai, C.-E.; Kelsey, G.; Cattanach, B.M.; Ferguson-Smith, A.C. Epigenetic Analysis of the Dlk1-Gtl2 Imprinted Domain on Mouse Chromosome 12: Implications for Imprinting Control from Comparison with Igf2-H19 *Hum. Mol. Genet.* **2002**, *11*, 77–86. [CrossRef] [PubMed]
34. Schmidt, J.V.; Matteson, P.G.; Jones, B.K.; Guan, X.J.; Tilghman, S.M. The Dlk1 and Gtl2 Genes Are Linked and Reciprocally Imprinted. *Genes Dev.* **2000**, *14*, 1997–2002. [PubMed]
35. Liu, L.; Luo, G.-Z.; Yang, W.; Zhao, X.; Zheng, Q.; Lv, Z.; Li, W.; Wu, H.-J.; Wang, L.; Wang, X.-J.; et al. Activation of the Imprinted Dlk1-Dio3 Region Correlates with Pluripotency Levels of Mouse Stem Cells. *J. Biol. Chem.* **2010**, *285*, 19483–19490. [CrossRef] [PubMed]
36. Kameswaran, V.; Bramswig, N.C.; McKenna, L.B.; Penn, M.; Schug, J.; Hand, N.J.; Chen, Y.; Choi, I.; Vourekas, A.; Won, K.-J.; et al. Epigenetic Regulation of the DLK1-MEG3 MicroRNA Cluster in Human Type 2 Diabetic Islets. *Cell Metab.* **2014**, *19*, 135–145. [CrossRef] [PubMed]
37. Guo, W.; Dong, Z.; Liu, S.; Qiao, Y.; Kuang, G.; Guo, Y.; Shen, S.; Liang, J. Promoter Hypermethylation-Mediated Downregulation of MiR-770 and Its Host Gene MEG3, a Long Non-Coding RNA, in the Development of Gastric Cardia Adenocarcinoma. *Mol. Carcinog.* **2017**, *56*, 1924–1934. [CrossRef]
38. González-Vallinas, M.; Rodríguez-Paredes, M.; Albrecht, M.; Sticht, C.; Stichel, D.; Gutekunst, J.; Pitea, A.; Sass, S.; Sánchez-Rivera, F.J.; Lorenzo-Bermejo, J.; et al. Epigenetically Regulated Chromosome 14q32 MiRNA Cluster Induces Metastasis and Predicts Poor Prognosis in Lung Adenocarcinoma Patients. *Mol. Cancer Res.* **2018**, *16*, 390–402. [CrossRef]
39. Moradi, S.; Sharifi-Zarchi, A.; Ahmadi, A.; Mollamohammadi, S.; Stubenvoll, A.; Günther, S.; Salekdeh, G.H.; Asgari, S.; Braun, T.; Baharvand, H. Small RNA Sequencing Reveals Dlk1-Dio3 Locus-Embedded MicroRNAs as Major Drivers of Ground-State Pluripotency. *Stem Cell Rep.* **2017**, *9*, 2081–2096. [CrossRef]
40. Xi, S.; Xu, H.; Shan, J.; Tao, Y.; Hong, J.A.; Inchauste, S.; Zhang, M.; Kunst, T.F.; Mercedes, L.; Schrump, D.S. Cigarette Smoke Mediates Epigenetic Repression of MiR-487b during Pulmonary Carcinogenesis. *J. Clin. Investig.* **2013**, *123*, 1241–1261. [CrossRef]
41. Oshima, G.; Poli, E.C.; Bolt, M.J.; Chlenski, A.; Forde, M.; Jutzy, J.M.S.; Biyani, N.; Posner, M.C.; Pitroda, S.P.; Weichselbaum, R.R.; et al. DNA Methylation Controls Metastasis-Suppressive 14q32-Encoded MiRNAs. *Cancer Res.* **2019**, *79*, 650–662. [CrossRef]
42. Kelly, L.; Bryan, K.; Kim, S.Y.; Janeway, K.A.; Killian, J.K.; Schildhaus, H.-U.; Miettinen, M.; Helman, L.; Meltzer, P.S.; van de Rijn, M.; et al. Post-Transcriptional Dysregulation by MiRNAs Is Implicated in the Pathogenesis of Gastrointestinal Stromal Tumor [GIST]. *PLoS ONE* **2013**, *8*, e64102. [CrossRef]
43. Anaya-Ruiz, M.; Bandala, C.; Perez-Santos, J.L.M. MiR-485 Acts as a Tumor Suppressor by Inhibiting Cell Growth and Migration in Breast Carcinoma T47D Cells. *Asian Pac. J. Cancer Prev.* **2013**, *14*, 3757–3760. [CrossRef]
44. Qin, R.; Chen, Z.; Ding, Y.; Hao, J.; Hu, J.; Guo, F. Long Non-Coding RNA MEG3 Inhibits the Proliferation of Cervical Carcinoma Cells through the Induction of Cell Cycle Arrest and Apoptosis. *Neoplasma* **2013**, *60*, 486–492. [CrossRef]
45. Ying, L.; Huang, Y.; Chen, H.; Wang, Y.; Xia, L.; Chen, Y.; Liu, Y.; Qiu, F. Downregulated MEG3 Activates Autophagy and Increases Cell Proliferation in Bladder Cancer. *Mol. Biosyst.* **2013**, *9*, 407–411. [CrossRef]
46. Formosa, A.; Markert, E.K.; Lena, A.M.; Italiano, D.; Finazzi-Agrò, E.; Levine, A.J.; Bernardini, S.; Garabadgiu, A.V.; Melino, G.; Candi, E. MicroRNAs, MiR-154, MiR-299-5p, MiR-376a, MiR-376c, MiR-377, MiR-381, MiR-487b, MiR-485-3p, MiR-495 and

MiR-654-3p, Mapped to the 14q32.31 Locus, Regulate Proliferation, Apoptosis, Migration and Invasion in Metastatic Prostate Cancer Cells. *Oncogene* **2014**, *33*, 5173–5182. [CrossRef]
47. Sarver, A.L.; Thayanithy, V.; Scott, M.C.; Cleton-Jansen, A.-M.; Hogendoorn, P.C.; Modiano, J.F.; Subramanian, S. MicroRNAs at the Human 14q32 Locus Have Prognostic Significance in Osteosarcoma. *Orphanet J. Rare Dis.* **2013**, *8*, 7. [CrossRef] [PubMed]
48. Zehavi, L.; Avraham, R.; Barzilai, A.; Bar-Ilan, D.; Navon, R.; Sidi, Y.; Avni, D.; Leibowitz-Amit, R. Silencing of a Large MicroRNA Cluster on Human Chromosome 14q32 in Melanoma: Biological Effects of Mir-376a and Mir-376c on Insulin Growth Factor 1 Receptor. *Mol. Cancer* **2012**, *11*, 44. [CrossRef] [PubMed]
49. Niu, C.S.; Yang, Y.; Cheng, C.-D. MiR-134 Regulates the Proliferation and Invasion of Glioblastoma Cells by Reducing Nanog Expression. *Int. J. Oncol.* **2013**, *42*, 1533–1540. [CrossRef] [PubMed]
50. Chen, L.; Zhang, J.; Feng, Y.; Li, R.; Sun, X.; Du, W.; Piao, X.; Wang, H.; Yang, D.; Sun, Y.; et al. MiR-410 Regulates MET to Influence the Proliferation and Invasion of Glioma. *Int. J. Biochem. Cell Biol.* **2012**, *44*, 1711–1717. [CrossRef] [PubMed]
51. Lim, L.; Balakrishnan, A.; Huskey, N.; Jones, K.D.; Jodari, M.; Ng, R.; Song, G.; Riordan, J.; Anderton, B.; Cheung, S.-T.; et al. MicroRNA-494 within an Oncogenic MicroRNA Megacluster Regulates G1/S Transition in Liver Tumorigenesis through Suppression of Mutated in Colorectal Cancer. *Hepatology* **2014**, *59*, 202–215. [CrossRef]
52. Chien, W.W.; Domenech, C.; Catallo, R.; Kaddar, T.; Magaud, J.-P.; Salles, G.; Ffrench, M. Cyclin-Dependent Kinase 1 Expression Is Inhibited by P16 INK4a at the Post-Transcriptional Level through the MicroRNA Pathway. *Oncogene* **2011**, *30*, 1880–1891. [CrossRef]
53. Dixon-McIver, A.; East, P.; Mein, C.A.; Cazier, J.-B.; Molloy, G.; Chaplin, T.; Andrew Lister, T.; Young, B.D.; Debernardi, S. Distinctive Patterns of MicroRNA Expression Associated with Karyotype in Acute Myeloid Leukaemia. *PLoS ONE* **2008**, *3*, e2141. [CrossRef] [PubMed]
54. Chabre, O.; Libé, R.; Assie, G.; Barreau, O.; Bertherat, J.; Bertagna, X.; Feige, J.-J.; Cherradi, N. Serum MiR-483-5p and MiR-195 Are Predictive of Recurrence Risk in Adrenocortical Cancer Patients. *Endocr. Relat. Cancer* **2013**, *20*, 579–594. [CrossRef] [PubMed]
55. Lussier, Y.A.; Xing, H.R.; Salama, J.K.; Khodarev, N.N.; Huang, Y.; Zhang, Q.; Khan, S.A.; Yang, X.; Hasselle, M.D.; Darga, T.E.; et al. MicroRNA Expression Characterizes Oligometastasis(Es). *PLoS ONE* **2011**, *6*, e28650. [CrossRef] [PubMed]
56. Lussier, Y.A.; Khodarev, N.N.; Regan, K.; Corbin, K.; Li, H.; Ganai, S.; Khan, S.A.; Gnerlich, J.L.; Gnerlich, J.; Darga, T.E.; et al. Oligo- and Polymetastatic Progression in Lung Metastasis(Es) Patients Is Associated with Specific MicroRNAs. *PLoS ONE* **2012**, *7*, e50141. [CrossRef]
57. Yoshimoto, T.; Motoi, N.; Yamamoto, N.; Nagano, H.; Ushijima, M.; Matsuura, M.; Okumura, S.; Yamaguchi, T.; Fukayama, M.; Ishikawa, Y. Pulmonary Carcinoids and Low-Grade Gastrointestinal Neuroendocrine Tumors Show Common MicroRNA Expression Profiles, Different from Adenocarcinomas and Small Cell Carcinomas. *Neuroendocrinology* **2018**, *106*, 47–57. [CrossRef] [PubMed]
58. Mairinger, F.D.; Ting, S.; Werner, R.; Walter, R.F.H.; Hager, T.; Vollbrecht, C.; Christoph, D.; Worm, K.; Mairinger, T.; Sheu-Grabellus, S.-Y.; et al. Different Micro-RNA Expression Profiles Distinguish Subtypes of Neuroendocrine Tumors of the Lung: Results of a Profiling Study. *Mod. Pathol.* **2014**, *27*, 1632–1640. [CrossRef]
59. Deng, B.; Molina, J.; Aubry, M.C.; Sun, Z.; Wang, L.; Eckloff, B.W.; Vasmatzis, G.; You, M.; Wieben, E.D.; Jen, J.; et al. Clinical Biomarkers of Pulmonary Carcinoid Tumors in Never Smokers via Profiling MiRNA and Target MRNA. *Cell Biosci.* **2014**, *4*, 35. [CrossRef] [PubMed]
60. Rapa, I.; Votta, A.; Felice, B.; Righi, L.; Giorcelli, J.; Scarpa, A.; Speel, E.-J.M.; Scagliotti, G.V.; Papotti, M.; Volante, M. Identification of MicroRNAs Differentially Expressed in Lung Carcinoid Subtypes and Progression. *Neuroendocrinology* **2015**, *101*, 246–255. [CrossRef]
61. Wong, J.J.M.; Ginter, P.S.; Tyryshkin, K.; Yang, X.; Nanayakkara, J.; Zhou, Z.; Tuschl, T.; Chen, Y.-T.; Renwick, N. Classifying Lung Neuroendocrine Neoplasms through MicroRNA Sequence Data Mining. *Cancers* **2020**, *12*, 2653. [CrossRef] [PubMed]
62. Zimmermann, N.; Knief, J.; Kacprowski, T.; Lazar-Karsten, P.; Keck, T.; Billmann, F.; Schmid, S.; Luley, K.; Lehnert, H.; Brabant, G.; et al. MicroRNA Analysis of Gastroenteropancreatic Neuroendocrine Tumors and Metastases. *Oncotarget* **2018**, *9*, 28379–28390. [CrossRef] [PubMed]
63. Roldo, C.; Missiaglia, E.; Hagan, J.P.; Falconi, M.; Capelli, P.; Bersani, S.; Calin, G.A.; Volinia, S.; Liu, C.-G.; Scarpa, A.; et al. MicroRNA Expression Abnormalities in Pancreatic Endocrine and Acinar Tumors Are Associated with Distinctive Pathologic Features and Clinical Behavior. *J. Clin. Oncol.* **2006**, *24*, 4677–4684. [CrossRef] [PubMed]
64. Jiang, X.; Shan, A.; Su, Y.; Cheng, Y.; Gu, W.; Wang, W.; Ning, G.; Cao, Y. MiR-144/451 Promote Cell Proliferation via Targeting PTEN/AKT Pathway in Insulinomas. *Endocrinology* **2015**, *156*, 2429–2439. [CrossRef]
65. Zhou, H.-Q.; Chen, Q.-C.; Qiu, Z.-T.; Tan, W.-L.; Mo, C.-Q.; Gao, S.-W. Integrative MicroRNA-MRNA and Protein-Protein Interaction Analysis in Pancreatic Neuroendocrine Tumors. *Eur. Rev. Med. Pharmacol. Sci.* **2016**, *20*, 2842–2852. [PubMed]
66. Lee, Y.S.; Kim, H.; Kim, H.W.; Lee, J.-C.; Paik, K.-H.; Kang, J.; Kim, J.; Yoon, Y.-S.; Han, H.-S.; Sohn, I.; et al. High Expression of MicroRNA-196a Indicates Poor Prognosis in Resected Pancreatic Neuroendocrine Tumor. *Medicine* **2015**, *94*, e2224. [CrossRef]
67. Arvidsson, Y.; Rehammar, A.; Bergström, A.; Andersson, E.; Altiparmak, G.; Swärd, C.; Wängberg, B.; Kristiansson, E.; Nilsson, O. MiRNA Profiling of Small Intestinal Neuroendocrine Tumors Defines Novel Molecular Subtypes and Identifies MiR-375 as a Biomarker of Patient Survival. *Mod. Pathol.* **2018**, *31*, 1302–1317. [CrossRef]
68. Li, S.-C.; Essaghir, A.; Martijn, C.; Lloyd, R.V.; Demoulin, J.-B.; Oberg, K.; Giandomenico, V. Global MicroRNA Profiling of Well-Differentiated Small Intestinal Neuroendocrine Tumors. *Mod. Pathol.* **2013**, *26*, 685–696. [CrossRef]

69. Castro-Vega, L.J.; Letouzé, E.; Burnichon, N.; Buffet, A.; Disderot, P.-H.; Khalifa, E.; Loriot, C.; Elarouci, N.; Morin, A.; Menara, M.; et al. Multi-Omics Analysis Defines Core Genomic Alterations in Pheochromocytomas and Paragangliomas. *Nat. Commun.* **2015**, *6*, 6044. [CrossRef]
70. Tömböl, Z.; Eder, K.; Kovács, A.; Szabó, P.M.; Kulka, J.; Likó, I.; Zalatnai, A.; Rácz, G.; Tóth, M.; Patócs, A.; et al. MicroRNA Expression Profiling in Benign (Sporadic and Hereditary) and Recurring Adrenal Pheochromocytomas. *Mod. Pathol.* **2010**, *23*, 1583–1595. [CrossRef] [PubMed]
71. De Cubas, A.A.; Leandro-García, L.J.; Schiavi, F.; Mancikova, V.; Comino-Méndez, I.; Inglada-Pérez, L.; Perez-Martinez, M.; Ibarz, N.; Ximénez-Embún, P.; López-Jiménez, E.; et al. Integrative Analysis of MiRNA and MRNA Expression Profiles in Pheochromocytoma and Paraganglioma Identifies Genotype-Specific Markers and Potentially Regulated Pathways. *Endocr. Relat. Cancer* **2013**, *20*, 477–493. [CrossRef]
72. Meyer-Rochow, G.Y.; Jackson, N.E.; Conaglen, J.V.; Whittle, D.E.; Kunnimalaiyaan, M.; Chen, H.; Westin, G.; Sandgren, J.; Stålberg, P.; Khanafshar, E.; et al. MicroRNA Profiling of Benign and Malignant Pheochromocytomas Identifies Novel Diagnostic and Therapeutic Targets. *Endocr. Relat. Cancer* **2010**, *17*, 835–846. [CrossRef]
73. Calsina, B.; Castro-Vega, L.J.; Torres-Pérez, R.; Inglada-Pérez, L.; Currás-Freixes, M.; Roldán-Romero, J.M.; Mancikova, V.; Letón, R.; Remacha, L.; Santos, M.; et al. Integrative Multi-Omics Analysis Identifies a Prognostic MiRNA Signature and a Targetable MiR-21-3p/TSC2/MTOR Axis in Metastatic Pheochromocytoma/Paraganglioma. *Theranostics* **2019**, *9*, 4946–4958. [CrossRef] [PubMed]
74. Tömböl, Z.; Szabó, P.M.; Molnár, V.; Wiener, Z.; Tölgyesi, G.; Horányi, J.; Riesz, P.; Reismann, P.; Patócs, A.; Likó, I.; et al. Integrative Molecular Bioinformatics Study of Human Adrenocortical Tumors: MicroRNA, Tissue-Specific Target Prediction, and Pathway Analysis. *Endocr. Relat. Cancer* **2009**, *16*, 895–906. [CrossRef] [PubMed]
75. Özata, D.M.; Caramuta, S.; Velázquez-Fernández, D.; Akçakaya, P.; Xie, H.; Höög, A.; Zedenius, J.; Bäckdahl, M.; Larsson, C.; Lui, W.-O. The Role of MicroRNA Deregulation in the Pathogenesis of Adrenocortical Carcinoma. *Endocr. Relat. Cancer* **2011**, *18*, 643–655. [CrossRef]
76. Assié, G.; Letouzé, E.; Fassnacht, M.; Jouinot, A.; Luscap, W.; Barreau, O.; Omeiri, H.; Rodriguez, S.; Perlemoine, K.; René-Corail, F.; et al. Integrated Genomic Characterization of Adrenocortical Carcinoma. *Nat. Genet.* **2014**, *46*, 607–612. [CrossRef] [PubMed]
77. Mao, Z.-G.; He, D.-S.; Zhou, J.; Yao, B.; Xiao, W.-W.; Chen, C.-H.; Zhu, Y.-H.; Wang, H.-J. Differential Expression of MicroRNAs in GH-Secreting Pituitary Adenomas. *Diagn. Pathol.* **2010**, *5*, 79. [CrossRef]
78. D'Angelo, D.; Palmieri, D.; Mussnich, P.; Roche, M.; Wierinckx, A.; Raverot, G.; Fedele, M.; Croce, C.M.; Trouillas, J.; Fusco, A. Altered MicroRNA Expression Profile in Human Pituitary GH Adenomas: Down-Regulation of MiRNA Targeting HMGA1, HMGA2, and E2F1. *J. Clin. Endocrinol. Metab.* **2012**, *97*, E1128–E1138. [CrossRef]
79. Bottoni, A.; Zatelli, M.C.; Ferracin, M.; Tagliati, F.; Piccin, D.; Vignali, C.; Calin, G.A.; Negrini, M.; Croce, C.M.; Degli Uberti, E.C. Identification of Differentially Expressed MicroRNAs by Microarray: A Possible Role for MicroRNA Genes in Pituitary Adenomas. *J. Cell. Physiol.* **2007**, *210*, 370–377. [CrossRef] [PubMed]
80. Butz, H.; Liko, I.; Boyle, B.; Czirjak, S.; Igaz, P.; Patocs, A.; Racz, K. MicroRNA Expression in Human Sporadic Pituitary Adenomas. *Endocr. Abstr.* **2009**, *20*, 566.
81. Cheunsuchon, P.; Zhou, Y.; Zhang, X.; Lee, H.; Chen, W.; Nakayama, Y.; Rice, K.A.; Tessa Hedley-Whyte, E.; Swearingen, B.; Klibanski, A. Silencing of the Imprinted DLK1-MEG3 Locus in Human Clinically Nonfunctioning Pituitary Adenomas. *Am. J. Pathol.* **2011**, *179*, 2120–2130. [CrossRef]
82. He, Z.; Chen, L.; Hu, X.; Tang, J.; He, L.; Hu, J.; Fei, F.; Wang, Q. Next-Generation Sequencing of MicroRNAs Reveals a Unique Expression Pattern in Different Types of Pituitary Adenomas. *Endocr. J.* **2019**, *66*, 709–722. [CrossRef]
83. Gentilin, E.; Tagliati, F.; Filieri, C.; Molè, D.; Minoia, M.; Rosaria Ambrosio, M.; Degli Uberti, E.C.; Zatelli, M.C. MiR-26a Plays an Important Role in Cell Cycle Regulation in ACTH-Secreting Pituitary Adenomas by Modulating Protein Kinase Cδ. *Endocrinology* **2013**, *154*, 1690–1700. [CrossRef]
84. Amaral, F.C.; Torres, N.; Saggioro, F.; Neder, L.; Machado, H.R.; Silva, W.A.; Moreira, A.C.; Castro, M. MicroRNAs Differentially Expressed in ACTH-Secreting Pituitary Tumors. *J. Clin. Endocrinol. Metab.* **2009**, *94*, 320–323. [CrossRef] [PubMed]
85. Stilling, G.; Sun, Z.; Zhang, S.; Jin, L.; Righi, A.; Kovācs, G.; Korbonits, M.; Scheithauer, B.W.; Kovacs, K.; Lloyd, R.V. MicroRNA Expression in ACTH-Producing Pituitary Tumors: Up-Regulation of MicroRNA-122 and -493 in Pituitary Carcinomas. *Endocrine* **2010**, *38*, 67–75. [CrossRef] [PubMed]
86. Müssnich, P.; Raverot, G.; Jaffrain-Rea, M.-L.; Fraggetta, F.; Wierinckx, A.; Trouillas, J.; Fusco, A.; D'Angelo, D. Downregulation of MiR-410 Targeting the Cyclin B1 Gene Plays a Role in Pituitary Gonadotroph Tumors. *Cell Cycle* **2015**, *14*, 2590–2597. [CrossRef] [PubMed]
87. Chen, Y.; Li, Q.; Wang, C.; Su, Z.; Li, W.; Chen, X.; Wu, Z. Differential expression analysis of prolactinoma-related microRNAs. *Zhonghua Yi Xue Za Zhi* **2012**, *92*, 320–323. [PubMed]
88. Darvasi, O.; Szabo, P.M.; Nemeth, K.; Szabo, K.; Spisak, S.; Liko, I.; Czirjak, S.; Racz, K.; Igaz, P.; Patocs, A.; et al. Limitations of High Throughput Methods for MiRNA Expression Profiles in Non-Functioning Pituitary Adenomas. *Pathol. Oncol. Res.* **2019**, *25*, 169–182. [CrossRef] [PubMed]
89. Butz, H.; Likó, I.; Czirják, S.; Igaz, P.; Korbonits, M.; Rácz, K.; Patócs, A. MicroRNA Profile Indicates Downregulation of the TGFβ Pathway in Sporadic Non-Functioning Pituitary Adenomas. *Pituitary* **2011**, *14*, 112–124. [CrossRef]

90. Liang, H.; Wang, R.; Diao, C.; Li, J.; Su, J.; Zhang, S. The PTTG1-Targeting MiRNAs MiR-329, MiR-300, MiR-381, and MiR-655 Inhibit Pituitary Tumor Cell Tumorigenesis and Are Involved in a P53/PTTG1 Regulation Feedback Loop. *Oncotarget* **2015**, *6*, 29413–29427. [CrossRef]
91. Krokker, L.; Nyírő, G.; Reiniger, L.; Darvasi, O.; Szücs, N.; Czirják, S.; Tóth, M.; Igaz, P.; Patócs, A.; Butz, H. Differentially Expressed MiRNAs Influence Metabolic Processes in Pituitary Oncocytoma. *Neurochem. Res.* **2019**. [CrossRef]
92. Lassalle, S.; Zangari, J.; Popa, A.; Ilie, M.; Hofman, V.; Long, E.; Patey, M.; Tissier, F.; Belléannée, G.; Trouette, H.; et al. MicroRNA-375/SEC23A as Biomarkers of the in Vitro Efficacy of Vandetanib. *Oncotarget* **2016**, *7*, 30461–30478. [CrossRef] [PubMed]
93. Hudson, J.; Duncavage, E.; Tamburrino, A.; Salerno, P.; Xi, L.; Raffeld, M.; Moley, J.; Chernock, R.D. Over Expression of MiR-10a and MiR-375 and down Regulation of YAP1 in Medullary Thyroid Carcinoma. *Exp. Mol. Pathol.* **2013**, *95*, 62–67. [CrossRef] [PubMed]
94. Nikiforova, M.N.; Tseng, G.C.; Steward, D.; Diorio, D.; Nikiforov, Y.E. MicroRNA Expression Profiling of Thyroid Tumors: Biological Significance and Diagnostic Utility. *J. Clin. Endocrinol. Metab.* **2008**, *93*, 1600–1608. [CrossRef]
95. Geraldo, M.V.; Nakaya, H.I.; Kimura, E.T. Down-Regulation of 14q32-Encoded MiRNAs and Tumor Suppressor Role for MiR-654-3p in Papillary Thyroid Cancer. *Oncotarget* **2017**, *8*, 9597–9607. [CrossRef]
96. Rosignolo, F.; Sponziello, M.; Giacomelli, L.; Russo, D.; Pecce, V.; Biffoni, M.; Bellantone, R.; Lombardi, C.P.; Lamartina, L.; Grani, G.; et al. Identification of Thyroid-Associated Serum MicroRNA Profiles and Their Potential Use in Thyroid Cancer Follow-Up. *J. Endocr. Soc.* **2017**, *1*, 3–13. [CrossRef]
97. Tetzlaff, M.T.; Liu, A.; Xu, X.; Master, S.R.; Baldwin, D.A.; Tobias, J.W.; Livolsi, V.A.; Baloch, Z.W. Differential Expression of MiRNAs in Papillary Thyroid Carcinoma Compared to Multinodular Goiter Using Formalin Fixed Paraffin Embedded Tissues. *Endocr. Pathol.* **2007**, *18*, 163–173. [CrossRef]
98. Linwah, Y.; Kelly, L.; Yongli, S.; Armstrong, M.J.; Nikiforov, Y.E.; Carty, S.E.; Nikiforova, M.N. MicroRNA Signature Distinguishes the Degree of Aggressiveness of Papillary Thyroid Carcinoma. *Ann. Surg. Oncol.* **2011**, *18*, 2035–2041. [CrossRef]
99. Jacques, C.; Guillotin, D.; Fontaine, J.-F.; Franc, B.; Mirebeau-Prunier, D.; Fleury, A.; Malthiery, Y.; Savagner, F. DNA Microarray and MiRNA Analyses Reinforce the Classification of Follicular Thyroid Tumors. *J. Clin. Endocrinol. Metab.* **2013**, *98*, E981–E989. [CrossRef] [PubMed]
100. Lassalle, S.; Hofman, V.; Ilie, M.; Bonnetaud, C.; Puisségur, M.-P.; Brest, P.; Loubatier, C.; Guevara, N.; Bordone, O.; Cardinaud, B.; et al. Can the MicroRNA Signature Distinguish between Thyroid Tumors of Uncertain Malignant Potential and Other Well-Differentiated Tumors of the Thyroid Gland? *Endocr. Relat. Cancer* **2011**, *18*, 579–594. [CrossRef]
101. Mancikova, V.; Castelblanco, E.; Pineiro-Yanez, E.; Perales-Paton, J.; de Cubas, A.A.; Inglada-Perez, L.; Matias-Guiu, X.; Capel, I.; Bella, M.; Lerma, E.; et al. MicroRNA Deep-Sequencing Reveals Master Regulators of Follicular and Papillary Thyroid Tumors. *Mod. Pathol.* **2015**, *28*, 748–757. [CrossRef]
102. Peng, Y.; Li, C.; Luo, D.-C.; Ding, J.-W.; Zhang, W.; Pan, G. Expression Profile and Clinical Significance of MicroRNAs in Papillary Thyroid Carcinoma. *Molecules* **2014**, *19*, 11586–11599. [CrossRef] [PubMed]
103. Riesco-Eizaguirre, G.; Wert-Lamas, L.; Perales-Patón, J.; Sastre-Perona, A.; Fernández, L.P.; Santisteban, P. The MiR-146b-3p/PAX8/NIS Regulatory Circuit Modulates the Differentiation Phenotype and Function of Thyroid Cells during Carcinogenesis. *Cancer Res.* **2015**, *75*, 4119–4130. [CrossRef]
104. Saiselet, M.; Gacquer, D.; Spinette, A.; Craciun, L.; Decaussin-Petrucci, M.; Andry, G.; Detours, V.; Maenhaut, C. New Global Analysis of the MicroRNA Transcriptome of Primary Tumors and Lymph Node Metastases of Papillary Thyroid Cancer. *BMC Genom.* **2015**, *16*. [CrossRef] [PubMed]
105. Swierniak, M.; Wojcicka, A.; Czetwertynska, M.; Stachlewska, E.; Maciag, M.; Wiechno, W.; Gornicka, B.; Bogdanska, M.; Koperski, L.; de la Chapelle, A.; et al. In-Depth Characterization of the MicroRNA Transcriptome in Normal Thyroid and Papillary Thyroid Carcinoma. *J. Clin. Endocrinol. Metab.* **2013**, *98*, E1401–E1409. [CrossRef] [PubMed]
106. Rossing, M.; Borup, R.; Henao, R.; Winther, O.; Vikesaa, J.; Niazi, O.; Godballe, C.; Krogdahl, A.; Glud, M.; Hjort-Sørensen, C.; et al. Down-Regulation of MicroRNAs Controlling Tumourigenic Factors in Follicular Thyroid Carcinoma. *J. Mol. Endocrinol.* **2012**, *48*, 11–23. [CrossRef] [PubMed]
107. Dettmer, M.; Vogetseder, A.; Durso, M.B.; Moch, H.; Komminoth, P.; Perren, A.; Nikiforov, Y.E.; Nikiforova, M.N. MicroRNA Expression Array Identifies Novel Diagnostic Markers for Conventional and Oncocytic Follicular Thyroid Carcinomas. *J. Clin. Endocrinol. Metab.* **2013**, *98*, E1–E7. [CrossRef]
108. Wojtas, B.; Ferraz, C.; Stokowy, T.; Hauptmann, S.; Lange, D.; Dralle, H.; Musholt, T.; Jarzab, B.; Paschke, R.; Eszlinger, M. Differential MiRNA Expression Defines Migration and Reduced Apoptosis in Follicular Thyroid Carcinomas. *Mol. Cell. Endocrinol.* **2014**, *388*, 1–9. [CrossRef]
109. Hébrant, A.; Floor, S.; Saiselet, M.; Antoniou, A.; Desbuleux, A.; Snyers, B.; La, C.; de Saint Aubain, N.; Leteurtre, E.; Andry, G.; et al. MiRNA Expression in Anaplastic Thyroid Carcinomas. *PLoS ONE* **2014**, *9*. [CrossRef]
110. Visone, R.; Pallante, P.; Vecchione, A.; Cirombella, R.; Ferracin, M.; Ferraro, A.; Volinia, S.; Coluzzi, S.; Leone, V.; Borbone, E.; et al. Specific MicroRNAs Are Downregulated in Human Thyroid Anaplastic Carcinomas. *Oncogene* **2007**, *26*, 7590–7595. [CrossRef]
111. Boufraqech, M.; Nilubol, N.; Zhang, L.; Gara, S.K.; Sadowski, S.M.; Mehta, A.; He, M.; Davis, S.; Dreiling, J.; Copland, J.A.; et al. MiR30a Inhibits LOX Expression and Anaplastic Thyroid Cancer Progression. *Cancer Res.* **2015**, *75*, 367–377. [CrossRef]
112. Braun, J.; Hoang-Vu, C.; Dralle, H.; Hüttelmaier, S. Downregulation of MicroRNAs Directs the EMT and Invasive Potential of Anaplastic Thyroid Carcinomas. *Oncogene* **2010**, *29*, 4237–4244. [CrossRef]

113. Yao, J.C.; Hassan, M.; Phan, A.; Dagohoy, C.; Leary, C.; Mares, J.E.; Abdalla, E.K.; Fleming, J.B.; Vauthey, J.-N.; Rashid, A.; et al One Hundred Years after "Carcinoid": Epidemiology of and Prognostic Factors for Neuroendocrine Tumors in 35,825 Cases in the United States. *J. Clin. Oncol.* **2008**, *26*, 3063–3072. [CrossRef]
114. Hendifar, A.E.; Marchevsky, A.M.; Tuli, R. Neuroendocrine Tumors of the Lung: Current Challenges and Advances in the Diagnosis and Management of Well-Differentiated Disease. *J. Thorac. Oncol.* **2017**, *12*, 425–436. [CrossRef] [PubMed]
115. Malczewska, A.; Kidd, M.; Matar, S.; Kos-Kudla, B.; Modlin, I.M. A Comprehensive Assessment of the Role of MiRNAs as Biomarkers in Gastroenteropancreatic Neuroendocrine Tumors. *Neuroendocrinology* **2018**, *107*, 73–90. [CrossRef]
116. Zatelli, M.C.; Grossrubatscher, E.M.; Guadagno, E.; Sciammarella, C.; Faggiano, A.; Colao, A. Circulating Tumor Cells and MiRNAs as Prognostic Markers in Neuroendocrine Neoplasms. *Endocr. Relat. Cancer* **2017**, *24*, R223–R237. [CrossRef] [PubMed]
117. Abuhatzira, L.; Xu, H.; Tahhan, G.; Boulougoura, A.; Schäffer, A.A.; Notkins, A.L. Multiple microRNAs within the 14q32 cluster target the mRNAs of major type 1 diabetes autoantigens IA-2, IA-2β, and GAD65. *FASEB J.* **2015**, *29*, 4374–4383 [CrossRef] [PubMed]
118. Miller, H.C.; Frampton, A.E.; Malczewska, A.; Ottaviani, S.; Stronach, E.A.; Flora, R.; Kaemmerer, D.; Schwach, G.; Pfragner, R. Faiz, O.; et al. MicroRNAs Associated with Small Bowel Neuroendocrine Tumours and Their Metastases. *Endocr. Relat. Cancer* **2016**, *23*, 711–726. [CrossRef] [PubMed]
119. Ruebel, K.; Leontovich, A.A.; Stilling, G.A.; Zhang, S.; Righi, A.; Jin, L.; Lloyd, R.V. MicroRNA Expression in Ileal Carcinoid Tumors: Downregulation of MicroRNA-133a with Tumor Progression. *Mod. Pathol.* **2010**, *23*, 367–375. [CrossRef] [PubMed]
120. Heverhagen, A.E.; Legrand, N.; Wagner, V.; Fendrich, V.; Bartsch, D.K.; Slater, E.P. Overexpression of MicroRNA MiR-7-5p Is a Potential Biomarker in Neuroendocrine Neoplasms of the Small Intestine. *Neuroendocrinology* **2018**, *106*, 312–317 [CrossRef] [PubMed]
121. Hamfjord, J.; Stangeland, A.M.; Hughes, T.; Skrede, M.L.; Tveit, K.M.; Ikdahl, T.; Kure, E.H. Differential Expression of MiRNAs in Colorectal Cancer: Comparison of Paired Tumor Tissue and Adjacent Normal Mucosa Using High-Throughput Sequencing. *PLoS ONE* **2012**, *7*, e34150. [CrossRef] [PubMed]
122. Castro-Vega, L.J.; Lepoutre-Lussey, C.; Gimenez-Roqueplo, A.-P.; Favier, J. Rethinking Pheochromocytomas and Paragangliomas from a Genomic Perspective. *Oncogene* **2016**, *35*, 1080–1089. [CrossRef] [PubMed]
123. Pereira, B.D.; Luiz, H.V.; Ferreira, A.G.; Portugal, J. Genetics of Pheochromocytoma and Paraganglioma. In *Paraganglioma: A Multidisciplinary Approach*; Mariani-Costantini, R., Ed.; Codon Publications: Brisbane, Austrilia, 2019; ISBN 978-0-9944381-7-1.
124. Patterson, E.; Webb, R.; Weisbrod, A.; Bian, B.; He, M.; Zhang, L.; Holloway, A.K.; Krishna, R.; Nilubol, N.; Pacak, K.; et al. The MicroRNA Expression Changes Associated with Malignancy and SDHB Mutation in Pheochromocytoma. *Endocr. Relat. Cancer* **2012**, *19*, 157–166. [CrossRef] [PubMed]
125. Igaz, P.; Igaz, I.; Nagy, Z.; Nyírő, G.; Szabó, P.M.; Falus, A.; Patócs, A.; Rácz, K. MicroRNAs in Adrenal Tumors: Relevance for Pathogenesis, Diagnosis, and Therapy. *Cell. Mol. Life Sci.* **2015**, *72*, 417–428. [CrossRef] [PubMed]
126. Robertson, S.; MacKenzie, S.M.; Alvarez-Madrazo, S.; Diver, L.A.; Lin, J.; Stewart, P.M.; Fraser, R.; Connell, J.M.; Davies, E. MicroRNA-24 Is a Novel Regulator of Aldosterone and Cortisol Production in the Human Adrenal Cortex. *Hypertension* **2013**, *62*, 572–578. [CrossRef]
127. MacKenzie, S.M.; van Kralingen, J.; Martin, H.; Davies, E. MicroRNAs in Aldosterone Production and Action. *Aldosterone Miner. Recept. Cell Biol. Transl. Med.* **2019**. [CrossRef]
128. He, J.; Cao, Y.; Su, T.; Jiang, Y.; Jiang, L.; Zhou, W.; Zhang, C.; Wang, W.; Ning, G. Downregulation of MiR-375 in Aldosterone-Producing Adenomas Promotes Tumour Cell Growth via MTDH. *Clin. Endocrinol.* **2015**, *83*, 581–589. [CrossRef]
129. Peng, K.-Y.; Chang, H.-M.; Lin, Y.-F.; Chan, C.-K.; Chang, C.-H.; Chueh, S.-C.J.; Yang, S.-Y.; Huang, K.-H.; Lin, Y.-H.; Wu, V.-C.; et al. MiRNA-203 Modulates Aldosterone Levels and Cell Proliferation by Targeting Wnt5a in Aldosterone-Producing Adenomas. *J. Clin. Endocrinol. Metab.* **2018**, *103*, 3737–3747. [CrossRef]
130. Nakano, Y.; Yoshimoto, T.; Watanabe, R.; Murakami, M.; Fukuda, T.; Saito, K.; Fujii, Y.; Akashi, T.; Tanaka, T.; Yamada, T.; et al. MiRNA299 Involvement in CYP11B2 Expression in Aldosterone-Producing Adenoma. *Eur. J. Endocrinol.* **2019**, *181*, 69–78. [CrossRef]
131. Patterson, E.E.; Holloway, A.K.; Weng, J.; Fojo, T.; Kebebew, E. MicroRNA Profiling of Adrenocortical Tumors Reveals MiR-483 as a Marker of Malignancy. *Cancer* **2011**, *117*, 1630–1639. [CrossRef]
132. Kwok, G.T.Y.; Zhao, J.T.; Glover, A.R.; Gill, A.J.; Clifton-Bligh, R.; Robinson, B.G.; Ip, J.C.Y.; Sidhu, S.B. MicroRNA-431 as a Chemosensitizer and Potentiator of Drug Activity in Adrenocortical Carcinoma. *Oncologist* **2019**, *24*, e241–e250. [CrossRef] [PubMed]
133. Turányi, E.; Dezso, K.; Paku, S.; Nagy, P. DLK Is a Novel Immunohistochemical Marker for Adrenal Gland Tumors. *Virchows Arch.* **2009**, *455*, 295–299. [CrossRef] [PubMed]
134. Aflorei, E.D.; Korbonits, M. Epidemiology and Etiopathogenesis of Pituitary Adenomas. *J. Neurooncol.* **2014**, *117*, 379–394. [CrossRef] [PubMed]
135. Feng, Y.; Mao, Z.-G.; Wang, X.; Du, Q.; Jian, M.; Zhu, D.; Xiao, Z.; Wang, H.-J.; Zhu, Y.-H. MicroRNAs and Target Genes in Pituitary Adenomas. *Horm. Metab. Res.* **2018**, *50*, 179–192. [CrossRef]
136. Wierinckx, A.; Roche, M.; Legras-Lachuer, C.; Trouillas, J.; Raverot, G.; Lachuer, J. MicroRNAs in Pituitary Tumors. *Mol. Cell. Endocrinol.* **2017**, *456*, 51–61. [CrossRef]
137. Vicchio, T.M.; Aliquò, F.; Ruggeri, R.M.; Ragonese, M.; Giuffrida, G.; Cotta, O.R.; Spagnolo, F.; Torre, M.L.; Alibrandi, A.; Asmundo, A.; et al. MicroRNAs Expression in Pituitary Tumors: Differences Related to Functional Status, Pathological Features, and Clinical Behavior. *J. Endocrinol. Investig.* **2020**, *43*, 947–958. [CrossRef] [PubMed]

138. Palumbo, T.; Faucz, F.R.; Azevedo, M.; Xekouki, P.; Iliopoulos, D.; Stratakis, C.A. Functional Screen Analysis Reveals MiR-26b and MiR-128 as Central Regulators of Pituitary Somatomammotrophic Tumor Growth through Activation of the PTEN–AKT Pathway. *Oncogene* **2013**, *32*, 1651–1659. [CrossRef]
139. Beylerli, O.; Beeraka, N.M.; Gareev, I.; Pavlov, V.; Yang, G.; Liang, Y.; Aliev, G. MiRNAs as Noninvasive Biomarkers and Therapeutic Agents of Pituitary Adenomas. *Int. J. Mol. Sci.* **2020**, *21*, 7287. [CrossRef]
140. Gejman, R.; Batista, D.L.; Zhong, Y.; Zhou, Y.; Zhang, X.; Swearingen, B.; Stratakis, C.A.; Hedley-Whyte, E.T.; Klibanski, A. Selective Loss of MEG3 Expression and Intergenic Differentially Methylated Region Hypermethylation in the MEG3/DLK1 Locus in Human Clinically Nonfunctioning Pituitary Adenomas. *J. Clin. Endocrinol. Metab.* **2008**, *93*, 4119–4125. [CrossRef]
141. Zhao, J.; Dahle, D.; Zhou, Y.; Zhang, X.; Klibanski, A. Hypermethylation of the Promoter Region Is Associated with the Loss of MEG3 Gene Expression in Human Pituitary Tumors. *J. Clin. Endocrinol. Metab.* **2005**, *90*, 2179–2186. [CrossRef] [PubMed]
142. Jin, L.; Riss, D.; Ruebel, K.; Kajita, S.; Scheithauer, B.W.; Horvath, E.; Kovacs, K.; Lloyd, R.V. Galectin-3 Expression in Functioning and Silent ACTH-Producing Adenomas. *Endocr. Pathol.* **2005**, *16*, 107–114. [CrossRef] [PubMed]
143. Riss, D.; Jin, L.; Qian, X.; Bayliss, J.; Scheithauer, B.W.; Young, W.F.; Vidal, S.; Kovacs, K.; Raz, A.; Lloyd, R.V. Differential Expression of Galectin-3 in Pituitary Tumors. *Cancer Res.* **2003**, *63*, 2251–2255.
144. Ruebel, K.H.; Jin, L.; Qian, X.; Scheithauer, B.W.; Kovacs, K.; Nakamura, N.; Zhang, H.; Raz, A.; Lloyd, R.V. Effects of DNA Methylation on Galectin-3 Expression in Pituitary Tumors. *Cancer Res.* **2005**, *65*, 1136–1140. [CrossRef]
145. Zhang, H.-Y.; Jin, L.; Stilling, G.A.; Ruebel, K.H.; Coonse, K.; Tanizaki, Y.; Raz, A.; Lloyd, R.V. RUNX1 and RUNX2 Upregulate Galectin-3 Expression in Human Pituitary Tumors. *Endocrine* **2009**, *35*, 101–111. [CrossRef]
146. Raue, F.; Frank-Raue, K. Epidemiology and Clinical Presentation of Medullary Thyroid Carcinoma. *Recent Results Cancer Res.* **2015**, *204*, 61–90. [CrossRef] [PubMed]
147. Malek, A.; Sleptsov, I.E.; Cheburkin, Y.V.; Roman, S.; Kolesnikov, N. MiRNA as Potential Tool for Thyroid Cancer Diagnostics and Follow up: Practical Considerations. *JSM Thyroid Disord. Manag.* **2017**, *2*, 1007.
148. Romeo, P.; Colombo, C.; Granata, R.; Calareso, G.; Gualeni, A.V.; Dugo, M.; De Cecco, L.; Rizzetti, M.G.; Zanframundo, A.; Aiello, A.; et al. Circulating MiR-375 as a Novel Prognostic Marker for Metastatic Medullary Thyroid Cancer Patients. *Endocr. Relat. Cancer* **2018**, *25*, 217–231. [CrossRef]
149. Saiselet, M.; Pita, J.M.; Augenlicht, A.; Dom, G.; Tarabichi, M.; Fimereli, D.; Dumont, J.E.; Detours, V.; Maenhaut, C. MiRNA Expression and Function in Thyroid Carcinomas: A Comparative and Critical Analysis and a Model for Other Cancers. *Oncotarget* **2016**, *7*, 52475–52492. [CrossRef]
150. Boufraqech, M.; Klubo-Gwiezdzinska, J.; Kebebew, E. MicroRNAs in the Thyroid. *Best Pract. Res. Clin. Endocrinol. Metab.* **2016**, *30*, 603–619. [CrossRef]
151. Fiore, R.; Khudayberdiev, S.; Christensen, M.; Siegel, G.; Flavell, S.W.; Kim, T.-K.; Greenberg, M.E.; Schratt, G. Mef2-Mediated Transcription of the MiR379-410 Cluster Regulates Activity-Dependent Dendritogenesis by Fine-Tuning Pumilio2 Protein Levels. *EMBO J.* **2009**, *28*, 697–710. [CrossRef] [PubMed]
152. Aavik, E.; Lumivuori, H.; Leppänen, O.; Wirth, T.; Häkkinen, S.-K.; Bräsen, J.-H.; Beschorner, U.; Zeller, T.; Braspenning, M.; van Criekinge, W.; et al. Global DNA Methylation Analysis of Human Atherosclerotic Plaques Reveals Extensive Genomic Hypomethylation and Reactivation at Imprinted Locus 14q32 Involving Induction of a MiRNA Cluster. *Eur. Heart J.* **2015**, *36*, 993–1000. [CrossRef] [PubMed]
153. Romitti, M.; Wajner, S.M.; Zennig, N.; Goemann, I.M.; Bueno, A.L.; Meyer, E.L.S.; Maia, A.L. Increased Type 3 Deiodinase Expression in Papillary Thyroid Carcinoma. *Thyroid* **2012**, *22*, 897–904. [CrossRef] [PubMed]
154. Greife, A.; Knievel, J.; Ribarska, T.; Niegisch, G.; Schulz, W.A. Concomitant Downregulation of the Imprinted Genes DLK1 and MEG3 at 14q32.2 by Epigenetic Mechanisms in Urothelial Carcinoma. *Clin. Epigenetics* **2014**, *6*, 29. [CrossRef]
155. Jishnu, P.V.; Jayaram, P.; Shukla, V.; Varghese, V.K.; Pandey, D.; Sharan, K.; Chakrabarty, S.; Satyamoorthy, K.; Kabekkodu, S.P. Prognostic Role of 14q32.31 MiRNA Cluster in Various Carcinomas: A Systematic Review and Meta-Analysis. *Clin. Exp. Metastasis* **2020**, *37*, 31–46. [CrossRef] [PubMed]
156. Shu, J.; Li, L.; Sarver, A.E.; Pope, E.A.; Varshney, J.; Thayanithy, V.; Spector, L.; Largaespada, D.A.; Steer, C.J.; Subramanian, S. Imprinting Defects at Human 14q32 Locus Alters Gene Expression and Is Associated with the Pathobiology of Osteosarcoma. *Oncotarget* **2016**, *7*, 21298–21314. [CrossRef] [PubMed]
157. Manodoro, F.; Marzec, J.; Chaplin, T.; Miraki-Moud, F.; Moravcsik, E.; Jovanovic, J.V.; Wang, J.; Iqbal, S.; Taussig, D.; Grimwade, D.; et al. Loss of Imprinting at the 14q32 Domain Is Associated with MicroRNA Overexpression in Acute Promyelocytic Leukemia. *Blood* **2014**, *123*, 2066–2074. [CrossRef]
158. Romitti, M.; Wajner, S.M.; Ceolin, L.; Ferreira, C.V.; Ribeiro, R.V.P.; Rohenkohl, H.C.; Weber, S.D.S.; Lopez, P.L.D.C.; Fuziwara, C.S.; Kimura, E.T.; et al. MAPK and SHH Pathways Modulate Type 3 Deiodinase Expression in Papillary Thyroid Carcinoma. *Endocr. Relat. Cancer* **2016**, *23*, 135–146. [CrossRef]
159. Gardiner, E.; Beveridge, N.J.; Wu, J.Q.; Carr, V.; Scott, R.J.; Tooney, P.A.; Cairns, M.J. Imprinted DLK1-DIO3 Region of 14q32 Defines a Schizophrenia-Associated MiRNA Signature in Peripheral Blood Mononuclear Cells. *Mol. Psychiatry* **2012**, *17*, 827–840. [CrossRef]
160. Stark, K.L.; Xu, B.; Bagchi, A.; Lai, W.-S.; Liu, H.; Hsu, R.; Wan, X.; Pavlidis, P.; Mills, A.A.; Karayiorgou, M.; et al. Altered Brain MicroRNA Biogenesis Contributes to Phenotypic Deficits in a 22q11-Deletion Mouse Model. *Nat. Genet.* **2008**, *40*, 751–760. [CrossRef]

161. Schaefer, A.; Im, H.-I.; Venø, M.T.; Fowler, C.D.; Min, A.; Intrator, A.; Kjems, J.; Kenny, P.J.; O'Carroll, D.; Greengard, P. Argonaute 2 in Dopamine 2 Receptor–Expressing Neurons Regulates Cocaine Addiction. *J. Exp. Med.* **2010**, *207*, 1843–1851. [CrossRef]
162. Diederichs, S.; Haber, D.A. Dual Role for Argonautes in MicroRNA Processing and Posttranscriptional Regulation of MicroRNA Expression. *Cell* **2007**, *131*, 1097–1108. [CrossRef] [PubMed]
163. O'Carroll, D.; Mecklenbrauker, I.; Das, P.P.; Santana, A.; Koenig, U.; Enright, A.J.; Miska, E.A.; Tarakhovsky, A. A Slicer-Independent Role for Argonaute 2 in Hematopoiesis and the MicroRNA Pathway. *Genes Dev.* **2007**, *21*, 1999–2004. [CrossRef]
164. Welten, S.M.J.; de Vries, M.R.; Peters, E.A.B.; Agrawal, S.; Quax, P.H.A.; Nossent, A.Y. Inhibition of Mef2a Enhances Neovascularization via Post-Transcriptional Regulation of 14q32 MicroRNAs MiR-329 and MiR-494. *Mol. Ther. Nucleic Acids* **2017**, *7*, 61–70. [CrossRef] [PubMed]
165. Downie Ruiz Velasco, A.; Welten, S.M.J.; Goossens, E.A.C.; Quax, P.H.A.; Rappsilber, J.; Michlewski, G.; Nossent, A.Y. Posttranscriptional Regulation of 14q32 MicroRNAs by the CIRBP and HADHB during Vascular Regeneration after Ischemia. *Mol. Ther. Nucleic Acids* **2019**, *14*, 329–338. [CrossRef] [PubMed]
166. Rosa, A.L.; Wu, Y.-Q.; Kwabi-Addo, B.; Coveler, K.J.; Reid Sutton, V.; Shaffer, L.G. Allele-Specific Methylation of a Functional CTCF Binding Site Upstream of MEG3 in the Human Imprinted Domain of 14q32. *Chromosome Res.* **2005**, *13*, 809–818. [CrossRef] [PubMed]
167. Bell, A.C.; West, A.G.; Felsenfeld, G. The Protein CTCF Is Required for the Enhancer Blocking Activity of Vertebrate Insulators. *Cell* **1999**, *98*, 387–396. [CrossRef]
168. Szabó, P.E.; Tang, S.-H.E.; Rentsendorj, A.; Pfeifer, G.P.; Mann, J.R. Maternal-Specific Footprints at Putative CTCF Sites in the H19 Imprinting Control Region Give Evidence for Insulator Function. *Curr. Biol.* **2000**, *10*, 607–610. [CrossRef]
169. Welten, S.M.J.; Bastiaansen, A.J.N.M.; de Jong, R.C.M.; de Vries, M.R.; Peters, E.A.B.; Boonstra, M.C.; Sheikh, S.P.; La Monica, N.; Kandimalla, E.R.; Quax, P.H.A.; et al. Inhibition of 14q32 MicroRNAs MiR-329, MiR-487b, MiR-494, and MiR-495 Increases Neovascularization and Blood Flow Recovery after Ischemia. *Circ. Res.* **2014**, *115*, 696–708. [CrossRef]
170. Gupta, G.P.; Massagué, J. Cancer Metastasis: Building a Framework. *Cell* **2006**, *127*, 679–695. [CrossRef]
171. Uppal, A.; Ferguson, M.K.; Posner, M.C.; Hellman, S.; Khodarev, N.N.; Weichselbaum, R.R. Towards a Molecular Basis of Oligometastatic Disease: Potential Role of Micro-RNAs. *Clin. Exp. Metastasis* **2014**, *31*, 735–748. [CrossRef]
172. Qian, P.; He, X.C.; Paulson, A.; Li, Z.; Tao, F.; Perry, J.M.; Guo, F.; Zhao, M.; Zhi, L.; Venkatraman, A.; et al. The Dlk1-Gtl2 Locus Preserves LT-HSC Function by Inhibiting the PI3K-MTOR Pathway to Restrict Mitochondrial Metabolism. *Cell Stem Cell* **2016**, *18*, 214–228. [CrossRef] [PubMed]
173. Thayanithy, V.; Sarver, A.L.; Kartha, R.V.; Li, L.; Angstadt, A.Y.; Breen, M.; Steer, C.J.; Modiano, J.F.; Subramanian, S. Perturbation of 14q32 MiRNAs-CMYC Gene Network in Osteosarcoma. *Bone* **2012**, *50*, 171–181. [CrossRef] [PubMed]
174. Feldmann, G.; Mishra, A.; Hong, S.-M.; Bisht, S.; Strock, C.J.; Ball, D.W.; Goggins, M.; Maitra, A.; Nelkin, B.D. Inhibiting the Cyclin-Dependent Kinase CDK5 Blocks Pancreatic Cancer Formation and Progression through the Suppression of Ras-Ral Signaling. *Cancer Res.* **2010**, *70*, 4460–4469. [CrossRef]
175. Reis, E.M.; Nakaya, H.I.; Louro, R.; Canavez, F.C.; Flatschart, A.V.F.; Almeida, G.T.; Egidio, C.M.; Paquola, A.C.; Machado, A.A.; Festa, F.; et al. Antisense Intronic Non-Coding RNA Levels Correlate to the Degree of Tumor Differentiation in Prostate Cancer. *Oncogene* **2004**, *23*, 6684–6692. [CrossRef] [PubMed]
176. Konoplev, S.N.; Fritsche, H.A.; O'Brien, S.; Wierda, W.G.; Keating, M.J.; Gornet, T.G.; St Romain, S.; Wang, X.; Inamdar, K.; Johnson, M.R.; et al. High Serum Thymidine Kinase 1 Level Predicts Poorer Survival in Patients with Chronic Lymphocytic Leukemia. *Am. J. Clin. Pathol.* **2010**, *134*, 472–477. [CrossRef]
177. Xu, Y.; Shi, Q.-L.; Ma, H.; Zhou, H.; Lu, Z.; Yu, B.; Zhou, X.; Eriksson, S.; He, E.; Skog, S. High Thymidine Kinase 1 (TK1) Expression Is a Predictor of Poor Survival in Patients with PT1 of Lung Adenocarcinoma. *Tumour Biol.* **2012**, *33*, 475–483. [CrossRef]
178. He, E.; Xu, X.H.; Guan, H.; Chen, Y.; Chen, Z.H.; Pan, Z.L.; Tang, L.L.; Hu, G.Z.; Li, Y.; Zhang, M.; et al. Thymidine Kinase 1 Is a Potential Marker for Prognosis and Monitoring the Response to Treatment of Patients with Breast, Lung, and Esophageal Cancer and Non-Hodgkin's Lymphoma. *Nucleosides Nucleotides Nucleic Acids* **2010**, *29*, 352–358. [CrossRef]
179. Pathak, S.; Meng, W.-J.; Nandy, S.K.; Ping, J.; Bisgin, A.; Helmfors, L.; Waldmann, P.; Sun, X.-F. Radiation and SN38 Treatments Modulate the Expression of MicroRNAs, Cytokines and Chemokines in Colon Cancer Cells in a P53-Directed Manner. *Oncotarget* **2015**, *6*, 44758–44780. [CrossRef]
180. Liu, D.; Zhang, X.; Yan, C.; Li, Y.; Tian, X.; Zhu, N.; Rong, J.; Peng, C.; Han, Y. MicroRNA-495 Regulates the Proliferation and Apoptosis of Human Umbilical Vein Endothelial Cells by Targeting Chemokine CCL2. *Thromb. Res.* **2015**, *135*, 146–154. [CrossRef]
181. Wilhelm, K.; Happel, K.; Eelen, G.; Schoors, S.; Oellerich, M.F.; Lim, R.; Zimmermann, B.; Aspalter, I.M.; Franco, C.A.; Boettger, T.; et al. FOXO1 Couples Metabolic Activity and Growth State in the Vascular Endothelium. *Nature* **2016**, *529*, 216–220. [CrossRef]
182. Yao, G.-D.; Zhang, Y.-F.; Chen, P.; Ren, X.-B. MicroRNA-544 Promotes Colorectal Cancer Progression by Targeting Forkhead Box O1. *Oncol. Lett.* **2018**, *15*, 991–997. [CrossRef]
183. Jeon, H.H.; Yu, Q.; Lu, Y.; Spencer, E.; Lu, C.; Milovanova, T.; Yang, Y.; Zhang, C.; Stepanchenko, O.; Vafa, R.P.; et al. FOXO1 Regulates VEGFA Expression and Promotes Angiogenesis in Healing Wounds. *J. Pathol.* **2018**, *245*, 258–264. [CrossRef]
184. Haig, D.; Mainieri, A. The Evolution of Imprinted MicroRNAs and Their RNA Targets. *Genes* **2020**, *11*, 1038. [CrossRef] [PubMed]
185. Haig, D. Kin Conflict in Seed Development: An Interdependent but Fractious Collective. *Annu. Rev. Cell Dev. Biol.* **2013**, *29*, 189–211. [CrossRef] [PubMed]
186. Haig, D. Imprinted Green Beards: A Little Less than Kin and More than Kind. *Biol. Lett.* **2013**, *9*. [CrossRef]

187. Patten, M.M.; Ross, L.; Curley, J.P.; Queller, D.C.; Bonduriansky, R.; Wolf, J.B. The Evolution of Genomic Imprinting: Theories, Predictions and Empirical Tests. *Heredity* **2014**, *113*, 119–128. [CrossRef]
188. Pallante, P.; Sepe, R.; Puca, F.; Fusco, A. High Mobility Group A Proteins as Tumor Markers. *Front. Med.* **2015**, *2*. [CrossRef]
189. Fusco, A.; Fedele, M. Roles of HMGA Proteins in Cancer. *Nat. Rev. Cancer* **2007**, *7*, 899–910. [CrossRef] [PubMed]

genes

MDPI

Review

Role of MicroRNAs in Acute Myeloid Leukemia

Aneta Wiśnik [1,2,*], Dariusz Jarych [3], Kinga Krawiec [1,2], Piotr Strzałka [1,2], Natalia Potocka [4], Magdalena Czemerska [1,2], Aleksandra Sałagacka-Kubiak [2], Agnieszka Pluta [1,2], Agnieszka Wierzbowska [1,2] and Izabela Zawlik [4,5,*]

1 Department of Hematology, Medical University of Lodz, 93-510 Lodz, Poland
2 Copernicus Memorial Multi-Specialist Oncology and Trauma Center, 93-510 Lodz, Poland
3 Laboratory of Virology, Institute of Medical Biology, Polish Academy of Sciences, 93-232 Lodz, Poland
4 Laboratory of Molecular Biology, Centre for Innovative Research in Medical and Natural Sciences, Collegium Medicum, University of Rzeszow, 35-959 Rzeszow, Poland
5 Department of General Genetics, Faculty of Medicine, Collegium Medicum, University of Rzeszow, 35-959 Rzeszow, Poland
* Correspondence: aneta.wisnik@stud.umed.lodz.pl (A.W.); izawlik@ur.edu.pl (I.Z.)

Abstract: MicroRNA (miRNA), a significant class of regulatory non-coding RNA (ncRNA), can regulate the expression of numerous protein-coding messenger RNAs (mRNAs). miRNA plays an important part in shaping the human transcriptome. So far, in the human genome, about 2500 miRNAs have been found. Acute myeloid leukemia (AML) belongs to a malignant clonal disorder of hematopoietic stem cells and is characterized by the uncontrolled clonal proliferation of abnormal progenitor cells in the bone marrow and blood. For the past several years, significant scientific attention has been attracted to the role of miRNAs in AML, since alterations in the expression levels of miRNAs may contribute to AML development. This review describes the main functions of non-coding RNA classes and presents miRNA biogenesis. This study aims to review recent reports about altered microRNA expression and their influence on AML cell survival, cell cycle, and apoptotic potential. Additionally, it summarizes the correlations between miRNAs and their target mRNAs in AML and outlines the role of particular miRNAs in AML subtypes according to ELN recommendations.

Keywords: microRNA; acute myeloid leukemia; non-coding RNA

1. Introduction

MicroRNAs (miRNAs), one of the ncRNA classes, are short molecules, typically composed of 19–25 nucleotides, which can regulate gene expression by reducing target genes' expression (either by degrading mRNA or inhibiting mRNA translation) [1]. miRNA can function as oncogenes (oncomiRs) or tumor suppressors. Due to their high stability in body fluids and tissues, miRNAs are promising biomarkers of various diseases, including cancers [2,3]. miRNA levels can be assessed by measuring their expression. The rapid development of molecular techniques and advancements in transcriptomics have expanded our understanding of miRNA and gene expression in oncology [4]. miRNA biogenesis is complex and involves several stages, beginning with transcription of DNA sequences to produce primary miRNA transcripts (pri-miRNAs). These are then processed into precursor miRNAs (pre-miRNAs) and, ultimately, into mature miRNAs [5,6]. Dysregulations of expression levels of various miRNAs are described in multiple diseases, including AML. Abnormal miRNA expression levels may serve as a valuable diagnostic or prognostic marker in AML [7]. In this review, we describe the different classes of ncRNA, summarize miRNA biogenesis, and examine alterations in miRNA expression in the context of AML.

2. Types of Non-Coding RNAs

It has been confirmed that less than 2% of the human genome undergoes the transcription and translation processes that lead to protein synthesis, while the rest consists of non-coding DNA. The ENCODE and FANTOM projects have shown that most non-coding DNA sequences are transcriptionally active. The transcribed but not translated sequences are known as non-coding RNAs (ncRNAs) [8–11]. ncRNA includes rRNAs (ribosomal RNAs), tRNAs (transfer RNAs), siRNAs (small interfering RNAs), miRNAs (microRNAs), lncRNAs (long non-coding RNAs), circRNAs (circular RNAs), piRNAs (piwi-interacting RNAs), snRNAs (small nuclear RNAs), snoRNAs (small nucleolar RNAs), eRNA (enhancer-derived RNA), paRNA (promoter-associated RNA), and others. In eukaryotic cells, we can distinguish regulatory and housekeeping types of ncRNA. The first group of ncRNAs plays an epigenetic function and regulates gene expression at the post-transcriptional and transcriptional stages, while the second group controls essential cellular functions. Structurally, non-coding RNAs are divided into short (fewer than 200 nucleotides) and long (more than 200 nucleotides) categories [12,13]. ncRNAs play a pivotal role in many biological processes, including translation (e.g., rRNA, tRNA, snoRNA, snRNA), control of epigenetic variations (e.g., lncRNA), alternative splicing (e.g., snRNA, snoRNA), gene expression regulation (e.g., miRNA, eRNA, snRNA, lncRNA, siRNA, piRNA, paRNA), and miRNA regulation (e.g., circRNA). The primary function of miRNAs is to regulate gene expression at the post-transcriptional level by destabilizing mRNA or inhibiting translation. miRNA targets over 60% of protein-coding genes, which can be responsible for essential cellular processes such as cell growth, apoptosis, differentiation, development, or the development of the disease process [14]. Over the past few decades, advancements in high-throughput molecular biology techniques, particularly those based on sequencing and complementary hybridization, have greatly expanded the amount of data available on ncRNAs. However, despite the growth in our knowledge regarding ncRNA, the functions of many remain unknown [15,16]. Our present understanding of ncRNAs and the genes they regulate is summarized in Table 1 below.

Table 1. Characteristics of non-coding RNA (ncRNA) according to GENCODE [17].

Abbreviation/ Numbers of Nucleotides	Full Name	Main Function	Type	Number of Genes by GENCODE *	Reference
lncRNA/>200	long non-coding RNA	epigenetic, transcriptional, post-transcriptional regulation	regulatory	35,934	[18]
snRNA/75–300	small nuclear RNA	spliceosome formation, which catalyzes the splicing of pre-mRNA	housekeeping	1901	[19,20]
miRNA/19–25	microRNA	gene expression regulation at the post-transcriptional level through the destabilization of mRNA or inhibition of translation	regulatory	~2500 **	[21–23]
snoRNA/60–300	small nucleolar RNA	regulation of spliceosomal and ribosomal functions, maintenance of the structure of rRNA	housekeeping	942	[24,25]
tRNA/70–80	transfer RNA	protein synthesis by codon–anticodon interactions during translation	housekeeping	416 **	[26,27]

Table 1. *Cont.*

Abbreviation/ Numbers of Nucleotides	Full Name	Main Function	Type	Number of Genes by GENCODE *	Reference
rRNA/up to~5000	ribosomal RNA	ribosome subunits formation, which take part in translation, indicating the precise positioning of ribosomal proteins within the ribosome	housekeeping	47	[28,29]
siRNA/21–23	small interfering RNA	suppression of genes expression by RNA interference (RNAi)	regulatory	N/A	[30]
circRNA/N/A	circular RNA	regulation of miRNA through the sponge effect	regulatory	~11,000 **	[31,32]
piRNA/24–32	piwi-interacting RNA	gene suppression by interactions with PIWI proteins	regulatory	~20,000 **	[33–35]
paRNA/200–500	promoter-associated RNA	scaffolding for proteins regulating gene expression e.g., during chromatin remodelling or transcription	regulatory	N/A	[36]
eRNA/50–2000	enhancer RNA	regulation of gene expression by modulating chromatin	regulatory	N/A	[37,38]

* Statistics about the current GENCODE Release (version 47); ** Data from article cited as a reference; N/A—not available.

3. MicroRNA Biogenesis

MicroRNAs are small, endogenous, single-stranded, approximately 19–25 nucleotides in length, highly evolutionarily conserved molecules, and their roles continue to be explored. These short regulatory RNAs mediate post-transcriptional silencing of genes, typically by complementary binding to target mRNAs in the location of their 3′ untranslated regions (UTRs) [39,40]. One miRNA has the potential to control the expression of numerous messenger RNAs (mRNAs), resulting in mRNAs' translational repression or degradation [41]. miRNAs play a role in regulating numerous cellular physiological processes, and alterations in their expression level can significantly influence the progression of numerous diseases [42].

The first miRNA, lin-4, was discovered in the nematode Caenorhabditis elegans in 1993. It was shown to be essential for postembryonic development by downregulating the expression of the protein LIN-14 [43].

In the human genome, two categories of miRNA are distinguished: intragenic and intergenic. These two groups differ in the transcription process and location in the genome. Intragenic miRNA is transcribed by RNA polymerase II (Pol II) and intergenic by RNA polymerase II or III (Pol III). In a significant proportion, the first group is transcribed together with their host genes by the same promoters, while the other group, using their own promoters, can be transcribed autonomously from host genes or co-transcribed with neighboring genes. Emerging studies have revealed that this co-transcription is not an absolute rule in the case of intragenic miRNAs. These two categories of miRNA are located differently in the genome—intragenic miRNA within protein-coding and non-coding genes, and intergenic between genes. Intragenic miRNAs can be further classified based on their position into intronic, exonic, exon-intron junction, and antisense gene strands [44]. Sometimes, miRNA genes are organized into clusters, where multiple miRNA genes are present at the same locus and can be transcribed simultaneously into one long primary transcript that encodes several miRNAs [45].

The production of miRNAs is a complicated process that involves several stages. Depending on the origin of the miRNA, its biogenesis is divided into two pathways: canonical and non-canonical. This study will focus on the canonical pathway, which is the most prevalent route of miRNA synthesis. Initially, DNA sequences are transcribed into primary miRNA transcripts (pri-miRNAs). These pri-miRNAs are then processed into precursor miRNAs (pre-miRNAs) and, ultimately, into mature miRNAs [46,47].

The canonical pathway starts with the transcription of miRNA genes into pri-miRNA by Pol II or III. These are long stem-loop structures composed of hundreds to thousands of nucleotides (nts), and the pri-miRNA carries the mature miRNA within the 3′ or 5′ arm (or both) [48,49]. The pri-miRNA undergoes splicing similarly to protein-coding mRNAs, and during polyadenylation, it acquires a 5′ m7G cap and a 3′ poly(A) tail [50]. Then, the pri-miRNA undergoes cleavage by the Microprocessor, a complex that includes one Drosha ribonuclease III (DROSHA) molecule and two molecules of its cofactor DiGeorge syndrome critical region 8 (DGCR8), a dsRNA binding protein. Microprocessor has a heterotrimeric structure. DROSHA cleaves approximately 11 base pairs from the ssRNA-dsRNA junction, which DGCR8 recognizes. Consequently, a smaller stem-loop precursor miRNA (pre-miRNA) is realized, about 80 nucleotides long, with one defined mature miRNA end [51–53]. The GHG motif in a stem part of pri-miRNA supports DROSHA in precisely recognizing the cleavage site [54].

The pre-miRNA is then transported to the cytoplasm by the RanGTP/Exportin-5 complex, which prevents nuclear degradation and enables transport [55]. RanGTP is subsequently hydrolyzed to RanGDP [56]. The pre-miRNA undergoes further processing in the cytoplasm by a complex which cleaves its loop. As a consequence, a 22-nt double-stranded miRNA duplex is realized. Duplexes have 2-nt 3′ overhangs at both ends, providing the second end of mature miRNA. This cleaving complex is composed of cytoplasmic ribonuclease III Dicer, transactivation response element RNA-binding protein (TRBP), and a protein activator of the double-stranded RNA-activated protein kinase (PACT) [57,58]. The duplex miRNA/*miRNA consists of the mature miRNA and the passenger strand (denoted by a star*) [59]. The duplex interacts with an Argonaut protein (AGO), and the duplex unwinds, leading to the destruction of the star strand. Along the course of one of the processes, when the central region of the duplex is complimentary, the star strand is cleaved by Argonaute 2 (AGO2), which is further degraded by the C3PO nuclease complex. The mature miRNA binds to the AGO protein, forming a miRNA-induced silencing complex (miRISC). It is possible that both strands of the precursor are incorporated into the RISC complex, with both strands serving as sources for mature miRNAs. When the mature miRNA derives from the 5′ strand of the stem-loop precursor, it is referred to as 5p, and when it derives from the 3′ strand, it is called 3p [60,61]. miRNA generally exhibit a complementarity with the 3′ UTR of their target mRNAs, through a crucial segment known as the "seed" region. This specific sequence, situated at the 2–7 nucleotide positions of the 5′ end of the mature miRNA, plays a vital role in the binding process. However, interactions with other regions, including the 5′ UTR, gene promoters, or coding sequences, have also been reported [4,62–64].

A non-canonical miRNA biogenesis pathway involves miRtrons, a class of miRNAs located within the introns of coding genes. MiRtrons form hairpin-containing lariat structures, which, after splicing, bypass Microprocessor cleavage. These molecules are debranched by the debranching enzyme 1 (DRB1) and are subsequently treated as pre-miRNA, ready for exporting to the cytoplasm by Exportin-5 [55].

After miRISC connects with its target, it triggers either mRNA decay or translation repression. The miRISC complex requires miRNA and Argonaute proteins (specifically AGO1-4 in humans) to become functional. However, it has been proved that miRISC consists of many other components: GW182 proteins (a family of proteins rich in tryptophan and glycine repeats, specifically TNRC6A-C in humans), the CCR4-NOT complex (carbon catabolite repression—negative on TATA-less complex), the PAN2/PAN3 complex (two subunits of the poly(A)-specific ribonuclease PAN complex), DCP1 (decapping protein 1), DDX6 (RNA helicase), and GIGYF2 (GRB10 interacting GYF protein 2). Several mechanisms of miRNA-induced silencing exist. GW182 proteins have been shown to be involved in the structural formation of the miRISC complex and in repressing target mRNA. They interact with AGO proteins through their N-terminal domain, while their C-terminal silencing domain is responsible for either transcriptional repression or mRNA degradation. Although their full function remains unclear, it has been observed that these proteins can silence the expression of target mRNAs independently of AGO proteins. Translation repression also occurs during the deadenylation process, which is mediated by the deadenylase CCR4-NOT complex. Additionally, the PAN2/PAN3 complex contributes to miRNA-induced deadenylation. In addition, the mechanism of translational repression entails decapping, which refers to the removal of the 5′ cap structure from mRNA by the decapping enzyme DCP2, activated by DCP1. The recruitment of DDX6 and GIGYF2 can also lead to transcriptional repression by inhibiting the eIF4F (eukaryotic translation Initiation Factor 4E) complex, which is responsible for translation initiation. If a miRNA matches targets complimentary in its central region (9–11 nucleotides), mRNA can be cleaved by the endonuclease activity of AGO2 [60,65–68]. The canonical pathway of microRNA biogenesis is illustrated in Figure 1.

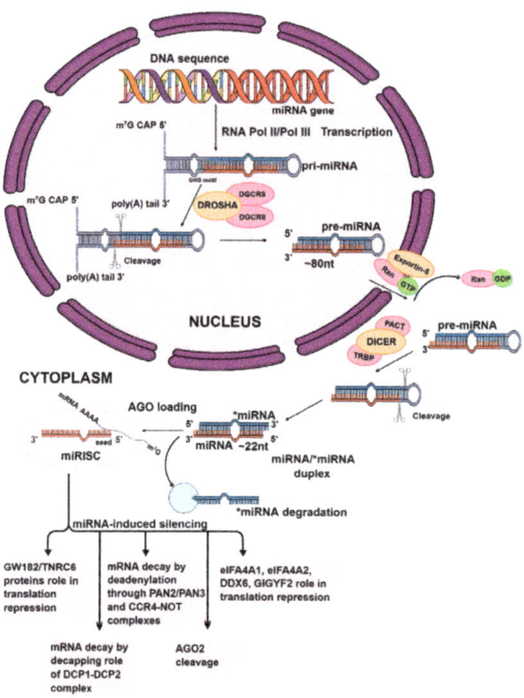

Figure 1. The canonical pathway of miRNA biogenesis. The initial stages of miRNA synthesis occur in the nucleus. The first stage is the transcription of the miRNA gene by RNA polymerase II or III. This process produces the primary miRNA transcript. The Microprocessor complex, consisting of

one Drosha molecule and two DGCR8 molecules, cleaves the pri-miRNA to form precursor miRNA. The precursor miRNA is then transported to the cytoplasm by the RanGTP/Exportin-5 complex. In the cytoplasm, the Dicer enzyme, with TRBP and PACT, cut the pre-miRNA, which generates miRNA/*miRNA duplexes. Mature miRNA, which is defined in one strand of the duplex, after its unwinding, as a single-stranded particle, is preserved within an active RISC complex, leading to the formation of miRISC, while the passenger strand is degraded. Subsequently, the expression of target mRNA is silenced by either cleaving the mRNA or inhibiting translation. The graphic was created using the program Mind the Graph (mindthegraph.com).

4. microRNA in AML

AML is a malignant clonal disorder of hematopoietic stem cells [69]. It is characterized by the uncontrolled clonal proliferation of abnormal progenitor cells in the bone marrow and blood [70]. Specific DNA alterations that affect the functions of normal bone marrow cells can cause them to transform into leukemia cells [71]. The majority of patients diagnosed with AML undergo a treatment that involves chemotherapy or undergo hematopoietic stem cell transplantation (HSCT), and certain subtypes of AML are treated with targeted therapies [72]. However, despite the advancements made in the past few years, the overall five-year survival rate continues to be disappointing [73,74]. According to the National Cancer Institute, the 5-year relative survival rate for AML was 31.9% during 2014–2020 [75]. Research has demonstrated that non-coding RNAs are intricately linked to the development and progression of AML, highlighting their significant role in the disease's pathogenesis [76,77].

Altered expression levels of miRNAs are recognized as a significant mechanism in the pathogenesis of AML [78]. A precise understanding of these small molecules' role is important for diagnosing, prognosticating, and developing therapies for AML patients [79]. This study reviews the role of miRNAs as potential molecular targets in AML. The articles for this review were selected based on the following inclusion criteria: (1) studies focused on AML; (2) examinations of miRNA expression levels using techniques such as reverse transcription quantitative polymerase chain reaction (RT-qPCR), sequencing, microarray technology, or data obtained from relevant databases; (3) the relationship between miRNAs and their target genes analysed using dual luciferase assays or databases (e.g., TargetScan, miRDB, ENCORI); (4) studies that included both a study and a control group; and (5) articles and abstracts published within the last ten years, from which data on miRNA alterations were collected [80–82]. The data presented in Tables 2 and 3 refer to articles describing the expression of miRNA and their target genes. Table 2 focuses on downregulated miRNAs in AML, and Table 3 refers to upregulated miRNAs in AML. In order to simplify the text, miRNA names were shortened, and the "hsa-" prefix referring to Homo sapiens was removed. The growth of molecular biology has been striking, and the number of articles on miRNA has significantly increased. The selection of relevant articles was facilitated using the PICO tool https://jefflibraries.libguides.com/PICO/WhyPICO (accessed on 28 March 2025).

Dysregulated miRNA expression can result from various genomic changes in AML, including deletions, insertions, translocations, and, rarely, somatic copy number alterations of miRNA genes [16,83]. Deletions or copy loss of miRNA genes can lead to decreased miRNA expression, while amplifications or copy gains of miRNA genes can result in increased miRNA expression, which is recognized to be involved in AML development. For example, 5q deletion, a frequent alteration in AML and MDS, has been correlated with significantly downregulated miR-146a expression [84]. Common alterations in AML, such as loss of 7q or *CEBPA* impairment, have been associated with suppressed miR-29b [85]. Additionally, CEBPA loss has been linked to reduced miR-34a expression, a tumour suppressor in AML [86]. Translocations may lead to dysregulated expression of miRNA genes located near the breakpoints. For instance, elevated miR-125b expression has been observed with coexisting translocation t(2;11)(p21;q23) [87]. Furthermore, oncoproteins arising from fusion genes, which are common chromosomal alterations in AML, can lead to changes in miRNA expression, as observed in cases of *MLL*-rearranged AML, with upregulated miR-17-92 cluster and miR-9 [88,89].

Other important mechanisms that alter miRNA expression include epigenetic modifications, such as DNA methylation. For instance, improper *EVI1* expression impedes myeloid cell differentiation, partly through DNA hypermethylation and downregulation of miR-9 [90]. The expression levels of miR-126/126* might be partially influenced by the process of methylation in AML [91]. Also, methylation deregulation of miR-193a is associated with myeloid leukemogenesis [92].

It is also suggested that altered miRNA expression may be linked to transcription factors. For example, E2F1 is proposed as the miR-223 gene's transcriptional inhibitor in AML [93]. Moreover, changes in chromatin accessibility, whether global or specific, can influence miRNA expression [94].

While mutations in the sequences of mature miRNAs are unlikely to directly alter their expression levels, such mutations could impact the specificity of target binding. For example, in the case of miR-142-3p, single-nucleotide variants (SNVs) that may affect target binding have been reported, although such occurrences are rare. On the other hand, polymorphisms in the 3′ UTR of miRNAs occur more frequently and can contribute to AML by regulating multiple genes [16]. Furthermore, the relationship between AML cell metabolism and miRNA expression could have an important impact on the advancement of the disease [95].

Understanding the role of miRNA in AML is crucial for selecting suitable miRNA candidates as potential therapeutic targets. When microRNAs that function as oncomiRs are overexpressed, they can contribute to the development of AML. Therefore, as a therapeutic approach, these miRNAs should be downregulated to prevent disease progression. Conversely, miRNAs that act as tumour suppressors, when downregulated, can also promote disease development. In this case, these miRNAs should be overexpressed to help prevent AML [96]. There are some possible strategies which can be used to modulate miRNA expression, as shown in Figure 2. The promise of miRNA-based therapies is gaining more interest in the fight against various cancers. As of now, specific RNA-based therapies have received approval for medical use, but none have been approved for AML yet. So far, such treatments have neither undergone phase III clinical trials nor been approved by the U.S. FDA for clinical use [97].

Many studies describe the expression of miRNA in AML. Some of them are well studied, and their function in AML is understood. Nevertheless, studies on miRNAs are ongoing, and new reports continue to arise in the meantime, which is why little is known about certain miRNAs in AML, and the role of some miRNAs is still being investigated. Sometimes, their function is confusing, so collecting data on miRNA is crucial for considering their role. miRNAs described below are characterized by altered expression in AML. This review presents their role mostly on AML cells and describes their function in that disease.

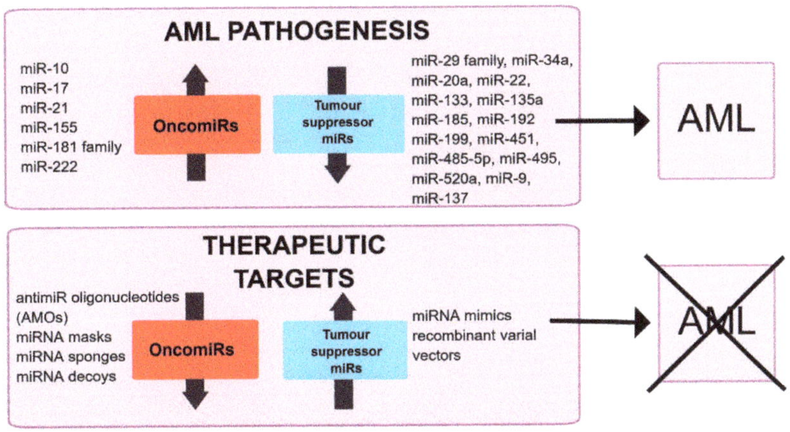

Figure 2. Schematic illustration of the potential role of miRNA in AML pathogenesis and their possible modulation in therapeutics.

Numerous papers have shown that the miR-29 family is significantly downregulated in AML, suggesting its role as a tumour suppressor in this disease. Wang et al. correlated negatively miR-29a relative expression with lncRNA XIST and *MYC* expression. They demonstrated that knocking down XIST led to an increased expression of miR-29a and a decreased expression of both MYC protein and mRNA, as well as reduced the tumorigenic ability of AML cells (KG-1) in vivo in mice [98]. Similarly, Tang et al. reported a negative correlation between expression levels of miR-29b-3p and both the mRNA and protein levels of HuR. The overexpression of miR-29b-3p led to a reduction in cell migration and cell proliferation, a decrease in colony-forming ability, and an increased apoptosis, along with K56a and U937 cell cycle arrest [99]. Furthermore, Randazzo et al. observed the co-occurrence of overexpressed miR-29a, miR-29c, and *DNMT3A* mutations in AML in contrast to patients without *DNMT3A* mutations [100]. Ngankeu et al. identified a germline polymorphism (a T base deletion) in the miR-29b-1/miR-29a cluster, which occurs more frequently in patients with core-binding factor AML (CBF-AML) [101].

miR-34a is generally considered a tumour suppressor in AML. Research has provided that the expression of miR-34a in AML is downregulated. Higher expression levels of miR-34a were present in AML patients with a platelet count lower than 34.5×10^9/L and in those who achieved earlier complete remission (CR), as showed Abdellateif et al. Conversely, lower expression levels of miR-34 were observed in AML patients who were refractory to treatment. No correlation was found between the expression level of miR-34a and overall survival (OS) or event-free survival (EFS) [102]. An inverse correlation has been noted between the expression of miR-34a and PD-L1 in AML. Upregulation of miR-34a led to reduced cell surface PD-L1 expression, inhibited PD-L1 surface expression induced by IFN-γ in AML cells (HL-60), inhibited apoptosis of PD-L1+/CD8+ T cells, and disrupted IL-10 production in AML cells treated with IFN-γ [103]. Furthermore, the expression level of miR-34a was found to be lower in AML patients with intermediate and poor prognoses compared to healthy controls and AML patients with good prognoses. Overexpressed miR-34a led to suppressed KG-1a cells, inhibited *DHAC2* expression and triggered the death of leukemic stem cells (LSCs). In vivo, upregulation of miR-34a was shown to prolong survival and inhibit weight loss in mice transfected with LSCs [104]. Higher expression of miR-34a has been correlated with a favourable cytogenetic risk, while lower expression was associated with the M5 AML subtype [105]. Wen et al. provided evidence for the pivotal function of LSCs eradication of miR-34c-5p. Mice treated with engineered exosomes containing miR-34c-5p demonstrated prolonged survival, accompanied by a decrease in tumor burden, compared to other treated groups, indicating that these engineered exosomes might potentially serve an inhibitory function in the AML progression [106]. While miR-34c is also noted as a downregulated miRNA in AML, its suppressive function is not as clear-cut as that of miR-34a. Peng et al. associated lower miR-34c-5p expression levels with an adverse prognosis in AML and negatively correlated expression of miR-34c-5p with RAB27B mRNA and protein expression levels [107]. Similarly, Yang et al. showed that lower miR-34c expression levels were associated with significantly shorter OS in AML patients, indicating that miR-34c has the potential to emerge as a valuable biomarker [108]. The restoration of miR-34a expression has been proposed as a potential therapeutic strategy to inhibit tumour growth in patients with solid tumours [109].

Several studies have shown that the expression of miR-142 is downregulated, suggesting its role as a tumour suppressor in AML. Zhang et al. negatively correlated miR-142-3p expression with the expression of both protein and mRNA of the target HMGB1. Additionally, the expression of miR-142-3p was lower in cell lines resistant to the drug (HL-60/ATRA and HL-60/ADR), compared to HL-60. Overexpression of miR-142-3p increased the sensitivity of HL-60/ATRA and HL-60/ADR cells, promoting apoptosis in these cells [110]. Elsewhere, miR-142-5p relative expression was found to be significantly reduced in THP-1, HL-60, TF-1, NB4, and U937 cell lines; it was also demonstrated that miR-142-5p is inversely correlated with *PFKP* (platelet isoform of phosphofructokinase) expression, and with lncRNA XIST expression, compared to the control cell line (HS-5). In their study, lncRNA knocking down increased the expression of miR-142-5p, which resulted in the downregulation of *PFKP* in AML cells (U937 and THP-1) and inhibited cell cycle, proliferation, viability and colony formation. This also induced apoptosis in AML cells. Conversely, knocking down miR-142-5p led to AML progression, highlighting the role of the lncRNA XIST/miR-142-5p/PFKP axis in this disease [111]. miR-142 was also found to be downregulated in pediatric samples compared to healthy samples, and its expression was negatively correlated with the expression of circ-004136 [112]. Moreover, miR-142 expression has been described as downregulated in chronic myeloid leukemia (CML), miR-142 knockdown in mice has been linked to the development of a blast crisis [113].

Research indicates that miR-9 expression is reduced in AML, suggesting its function as a tumour suppressor. Liu et al. showed that overexpression of miR-9 enhances daunorubicin sensitivity in AML cells. Additionally, Zhu et al. found a negative correlation between miR-9 expression and *CXCR4* expression. miR-9 has been shown to reduce the proliferation and mobility of AML cells while increasing their apoptotic rate [114–116].

Recent studies also suggest a suppressive role for miR-20 in AML. Studies show that the relative expression of miR-20a-5p is decreased in AML. Bao et al. reported that lower relative expression of miR-20a-5p, according to the median value, is associated with shorter OS in AML patients compared to higher expression levels. They also found an inverse correlation between expression levels of miR-20a-5p and protein phosphatase 6 catalytic subunit, *PPP6C*. Upregulation of miR-20a-5p led to the inhibition of proliferation in AML cells (THP-1, U937), induction of cell cycle arrest, and increased apoptosis. An in vivo study showed that miR-20a-5p upregulation in mice resulted in smaller tumour sizes [117]. Ping et al. reported that the upregulation of miR-20a-5p inhibits the growth of AML cells through circRNA circ_0009910 [118]. In addition, miR-20a, along with miR-106b and miR-125b, contributes to the degradation of the PML-RARA oncoprotein [119].

A recently published article found that miR-103a-2-5p has tumor suppressive properties. The relative expression level of miR-103a-2-5p was found to be lower in samples from AML patients and AML cell lines (OCI-AML3, MOLM-13, THP-1, OCI-AML2 and MV4-11) in contrast to samples from healthy donors and normal cells (HS-5). Additionally, there was a negative correlation between the level of miR-103a-2-5p and the expression of its target, LILRB3, at both the mRNA and protein levels. Higher expression levels of miR-103a-2-5p were associated with longer OS. miR-103a-2-5p overexpressed inhibited proliferation, promoted cell cycle arrest, significantly increased the apoptosis rate and reduced migration of AML cells, and attenuated their clonality in AML cells (THP-1, MV4-11). An in vivo study revealed slower tumour growth in mice injected with miR-103a-2-5p encapsulated in cationic liposomes (CLPs) [120].

Some studies suggest that miR-133 and miR-135a may play a suppressive role in AML. Research indicates that miR-133 and miR-135a expression levels are downregulated [121–124]. Cheng et al. negatively correlated the expression of these miRNAs with their target gene, *CDX2*. In HL-60 AML cells treated with all-trans-retinoic acid (ATRA), there was a significant induction in the expression of both miRNAs. Overexpression of these miRNAs led to reduced proliferation of AML cells (NB4, HL-60). Among various AML subtypes, in M1, M2, and M3, the expression levels of both miRNAs were observed to be higher compared to other subtypes. Furthermore, the study showed that patients who achieved complete remission (CR) exhibited higher relative expression of these miRNAs compared to those who did not, correlating with a better leukemia-free survival rate [125]. According to databases by Wang et al., the target genes for miR-133 include *ZC3H15* and *BCLAF1*. An in vitro study using NB4 cells revealed that introducing miR-133 reduced the capacity for proliferation and accelerated apoptosis [123]. In addition, patients with a favourable karyotype demonstrated elevated expression of miR-133, which correlated with longer OS and relapse-free survival (RFS) [121]. A negative correlation was also observed between expression levels of miR-135a and *HOXA10*. In their examination of AML blood samples and AML cell lines (HL-60, AML193, AML2, and AML5), they noted decreased relative expression of miR-135a compared to normal blood samples and the normal cell line HS-5. Higher expression levels of miR-135a were associated with a better prognosis, that is, with longer OS in AML patients. Overexpression of miR-135a significantly elevated cellular apoptosis and suppressed the proliferation and cell cycle of HL-60 and AML5 cells [124]. Liu et al. also showed that miR-133 targets the *CXCL12* gene, and inhibiting miR-133 expression can enhance the sensitivity of HL-60/ADR cells to daunorubicin [126].

miR-137 is reported as downregulated and suggested to act as a tumour suppressor in AML. miR-137 expression was found to be negatively correlated with *c-KIT* expression, and restoring miR-137 expression inhibited proliferation and promoted differentiation [127]. Additionally, Wang et al. inversely correlated the expression levels of miR-137 and *TRIM25*. Overexpressing miR-137 resulted in reduced proliferation, migration, and invasion of AML cells [128].

Similarly, miR-148 is also described as downregulated and acts as a tumour suppressor in AML [129,130]. Wang et al. reported that the expression of its direct target, *DNMT1*, was inversely correlated with miR-148 expression. Moreover, overexpression of miR-148 inhibited the proliferation and increased the apoptosis of AML cells [130].

Research indicates that miR-185 is downregulated in AML. Pang et al. highlighted the role of miR-185 as a tumour suppressor, reporting significantly lower levels of relative miR-185 expression in specimens collected from AML patients and AML cell lines compared to specimens from healthy individuals and normal cells. They found that *GPX1* expression was negatively correlated with miR-185-5p expression. Overexpression of miR-185-5p decreased the viability, proliferation, invasion, and apoptosis of AML cells (HL-60, KG-1) [131]. Similarly, Zhang et al. also described downregulated miR-185 expression in AML samples and cell lines compared to normal samples, noting its regulation by TUG1 lncRNA [132].

The expression of miR-192 is found to be downregulated in AML. A study conducted by Wu et al. [133] examined the role of the circ_0001602/miR-192-5p/ZBTB20 axis, revealing that circ_0001602 and *ZBTB20* expression levels were upregulated, while miR-192-5p expression levels were downregulated in AML cell lines (THP-1, HL-60, NB4) and patient samples compared to samples from healthy individuals and the HS-5 cell line. Knockdown of circ_0001602 resulted in reduced proliferation of AML cells and induced apoptosis and cell cycle arrest. However, a deficiency of miR-192-5p abolished this effect. Moreover, overexpression of miR-192-5p reduced cell viability, inhibited cell cycle progression, and induced cell apoptosis in AML cell lines; however, this effect was diminished after the addition of *ZBTB20*. This evidence suggests a sponge effect of circ_0001602 affecting this miRNA; the study also highlights the downstream influence of miRNA on its target gene, which overall indicates a pivotal role of the examined axis in the progression of AML. Additionally, the expression of miR-192 was downregulated in AML pediatric patients compared to healthy controls [134]. Ke et al. reported downregulated expression of miR-192 in AML patients compared to controls and its binding to *CCNT2*, indicating that miR-192 functions as a tumour suppressor. Overexpression of miR-192 induced cell cycle arrest, apoptosis, and cell differentiation [135].

Research shows that miR-451 is downregulated in AML and suggests its role as a tumour suppressor. Krakowsky et al. found an inverse correlation between the expression of miR-451 and *MDR1* expression, suggesting that miR-451 has the potential to enhance the chemotherapy sensitivity of *FLT3*-ITD+ cells by targeting *MDR1*. Li et al. found that miR-451a was one of the significantly decreased miRNAs, alongside miR-185-5p, miR-443b-3p, miR-199a-5p, and miR-151a-3p among the examined miRNAs. Furthermore, Song et al. suggested the role of the hnRNP A1/miR-451/c-Myc axis in AML development. Su et al. discovered that miR-451 targets *YWHAZ* and suppresses YHWAZ-AKT signalling pathway in AML. Overexpression of miR-451 resulted in increased apoptosis rates of AML cells and inhibited their proliferation [136–140].

Recently, studies by Wu et al. and by Xie et al. have shown that miR-455-3p is downregulated in AML. The first study indicates that miR-455-3p functions as a tumour suppressor in AML [141,142]. The second study demonstrated a negative correlation between miR-455-3p and its target gene, *UBN2*. AML cells transfected with the miR-455-3p mimic exhibited decreased proliferation, inhibited cell viability, induced cell cycle arrest, and induced both apoptosis and autophagy. The study highlighted the critical regulatory role of miR-455-3p in AML through the mediation of peroxisome proliferator-activated receptor α (PPARα) [142].

miR-485-5p was identified as downregulated and suggested to function as a tumour suppressor in AML. Wang et al. found a negative correlation between miR-485-5p expression and its target gene, *SALL4*. The upregulation of miR-485-5p elevated the apoptotic rate and attenuated the proliferative ability of AML cells (AML5 and U937) [143,144].

Similarly, miR-495 was described as downregulated, and Wang et al. confirmed its suppressive role in AML [145–148]. Lei et al. reported that the relative expression levels of miR-495-3p and miR-543 were lower in AML bone marrow samples and AML cells (HL-60 and Kasumi-1) compared to control bone marrow samples and control cells (HS-5). They found that the expression of the target gene *PDK1* was negatively correlated with the levels of these miRNAs. In AML cell lines treated with matrine, the expression levels of miR-495-3p and miR-543 increased. Overexpression of these miRNAs led to attenuated viability, promoted apoptosis, induced cell cycle arrest, and inhibited glycolysis, indicating their regulatory role in AML under matrine treatment [146].

miR-520a is also downregulated in AML and is recognized as a tumour suppressor, according to studies by Chen et al. and Xiao et al. The first author correlated reversely the expressions of miR-520a-3p and *MUC1*, demonstrating that overexpression of miR-520a-3p inhibited proliferation and induced apoptosis of AML cells (THP-1) [149]. Xiao et al. similarly showed that the overexpression of miR-520a reduced proliferation and cell viability while inducing apoptosis. The functional study demonstrated that the anti-proliferative effects of miR-520a were primarily due to its ability to block the PI3K/AKT signalling pathway [150].

The expression of miR-30e-5p was also found to be downregulated and negatively correlated with *CYB561* expression. Overexpression of miR-30e-5p delayed the onset of leukemia in mice with *KMT2A::MLLT3*. Additionally, upregulation of miR-30e-5p also impaired the self-renewal abilities of LSCs. Notably, no significant correlation was reported in the case of miR-30e-3p in AML samples compared to healthy samples [151].

Ye Q et al. reported that the relative expression of miR-211-5p was downregulated in pediatric AML samples compared to control samples and was negatively correlated with the expression of its target gene, *JAK2*. Overexpression of miR-211-5p, induced by cryptanshinone (CPT), led to increased apoptosis of LSCs, decreased LSC proliferation and viability, and reduced concentrations of proinflammatory cytokines (IL-1β, TNF-α). Inhibition of miR-211-5p resulted in the opposite effect, highlighting its regulatory role in CPT-treated AML cells [152].

Xie et al. showed that miR-409-3p was downregulated in the THP-1 cell line compared to the normal cell line (HS-5). miR-409-3p negatively regulated its target gene, *RAB10*. Overexpression of miR-409-3p reduced the proliferation of THP-1 cells and increased apoptosis, thus exerting its suppressive effect in AML [153].

Wang et al. demonstrated that the expression of miR-454-3p was significantly decreased in AML cells, and the target for miR-454-3p is *ZEB2*. Overexpression of miR-454-3p in THP-1 cells resulted in reduced cell viability, induced cell cycle arrest, and promoted apoptosis and autophagy, confirming its role as a tumour suppressor in AML [154].

The circPLXNB2/miR-654-3p/CCND1 axis was examined in AML by Wang et al. They found that the relative expression of miR-654-3p in AML patients was downregulated, and *CCND1* was identified as its direct target. Elevated expression of miR-654-3p led to decreased proliferation of HL-60 and Kasumi-1 cells, increased apoptosis, and induced cell cycle arrest [155].

miR-1294 was downregulated in AML and negatively regulated by circFN1, as described by Wang et al. AML cells (HL-60) transfected with the miR-1294 mimic exhibited significantly lower cell proliferation and invasion abilities compared to those transfected with the negative control (NC) mimic. Moreover, apoptosis was higher in miR-1294 mimic-transfected cells than in the NC mimic-transfected cells. The study confirmed that miR-1294 negatively regulates its target gene, *ARHGEF10L*, thereby demonstrating the regulatory role of the circFN1/miR-1294/ARHGEF10L axis in AML [156].

Conversely, miR-10 is reported to be upregulated in AML, and its role is characterized as oncogenic. Zhi et al. found a correlation between higher expression levels of miR-10a-5p and shorter OS. Additionally, Wang et al. reported overexpressed miR-10b in pediatric specimens in contrast to the ITP group and was negatively correlated with *HOXD10* expression. Overexpression of miR-10b accelerated the proliferation of AML cells (HL-60), increased the number of cells in the S phase, and decreased the apoptosis rate [157–160].

miR-21 is described as upregulated and acting as an oncogene in AML. Li et al. negatively correlated miR-21 expression with *KLF5* expression. miR-21 inhibition decreased AML cell proliferation. Moreover, overexpression of miR-21 resulted in decreased *KLF5* expression, thereby promoting tumorigenesis in vivo [161]. Higher levels of miR-21 expression were associated with shorter OS and RFS in AML patients [162]. Notably, in patients with mutated *NPM1*, the expression of miR-21 was markedly increased [163].

Similarly, miR-17 was also found to be upregulated in AML [164–166]. Studies indicate that miR-17-5p is upregulated in AML cells (THP-1 and HL-60) compared to normal cells (HS-5). The expression levels of its target gene, *JAK1*, and the lncRNA SUCLG2-AS1 are negatively correlated with miR-17-5p, as shown by Liu et al. Their research demonstrated that SUCLG2-AS1 modulates the expression of *JAK1* and influences the leukemic (THP-1, HL-60) cells' apoptosis, proliferation, invasion, and migration via miR-17-5p. Furthermore, miR-17-5p modulates apoptosis, invasion, and migration of AML cells through *JAK1*, highlighting the regulatory role of the lncRNA SUCLG2-AS1/miR-17-5p/JAK1 axis in AML [165]. Wang et al. also reported an upregulation of miR-17-5p expression in AML samples compared to healthy volunteers, noting that higher expression of miR-17-5p correlated with shorter RFS and OS in AML patients. Additionally, the research indicates a negative correlation between miR-17-5p and *BECN1* expression. After treatment with vitamin D, miR-17-5p expression decreased and reduced AML cell (HL-60) proliferation [166].

According to Huang et al., miR-93 is upregulated in AML and activates the PI3K/AKT pathway via *DAB2*. Knocking down miR-93 inhibited AML cell proliferation. Moreover, miR-93 downregulation led to AML cells' apoptosis and cell cycle arrested. Conversely, overexpression of miR-93 led to an increase in tumour size in mice in vivo [167].

A recent study indicated that the relative expression of both miR-106b-5p and its target, Rab10 mRNA and protein, is upregulated in AML. In patients at the initial stage of AML, lower expression levels of miR-106b-5p and Rab10 protein were associated with higher survival rates [168].

miR-126 is upregulated in inv(16)(p12q22) AML, as demonstrated by Zhang et al. A decrease in miR-126 expression resulted in inhibited AML cell survival and activity in vivo [169].

miR-146 was also described as being upregulated in AML. Wang et al. found that the relative expression of miR-146a was significantly increased in pediatric AML patients compared to healthy children. Moreover, the miR-146a expression level was negatively correlated with the expression of its target, *CNTFR*. When miR-146a was downregulated, there was a significant decrease in proliferation, increased apoptosis, and reduced migration of leukemic cells (HL-60) [170]. Li et al. confirmed the upregulation of miR-146b-5p in examined patients with AML in contrast to healthy samples of RNA collected from extracellular vesicles (EVs) in serum. Additionally, their study identified other upregulated miRNAs, including miR-10a-5p, miR-155-5p, miR-100-5p, and let-7a-5p, while downregulated miRNAs included miR-185-5p, miR-4433b-3p, miR-199a-5p, miR-451a, and miR-151a-3p [171].

miR-155 is similarly recognized as upregulated in AML, suggesting an oncogenic role in AML [172–176]. Garavand et al. observed no correlation between miR-155-5p expression and either *KRAS* or *CREB* expression [172]. Xue et al. found that the relative expression of miR-155 was upregulated in samples from AML patients and cell lines compared to healthy volunteers and CML cell lines, with a negative correlation to relative *SHIP1* expression. After inhibition of miR-155, the proliferation of AML cells (THP-1, U937) decreased, and the apoptosis rate increased [173]. Hatem et al. also reported in AML an upregulation of miR-155 along with miR-10a and let-7a [176].

miR-181 was identified as upregulated in AML and is thought to function as an oncomiR [177–179]. A study of the miR-181 family by Sue et al. found that the expression of miR-181a, miR-181b, miR-181c, and miR-181d was higher in AML samples compared to controls. This study indicates a negative correlation between these miRNAs and genes *PRKCD*, *CTDSPL*, and *CAMKK1*. The study highlighted the role of the miR-181 family in granulocytic and macrophage-like differentiation [179].

According to Gao et al., miR-1306-5p is increased in AML and acts as a tumour oncogene. The relative expression of *PHF6* was negatively correlated with miR-1306-5p expression levels. Downregulation of miR-1306-5p led to decreased proliferation and increased apoptosis of AML cells [180].

The function of miR-22 remains unclear. Many studies suggest that it acts as a tumour suppressor and report the downregulation of miR-22 in AML. On the other hand, a singular study indicates upregulation of miR-22 expression in AML. For instance, Jiang et al. found that the relative expression of miR-22 was significantly lower in de novo AML and negatively correlated with *CRTC1* and *FLT3* expression. Restoring miR-22 expression inhibited in vitro AML colony formation, promoted apoptosis of AML cells, and decreased AML cell viability and growth in the MONOMAC-6 cell line. In an in vivo study, it was reported that the upregulation of miR-22 significantly inhibited leukemogenesis in mice, supporting its anti-tumour role in AML [181]. Similarly, Shen et al. reported downregulated expression of miR-22 in de novo AML, finding a reverse association between expression of miR-22 and *MECOM(EVI1)*, and a positive correlation with the expression level of transcription factor PU.1. The study proved that upregulation of miR-22 alleviated differentiation inhibition in AML bone marrow blasts, resulting in a notable reduction in the HL60 and THP1 cells growth [182]. Qu et al. also noted downregulated miR-22 expression in AML and associated it with worse OS and RFS in AML patients. Additionally, increased levels of miR-22 were observed in pre- and post-treated specimens and was linked with the achievement of CR [183]. On the other hand, Yao et al. presented upregulated expression of miR-22-3p and miR-22-5p in AML [184]. Although the expression of miR-22 is described differently in various research studies, the significant majority of them focus on the anti-tumour function of this miRNA.

The expression of miR-92a shows a dual role in studies on AML, appearing both upregulated and downregulated. For example, Gu et al. reported significantly lower relative levels of miR-92a expression in AML cell lines (HL-60 and THP-1) compared to a normal cell line (HS-5). They negatively correlated the expression of miR-92a with *MTHFD2* expression, suggesting a suppressive role for miR-92a in AML. Moreover, the upregulation of miR-92a inhibited AML cell proliferation and induced a surge in apoptosis by regulating *MTHFD2* [185]. On the other hand, Su et al. highlighted the role of the circ_0002232/miR-92a-3p/PTEN axis in AML, showing that in AML patients, miR-92a-3p was upregulated in contrast to healthy donors. Sue et al. correlated inversely the expressions of miR-92a and *PTEN*. [186]. Saadi et al. described increased expression of miR-92a and miR-181a in AML patients compared to the control group [187]. Additionally, Rashed et al. demonstrated that the post-induction level of miR-92a expression was significantly higher in patients who achieved CR compared to those who did not, indicating that miR-92a has the potential to be an effective biomarker [188].

miR-199 is predominantly recognized as a tumour suppressor in AML; however, there is some limited evidence suggesting that it may act as an onco-miR in a specific context. Most of the research indicates that the expression level of miR-199 is downregulated in AML. For example, Qi and Zhang found a correlation between decreased miR-199 expression and poor risk stratification in pediatric AML patients [189]. Likewise, Li et al. demonstrated that patients with relapsed/refractory AML had lower expression of miR-199a-5p than those who achieved CR. They also discovered that miR-199a-5p directly targets *DRAM1*, and overexpression of miR-199a-5p inhibited autophagy and reduced chemoresistance upon adriamycin treatment [190]. Ellson et al. associated miR-199 higher expressions with improved OS and EFS in AML patients [191]. Favreau et al. noted that lower expression of miR-199b correlated with shorter OS in AML patients and indicated its prognostic significance in the FAB-M5 subtype [192]. Conversely, Alemdehy et al. suggested that miR-199a-3p functions as an onco-miR in AML, showing that this miRNA caused AML in an in vivo mice model [193].

In contrast, the role of miR-222 is more frequently reported as oncogenic in AML studies; however, there is one article that indicates its potential suppressive function in AML. Yuan et al. reported decreased miR-222-3p expression in isolated exosomes secreted by bone marrow mesenchymal stem cells (BM-MSC) in AML patients compared to the exosomes from healthy donors. They found that upregulation of miR-222-3p elevated the Th1/Th2 ratio, confirmed in vivo, and induced apoptosis of AML cells (HL-60). The expression of miR-222-3p was also found to be negatively correlated with IRF2 expression [194]. Pei et al. reported an increase in the relative expression of both miR-221 and miR-222 in AML samples compared to healthy ones, noting a negative correlation between miRNA expression and *YOD1* expression. Downregulation of miR-221/222 increased the p53 protein level and reduced its ubiquitination [195]. Liu et al. observed higher relative expression of miR-222-3p in AML cell lines (NB4, U937, KG1a, THP1) compared to peripheral blood mononuclear cells (PBMCs) from healthy controls, identifying *AXIN2* as a target for miR-222-3p. Downregulation of miR-222-3p led to decreased viability and induced apoptosis of AML cells (NB4 and U937) [196]. Likewise, Pavlovic et al. also reported upregulation of miR-222 in AML, supporting its oncogenic function [197].

miR-361 has been reported as both downregulated and upregulated in AML. Xu et al. reported decreased relative expression of miR-361-3p in *KMT2A*-rearranged AML patients and AML cell lines (THP-1, HL-60, KG-1A, KO52) compared to control subjects and the cell line (HS-5). They confirmed a negative correlation between the expression of miR-361-3p and KMT2A. Overexpression of miR-361-3p decreased proliferation, migration, and invasion of AML cells (HL-60) [198]. Conversely, Liu et al. reported an increased relative expression of miR-361-3p in AML patients and the AML cell line (HL-60) compared to healthy donors, indicating its binding to *BTG2*. The downregulation of miR-361-3p expression reduced proliferation and increased cell apoptosis under 9s-HODE, a major active derivative of linoleic acid, suggesting an oncogenic role of miR-361-3p in AML [199].

Table 2. List of downregulated miRNAs in AML and their target genes; effect on AML cells while microRNA is overexpressed; type and size of microRNA expression study/control group.

↓-miRNA	Target	Effect on AML/Leukemic Cells When miRNA Is Overexpressed	Method (Study Group/Control Group or Cell Lines)	Reference
miR-9	*CXCR4*	Reduction of proliferation and mobility of AML cells. Increased apoptosis rate of AML cells.	RT-qPCR (36 AML BM/10 BM; NB4, HL-60, Kasumi-1, SKNO-1, KG-1a/normal CD34+ cells), miRNA mimics (Kasumi-1, SKNO-1)	[116]
miR-20a-5p	*PPP6C*	Inhibition of cell proliferation, induction of cell cycle arrest and apoptosis. Decrease in tumor size in vivo.	RT-qPCR (61 AML BM/61 BM; Kasumi-1, THP-1, U937, HL-60/HS-5), miRNA mimics (THP-1, U937)	[117]
miR-22	*CRTC1* *FLT3*	Inhibition of cell colony forming, viability, and growth. Inhibition of leukemogenesis in mice in vivo. Promotion of cell apoptosis.	RT-qPCR (42 AML MNC/5 MNC), miRNA mimics (MONOMAC-6, THP-1, KOCL-48)	[181]
	EVI1	Relieved blockage in the differentiation of bone marrow blasts. Inhibition of cell growth.	RT-qPCR (79 AML PB MNC/114 PB MNC; 41 AML BM MNC/8 BM MNC; 50 AML BM CD34+/10 BM CD34+), miRNA mimics (HL-60, THP1)	[182]

Table 2. *Cont.*

↓-miRNA	Target	Effect on AML/Leukemic Cells When miRNA Is Overexpressed	Method (Study Group/Control Group or Cell Lines)	Reference
miR-29a	*MYC*	Inhibition of cell tumorigenic ability in mice in vivo.	RT-qPCR (62 AML BM MNC/20 BM MNC; KG-1/BM MNC), miRNA mimics (KG-1)	[98]
miR-29b-3p	*HuR*	Inhibition of cell proliferation, colony-forming capability, and cell migration. Promotion of cell cycle arrest at G0/G1 phase and apoptosis.	RT-qPCR (K562, NB4, U937, K562/G01, Kasumi-1, HL60/PB MNC), miRNA mimics (K562, U937)	[99]
miR-30e-5p	*CYB561*	Impaired cell self-renewal. Inhibited onset of *KMT2A::MLLT3*-driven leukemia in mice in vivo.	RT-qPCR (29 AML BM MNC/6 BM MNC), miRNA mimics (*KMT2A::MLLT3* AML cells)	[151]
miR-34a	*PD-L1*	Reduction in cell surface PD-L1 expression. Reduction in INF-γ-induced PD-L1 surface expression, apoptosis of PD-L1/CD 8+ T cells and IL-10 production upon IFN-γ	RT-qPCR (13 AML BM/5 BM), miRNA mimics (HL-60, Kasumi-1)	[103]
	DHAC2	Suppressed proliferation. Induced LSC death. Prolonged survival in AML mice in vivo.	RT-qPCR (30 AML BM/10 BM), miRNA mimics (KG-1a)	[104]
miR-92a	*MTHFD2*	Inhibition of cell proliferation and promotion of apoptosis.	RT-qPCR (HL-60, THP-1/HS-5), miRNA mimics (HL-60, THP-1)	[185]
miR-103a-2-5p	*LILRB3*	Inhibition of cell proliferation, migration, and clonality. Promotion of apoptosis and cell cycle arrest.	RT-qPCR (30 AML BM MNC/BM MNC), miRNA mimics (THP-1, OCI-AML2, OCI-AML3, MV4-11)	[120]
miR-133	*ZC3H15* *BCLAF1*	Inhibition of AML cell proliferation. Acceleration of AML cell apoptosis.	miRNA sequencing (102 AML samples/CD34+) *, miRNA mimics (NB4)	[123]
miR-135a	*HOXA10*	Inhibition of AML cell proliferation and cell cycle. Acceleration of AML cell apoptosis.	RT-qPCR (29 AML PB/11 PB; HL-60, AML193, AML2, AML5/HS-5), miRNA mimics (HL-60, AML5)	[124]
miR-133a miR-135a	*CDX2*	Inhibition of AML cell proliferation.	RT-qPCR (59 AML BM/9 BM), miRNA mimics (HEK-293, NB4, HL-60)	[125]
miR-137	*C-kit*	Inhibition of proliferation and promotion differentiation of AML cells.	RT-qPCR (49 AML BM/57 BM), miRNA mimics (Kasumi-1, K562)	[127]
	TRIM25	Inhibited AML cells proliferation, migration and invasion.	RT-qPCR (45 AML BM/45 BM; HEL, Kasumi-1, HL-60, MEG01/HS-5), miRNA mimics (Ksaumi-1, HL-60)	[128]
miR-142-3p	*HMGB1*	Improvement of drug sensitivity in AML cells.	RT-qPCR (23 AML PB MNC/15 PB MNC), miRNA mimics (HL-60/ATRA, HL-60/ADR)	[110]
miR-142-5p	*PFKP*	Inhibition of AML cell proliferation, viability, cloning, and cycle. Induction of AML cell apoptosis.	RT-qPCR (THP-1, HL-60, TF-1, NB4, U937/HS-5), miRNA mimics (THP-1, U937)	[111]
miR-148	*DNMT1*	Inhibition of proliferation and induction of apoptosis of AML cells.	RT-qPCR (80 AML BM MNC/20 BM MNC; U937, THP-1, Kasumi-1/20 BM MNC), miRNA mimics (U937, Kasumi-1)	[130]

Table 2. *Cont.*

↓-miRNA	Target	Effect on AML/Leukemic Cells When miRNA Is Overexpressed	Method (Study Group/Control Group or Cell Lines)	Reference
miR-185-5p	*GPX1*	Inhibition of viability, proliferation, invasion, and promotion of apoptosis.	RT-qPCR (37 AML BM/37 BM; MOLM-14, HL-60, KG-1/HS-5), miRNA mimics (KG-1, HL-60)	[131]
miR-192-5p	*ZBTB20*	Inhibition of cell viability. Induction of cell apoptosis and cell cycle arrest.	RT-qPCR (52 AML BM/34 BM THP-1, HL-60, NB4/HS-5), miRNA mimics (THP-1, HL-60)	[134]
	CCNT2	Induction of cell cycle arrest, apoptosis, and cell differentiation.	RT-qPCR (10 AML BM/10 BM; NB4, HL-60/10 BM), miRNA mimics (NB4, HL-60)	[135]
miR-199a-5p	*DRAM1*	Reduction of chemoresistance upon ADM treatment. Inhibition of autophagy.	RT-qPCR (32 AML relapsed/refractor BM/11 complete remission BM), miRNA mimics (K562/ADM, K562)	[190]
miR-211-5p	*JAK2*	Reduction of AML cell proliferation, viability, and inflammation. Induction of AML cell apoptosis.	RT-qPCR (50 AML PB/50 PB), miRNA mimics (CPT treated LSCs)	[152]
miR-222-3p	*IRF2*	Increased Th1/Th2 ratio. Induction of cell apoptosis.	RT-qPCR (20 BM of AML patients and healthy donors), miRNA mimics (HL-60)	[194]
miR-361-3p	*KMT2A*	Reduction of cell proliferation, migration, and invasion.	RT-qPCR (30 AML PB/30 PB; HL-60, KG-1a, KO52, THP-1/HS-5), miRNA mimics (HL-60, H562)	[198]
miR-409-3p	*RAB10*	Inhibition of cell proliferation and induced apoptosis.	RT-qPCR (THP-1, NB4/HS-5), miRNA mimics (THP-1)	[153]
miR-451	*YWHAZ*	Suppression of AML cell proliferation. Increased AML cell apoptosis.	RT-qPCR (69 AML PB MNC/80 PB MNC; 56 AML BM MNC/9 BM MNC; 32 AML BM CD34+/9 BM CD34+), miRNA mimics (NB4, HL-60)	[140]
miR-454-3p	*ZEB2*	Inhibited viability and induced cell cycle arrest. Induction of apoptosis and autophagy.	RT-qPCR (NB4, THP-1, KG-1a, U937/PB MNC), miRNA mimics (THP-1)	[154]
miR-455-3p	*UBN2*	Inhibition of cell proliferation and viability. Induction of cell apoptosis, autophagy, and cell cycle arrest.	RT-qPCR (16 AML PB/16 PB; HL-60, Kasumi-1, KG1, THP-1, MV4-11/PB MNC), miRNA mimics (HL-60)	[142]
miR-485-5p	*SALL4*	Inhibition of cell proliferation and induction of cell apoptosis.	RT-qPCR (35 AML BM/35 BM; AML2, AML193, Kasumi-1, HL-60, AML5, U937/HS-5), miRNA mimics (AML5, U937)	[144]
miR-495-3p miR-543	*PDK1*	Inhibition of cell proliferation, cell glycolysis, viability under matrine. Induction of cell apoptosis and cell cycle arrest under matrine.	RT-qPCR (31 AML BM/31 BM; HL-60, Kasumi-1/HS-5), miRNA mimics (HL-60, Kasumi-1)	[147]
miR-520a-3p	*MUC1*	Inhibition of cell proliferation and induction of apoptosis.	RT-qPCR (25 AML PB/25 PB), miRNA mimics (THP-1)	[150]
miR-654-3p	*CCND1*	Inhibition of cell proliferation, induction of apoptosis and cell cycle arrest.	RT-qPCR (51 AML PB/51 PB; HL-60, Kasumi-1/HS-5), miRNA mimics (HL-60, Kasumi-1)	[155]
miR-1294	*ARHGEF10L*	Inhibition of AML cell proliferation and invasion. Induction of AML cell apoptosis.	RT-qPCR (16 plasma/16 plasma; HL-60/HS-5), miRNA mimics (HL-60)	[156]

*—data from GEO database; BM, bone marrow; MNC, mononuclear cells; PB, peripheral blood.

Table 3. List of upregulated miRNAs in AML and their target genes; effect on AML cells while microRNA is knockdown; type and size of microRNA expression study/control group.

↑-miRNA	Target	Effect on AML Cells While miRNA Is Downregulated	Method (Study Group/Control Group or Cell Lines)	Reference
miR-10b	*HOXD10*	ND	RT-qPCR (108 AML serum samples/25 serum samples)	[158]
miR-17-5p	*JAK1*	Inhibition of cell proliferation, migration, invasion, and promotion of apoptosis when lncRNA SUCLG2-AS1 overexpressed.	RT-qPCR (THP1, HL60/HS-5), miRNA mimics (THP-1, HL-60)	[165]
	BECN1	Inhibition of AML cell proliferation after treatment of vitamin D.	RT-qPCR (144 AML PB/45 PB), miRNA mimics (HL-60)	[166]
miR-21	*KLF5*	Reduction of AML cells proliferation.	RT-qPCR (SKM-1, HL-60/HS-5), miRNA mimics (SKM-1, HL-60)	[161]
miR-92a-3p	*PTEN*	ND	RT-qPCR (115 AML BM MNC/48 BM MNC)	[186]
miR-93	*DAB2*	Inhibition of AML cells proliferation. Promotion of cell cycle arrest and apoptosis of AML cells.	RT-qPCR (28 AML BM MNC/30 BM MNC), miRNA mimics (THP-1, HL-60, HS-5)	[167]
miR-106-5p	*RAB10*	ND	RT-qPCR (85 AML BM MNC/15 BM MNC)	[168]
miR-146a	*CNTFR*	Reduction of leukemic cells proliferation and migration. Increasing leukemic cells apoptosis.	RT-qPCR (11 AML BM/10 BM), miRNA mimics (HL-60)	[170]
miR-155	*SHIP1*	Inhibition of AML cell proliferation and promotion of apoptosis.	RT-qPCR (30 AML BM or PB MNC/10 BM MNC), miRNA mimics (U937, THP-1)	[173]
miR-181 family	*PRKCD* *CTDSPL* *CAMKK1*	Modulation of granulocytic and macrophage-like differentiation.	RT-qPCR (95 AML PB MNC/ 75 PB MNC; 36 AML BM CD34+/9 BM CD34+), miRNA mimics (HL-60)	[179]
miR-221 miR-222	*YOD1*	Increasing p53 protein level and reducing its ubiquitination.	RT-qPCR (18 AML PB/20 PB and 6 BM), miRNA mimics (U2OS, HCT116)	[195]
miR-222-3p	*AXIN2*	Decreased cell viability and induced apoptosis.	RT-qPCR (KG1a, NB4, U937, THP1/PB MNC), miRNA mimics (NB4, U937)	[196]
miR-361-3p	*BTG2*	Inhibition of cell proliferation and induction of apoptosis.	RT-qPCR (34 AML PB/5 PB; HL-60/5 PB), miRNA mimics (HL-60)	[199]
miR-1306-5p	*PHF6*	Inhibition of AML cell proliferation and induction of apoptosis rate.	RT-qPCR (48 AML BM/30 BM; HL-60, Kgla, K562, THP-1/HS-5), miRNA mimics (HL-60, K562)	[180]

BM, bone marrow; MNC, mononuclear cells; PB, peripheral blood; ND, no data.

miRNA Associated with Genetic Abnormalities

According to the 2022 ELN recommendations, AML is classified into numerous subtypes with respect to genetic abnormalities, including: AML with recurrent genetic abnormalities, AML with mutated *TP53*, AML with myelodysplasia-related cytogenetic abnormality, and AML not otherwise specified (AML NOS). AML with recurrent genetic abnormalities constitute a large group of AML subtypes, which comprises: APL with t(15;17)(q24.1;q21.2)/*PML::RARA*, AML with t(8;21)(q22;q22.1)/*RUNX1::RUNX1T1*, AML with inv(16)(p13.1q22) or t(16;16)(p13.1;q22)/*CBFB::MYH11*, AML with t(9;11)(p21.3;q23.3)/ *MLLT3::KMT2A*, AML with t(6;9)(p22.3;q34.1)/*DEK::NUP214*, AML with inv(3)(q21.3q26.2) or t(3;3)(q21.3;q26.2)/*GATA2, MECOM(EVI1)*, AML with other rare recurring translocations, AML with mutated *NPM1*, AML with in-frame bZIP mutated *CEBPA*, and AML with t(9;22)(q34.1;q11.2)/BCR::ABL1. Some studies indicate the significance of miRNAs in the development of leukemia, particularly regarding distinct subtypes of the disease, which may prove valuable clinically in the future. Data is collected in Table 4.

A few reports indicate particular miRNAs which may be relevant in the t(8;21) AML subtype; for instance, miRNA let-7b, when overexpressed, is reported to reduce AML1-ETO protein expression. Additionally, miRNA let-7b is noted to inhibit the proliferation of t(8;21) AML cell lines and induce differentiation, indicating that this miRNA may play a role in the leukemic characteristics associated with the t(8;21)(q22;q22.1) chromosomal translocation [200]. miR-223 is reported as downregulated in t(8;21) AML and is identified as a direct target of the AML1/ETO oncoprotein. AML1/ETO alters the expression level of miR-223, ultimately restoring cell differentiation in cases of AML with t(8;21)(q22;q22.1) [201]. miR-9-1 is showed as downregulated in AML associated with the t(8;21) chromosomal translocation. *RUNX1::RUNX1T1* is regulated by miR-9-1, and silencing miR-9-1 has been shown to increase the oncogenic activity of *RUNX1::RUNX1T1*. Additionally, the miR-9-1 overexpression reduces proliferation and promotes differentiation of t(8;21) cell lines [202]. Higher expression levels of miR-126 are reported to be associated with poor prognosis in patients with t(8;21) AML [203]. miR-130a expression is presented as increased in t(8;21) AML, which induce apoptosis and differentiation of t(8;21) AML cells. The level of miR-130a is suggested to be important for t(8;21) AML maintenance [204]. miR-383 is presented as upregulated in AML1-ETO positive AML, and AML1-ETO regulates its expression. In turn, miR-383 negatively regulates THAP10 expression, which functionally inhibits leukemogenesis of t(8;21) AML [205].

In APL with fusion gene *PML::RARA*, miR-15b is described to have an essential function in the proliferation and differentiation of acute promyelocytic leukemia. Also, miR-15b is indicated as a tumour suppressor in APL, and its target gene is CCNE1 [206]. miR-125b is described in APL, and its overexpression in t(15;17) APL expands PML-RARA-induced leukemogenesis in vivo [207]. micoRNA-382-5p is important in the differentiation of ATRA-induced APL, which targets PTEN [208].

In AML with fusion gene *CBFB::MYH11*, miR-126 was examined in vivo. The study indicated that deleting miR-126 prevented the development of leukemia in 50% of *CBFB::MYH11* knock-in mice and significantly extended their survival. The study also revealed that abnormal expression of miR-126 significantly contributes to the onset of leukemogenesis and maintains AML induced by *CBFB::MYH11* [169].

In the context of AML with fusion gene *MLLT3::KMT2A*, miR-30e is described to play a crucial role in developing this AML subtype. In vivo, the study showed that the overexpression of miR-30e delays the development of *KMT2A::MLLT3*-driven leukemia [151].

AML with mutated *NPM1* showed upregulated miR-10a compared to *NPM1* wild-type AML, indicating its potential role in *NPM1*-mutated AML by targeting *MDM4* [209]. Moreover, miR-21 is reported to be overexpressed in *NPM1*-mutated AML, where it targets

PDCD4. This abnormal expression is thought to contribute to the pathogenesis of *NPM1*-mutated AML [163]. Additionally, miR-215-5p is reported to be involved in downregulating *SMC1A* in *NPM1*-mutated AML. Expression of miR-215-5p was significantly higher, and expression of *SMC1A* was considerably lower in *NPM1*(transcript A)-mutated AML cells than in AML control cells [210].

In AML with mutated *TP53* and a complex karyotype, miR-34a is reported to be downregulated, while miR-100 is reported to be upregulated. miR-34a was correlated with resistance to chemotherapy and inferior survival. Among AML with biallelic TP53 and complex karyotype, miR-34a expression was divided into quartiles, and higher (the fourth quartile) expression of miR-34a was associated with better OS [211]. *TP53* encodes the TP53 protein, a well-known tumor suppressor. When activated, TP53 serves as a transcriptional activator for numerous genes that regulate cellular functions. Mutations in *TP53* can have various consequences. One effect is the loss of its tumour-suppressive function, which can occur due to point mutations (primarily affecting exons 5–8) or truncating variants (caused by indels) or *TP53* loss (caused by large deletions) that lead to a nonfunctional TP53 [212].

Chen et al. demonstrated that miR-363-3p can potentially trigger the *RUNX1*^{mut} AML onset [213]. Barreyro et al. showed that Hematopoietic Stem and Progenitor Cells (HSPCs) lacking miR-146a but with *RUNX1*^{mut} progresses to AML. These reports suggest a crucial role of both miR-363-3p and miR-146a in *RUNX1*^{mut} AML [214].

Table 4. miRNAs associated with genetic abnormalities in AML.

Genetic Abnormality	miRNA	Reference
t(8;21)(q22;q22.1)/ *RUNX1::RUNX1T1*	let-7b miR-223 miR-9-1 miR-126 miR-130a miR-383	[200] [201] [202] [203] [204] [205]
t(15;17)(q24.1;q21.2)/ *PML::RARA*	miR-15b miR-125b miR-382-5p	[206] [207] [208]
inv(16)(p13.1q22) or t(16;16)(p13.1;q22)/ *CBFB::MYH11*	miR-126	[169]
t(9;11)(p21.3;q23.3)/ *MLLT3::KMT2A*	miR-30e	[151]
NPM1	miR-10a miR-21 miR-215-5p	[209] [163] [210]
TP53	miR-34a miR-100	[211]
RUNX1	miR-363-3p miR-146a	[213] [214]

5. Conclusions

In this review, we summarized the downregulated and upregulated miRNAs observed in AML, highlighting the presence of both types. Research has demonstrated that many miRNA expression levels are downregulated in AML. Studies have also indicated that forced upregulation of some miRNAs can inhibit the tumorigenic potential of AML cells. The downregulation and suppressive role of the miR-29 family and miR-34a are well-established in numerous studies on AML. Additionally, various reports also highlight the suppressive role of miR-142 in this context. Recent studies have suggested the suppressive role of miR-20a-5p, miR-22, miR-133, miR-135a, miR-185, miR-192, miR-199, miR-451, miR-485-5p, miR-495, miR-520a, miR-9, and miR-137; however, further investigation is needed. There have also been isolated reports on the suppressive role of miR-103a-2-5p, miR-454, miR-455-3p, and miR-148, but little is known about their functions, and further studies are required. Additionally, miR-211-5p, miR-30e-5p, miR-409-3p, miR-654-3p, and miR-1294 appear to be downregulated in AML and require further studies.

Research has reported that the expression levels of many miRNAs are elevated in AML. Studies have shown that forced downregulation of certain miRNAs can inhibit the tumorigenic potential of AML cells. The upregulation and oncogenic functions of miR-10, miR-17, miR-21, miR-155, miR-181 family, and miR-222 have been documented, alongside isolated reports on the oncogenic function of miR-1306-5p. However, these also require further research. Additional upregulated microRNAs in AML include miR-93, miR-106b-5p, miR-126, and miR-146, all of which need more research to clarify their roles in AML. Some microRNAs have been described as upregulated and downregulated; for instance, miR-92a and miR-361, suggesting that their functions remain unclear and warrant furtherer studies.

The accumulating data increasingly supports the idea that miRNA expression could significantly influence clinical outcomes in AML, and their potential is actively being explored. Many miRNAs have been correlated with patient prognosis. Higher miR-34a expression levels were associated with achieving CR [115], overexpression of miR-9 improved daunorubicin sensitivity in AML cells [178], higher miR-133 expression level was linked with longer OS and RFS [130], higher miR-135a expression was associated with better prognosis [134], and lower miR-106b-5p expression and Rab10 protein expression at the initial stage were associated with longer survival rates [193]. Conversely, diminished levels of miR-34a expression have been associated with an unfavorable prognosis in AML [115], lower miR-142-3p expression levels were associated with drug-resistant AML cell lines [137], lower miR-20 expression was correlated with shorter OS [48], higher miR-10a-5p was associated with shorter OS [185], and higher miR-21 and miR-17-5p expressions were linked to worse OS and RFS [189,190].

In particular subtypes of AML, according to ELN recommendations, we can highlight the role of some miRNAs in their leukemogenesis. miR let-7b, miR-9-1, miR-126, miR-130a, miR-223, and miR-383 seem to play an important role in AML with t(8;21)(q22;q22.1)/*RUNX1::RUNX1T1*. miR-15b, miR-125b, and miR-382-5p seem to play an essential role in APL with t(15;17)(q24.1;q21.2)/*PML::RARA*. In AML with inv(16)(p13.1q22) or t(16;16)(p13.1;q22)/*CBFB::MYH11*, miR-126 seems to play a crucial role. In AML with fusion gene *MLLT3::KMT2A*, miR-30e appears to play an important role. In *NPM1*^{mut}AML, miR-10a, miR-21, and miR-215-5p appears essential. In *TP53*^{mut}AML, miR-34a plays a pivotal role. In turn, in *RUNX1*^{mut}AML, miR-146a and miR-363-3p may occur crucial.

Overall, the available evidence underscores the promising potential of miRNAs as novel therapeutic targets in AML. Changes in miRNA expression levels could serve as additional molecular markers for better assessment of AML patients' prognosis, especially when combined with other molecular and cytogenetic risk factors. This review highlights many promising genetic biomarkers that could aid in AML diagnosis, prognosis, and therapy response in the future; however, further studies are necessary. The use of miRNAs in therapy still faces many challenges, such as the capability of one miRNA to regulate multiple mRNAs simultaneously, leaving a vast number of possible miRNA-mRNA interactions yet to be explored. Over the past decades, studies on miRNAs have provided valuable insights into their biological functions, roles in disease development, and methods for effectively delivering them into organisms. It is possible that in the future, miRNA expression profiles might serve as markers for measurable residual disease in AML. It is essential to note that the outcomes of miRNA expression can differ depending on the analysis platforms employed and the various AML subtypes examined, which should be considered in future research.

Author Contributions: Conceptualization, I.Z. and A.W. (Aneta Wiśnik); methodology, A.W. (Aneta Wiśnik); software, I.Z.; investigation, A.W. (Aneta Wiśnik), A.S.-K. and M.C.; resources, A.W. (Agnieszka Wierzbowska); data curation, M.C. and A.P.; writing—original draft preparation, A.W. (Aneta Wiśnik); writing—review and editing, I.Z., D.J. and N.P.; visualization, K.K. and P.S.; supervision, I.Z. and A.S.-K.; project administration, I.Z. All authors have read and agreed to the published version of the manuscript.

Funding: This research received no external funding.

Institutional Review Board Statement: Not applicable.

Informed Consent Statement: Not applicable.

Data Availability Statement: Not applicable.

Conflicts of Interest: The authors declare no conflicts of interest.

References

1. Ranganathan, K.; Sivasankar, V. MicroRNAs—Biology and clinical applications. *J. Oral Maxillofac. Pathol.* **2014**, *18*, 229–234. [CrossRef] [PubMed]
2. Metcalf, G.A.D. MicroRNAs: Circulating biomarkers for the early detection of imperceptible cancers via biosensor and machine-learning advances. *Oncogene* **2024**, *43*, 2135–2142. [CrossRef]
3. Wang, J.; Chen, J.; Sen, S. MicroRNA as Biomarkers and Diagnostics. *J. Cell. Physiol.* **2016**, *231*, 25–30. [CrossRef]
4. Supplitt, S.; Karpinski, P.; Sasiadek, M.; Laczmanska, I. Current Achievements and Applications of Transcriptomics in Personalized Cancer Medicine. *Int. J. Mol. Sci.* **2021**, *22*, 1422. [CrossRef] [PubMed]
5. O'Brien, J.; Hayder, H.; Zayed, Y.; Peng, C. Overview of MicroRNA Biogenesis, Mechanisms of Actions, and Circulation. *Front. Endocrinol.* **2018**, *9*, 402. [CrossRef]
6. Bhaskaran, M.; Mohan, M. MicroRNAs: History, biogenesis, and their evolving role in animal development and disease. *Vet. Pathol.* **2014**, *51*, 759–774. [CrossRef] [PubMed]
7. Otmani, K.; Lewalle, P. Tumor Suppressor miRNA in Cancer Cells and the Tumor Microenvironment: Mechanism of Deregulation and Clinical Implications. *Front. Oncol.* **2021**, *11*, 708765. [CrossRef]
8. Poller, W.; Sahoo, S.; Hajjar, R.; Landmesser, U.; Krichevsky, A.M. Exploration of the Noncoding Genome for Human-Specific Therapeutic Targets-Recent Insights at Molecular and Cellular Level. *Cells* **2023**, *12*, 2660. [CrossRef]
9. Walter, N.G. Are non-protein coding RNAs junk or treasure?: An attempt to explain and reconcile opposing viewpoints of whether the human genome is mostly transcribed into non-functional or functional RNAs. *Bioessays* **2024**, *46*, e2300201. [CrossRef]
10. Kelly, R.C.; Morgan, R.A.; Brown, M.; Overton, I.; Hardiman, G. The Non-coding Genome and Network Biology. In *Systems Biology II Springer Medizin*; Springer Nature Switzerland: Cham, Switzerland, 2024; Volume 15, pp. 163–181. [CrossRef]
11. Nobusada, T.; Yip, C.W.; Agrawal, S.; Severin, J.; Abugessaisa, I.; Hasegawa, A.; Hon, C.C.; Ide, S.; Koido, M.; Kondo, A.; et al. Update of the FANTOM web resource: Enhancement for studying noncoding genomes. *Nucleic Acids Res.* **2025**, *53*, D419–D424. [CrossRef]
12. Loganathan, T.; Doss, G.P.C. Non-coding RNAs in human health and disease: Potential function as biomarkers and therapeutic targets. *Funct. Integr. Genom.* **2023**, *23*, 33. [CrossRef] [PubMed]
13. George, T.P.; Subramanian, S.; Supriya, M.H. A brief review of noncoding RNA. *Egypt. J. Med. Hum. Genet.* **2024**, *25*, 98. [CrossRef]
14. Wang, Z.; Wang, H.; Zhou, S.; Mao, J.; Zhan, Z.; Duan, S. miRNA interplay: Mechanisms and therapeutic interventions in cancer. *MedComm—Oncol.* **2024**, *3*, e93. [CrossRef]
15. Poliseno, L.; Lanza, M.; Pandolfi, P.P. Coding, or non-coding, that is the question. *Cell Res.* **2024**, *9*, 609–629. [CrossRef]
16. Li, S.; Hu, W.; Qian, L.; Sun, D. Insights into non-coding RNAS: Biogenesis, function and their potential regulatory roles in acute kidney disease and chronic kidney disease. *Mol. Cell. Biochem.* **2025**, *480*, 1287–1304. [CrossRef]
17. GENCODE. Available online: https://www.gencodegenes.org/human/stats.html (accessed on 27 March 2025).
18. Malgundkar, S.H.; Tamimi, Y. The pivotal role of long non-coding RNAs as potential biomarkers and modulators of chemoresistance in ovarian cancer (OC). *Hum. Genet.* **2024**, *143*, 107–124. [CrossRef]

19. Kasprzyk, M.E.; Kazimierska, M.; Podralska, M. *Navigating Non-Coding RNA from Biogenesis to Therapeutic Application*; Academic Press: Cambridge, MA, USA, 2023; Chapter 3; pp. 89–138.
20. Su, Y.; Wu, J.; Chen, W.; Shan, J.; Chen, D.; Zhu, G.; Ge, S.; Liu, Y. Spliceosomal snRNAs, the Essential Players in pre-mRNA Processing in Eukaryotic Nucleus: From Biogenesis to Functions and Spatiotemporal Characteristics. *Adv. Biol.* **2024**, *8*, 2400006. [CrossRef]
21. Wang, H.; Liu, J.; Fang, Y.; Shen, X.; Liu, H.; Yu, L.; Zeng, S.; Cai, S.; Zhou, J.; Li, Z. Design and analysis of self-priming extension DNA hairpin probe for miRNA detection based on a unified dynamic programming framework. *Anal. Chim. Acta* **2024**, *1303*, 342530. [CrossRef]
22. Venneri, M.; Passantino, A. MiRNA: What clinicians need to know. *Eur. J. Intern. Med.* **2023**, *113*, 6–9. [CrossRef]
23. Maji, R.K.; Leisegang, M.S.; Boon, R.A.; Schulz, M.H. Revealing microRNA regulation in single cells. *Trends Genet.* **2025**, S0168-9525(24)00317-2. [CrossRef]
24. Chauhan, W.; Sudharshan, S.J.; Kafle, S.; Zennadi, R. SnoRNAs: Exploring Their Implication in Human Diseases. *Int. J. Mol. Sci.* **2024**, *25*, 7202. [CrossRef] [PubMed]
25. Shen, L.P.; Zhang, W.C.; Deng, J.R.; Qi, Z.H.; Lin, Z.W.; Wang, Z.D. Advances in the mechanism of small nucleolar RNA and its role in DNA damage response. *Mil. Med. Res.* **2024**, *11*, 53. [CrossRef] [PubMed]
26. Smith, T.J.; Giles, R.N.; Koutmou, K.S. Anticodon stem-loop tRNA modifications influence codon decoding and frame maintenance during translation. *Semin. Cell Dev. Biol.* **2024**, *154*, 105–113. [CrossRef]
27. Wang, L.; Lin, S. Emerging functions of tRNA modifications in mRNA translation and diseases. *J. Genet. Genom.* **2023**, *50*, 223–232. [CrossRef] [PubMed]
28. News Medical Life Sciences. Available online: https://www.news-medical.net/life-sciences/-Types-of-RNA-mRNA-rRNA-and-tRNA.aspx#:~:text=In%20bacteria,%20the%20small%20and,1800%20and%205000%20nucleotides,%20respectively (accessed on 27 March 2025).
29. Rauscher, R.; Polacek, N. Ribosomal RNA expansion segments and their role in ribosome biology. *Biochem. Soc. Trans.* **2024**, *52*, 1317–1325. [CrossRef]
30. Li, R.; Zhu, M.; Hu, X.; Chen, J.; Yu, F.; Barth, S.; Sun, L.; He, H. Overcoming endosomal/lysosomal barriers: Advanced strategies for cytosolic siRNA delivery. *Chin. Chem. Lett.* **2024**, 110736. [CrossRef]
31. Saleem, A.; Khan, M.U.; Zahid, T.; Khurram, I.; Ghani, M.U.; Ullah, I.; Munir, R.; Calina, D.; Sharifi-Rad, J. Biological role and regulation of circular RNA as an emerging biomarker and potential therapeutic target for cancer. *Mol. Biol. Rep.* **2024**, *51*, 296. [CrossRef]
32. Drula, R.; Braicu, C.; Neagoe, I.B. Current advances in circular RNA detection and investigation methods: Are we running in circles? *Wiley Interdiscip. Rev.-RNA* **2024**, *15*, e1850. [CrossRef]
33. Werry, N.; Russell, S.J.; Sivakumar, R.; Miller, S.; Hickey, K.; Larmer, S.; Lohuis, M.; Librach, C.; LaMarre, J. piRNA expression patterns in high vs. low fertility bovine sperm. *Syst. Biol. Reprod. Med.* **2024**, *70*, 183–194. [CrossRef]
34. Claro-Linares, F.; Rojas-Ríos, P. PIWI proteins and piRNAs: Key regulators of stem cell biology. *Front. Cell Dev. Biol.* **2025**, *13*, 1540313. [CrossRef]
35. Jove. Available online: https://app.jove.com/science-education/v/11630/concepts/pirna-piwi-interacting-rnas (accessed on 27 March 2025).
36. Márton, É.; Varga, A.; Domoszlai, D.; Buglyó, G.; Balázs, A.; Penyige, A.; Balogh, I.; Nagy, B.; Szilágyi, M. Non-Coding RNAs in Cancer: Structure, Function, and Clinical Application. *Cancers* **2025**, *17*, 579. [CrossRef] [PubMed]
37. Colino-Sanguino, Y.; Clark, S.J.; Valdes-Mora, F. The H2A.Z-nucleosome code in mammals: Emerging functions. *Trends Genet.* **2022**, *38*, 273–289. [CrossRef]
38. Ma, S.; Wang, Z.; Xiong, Z.; Ge, Y.; Xu, M.-Y.; Zhang, J.; Peng, Y.; Zhang, Q.; Sun, J.; Xi, Z.; et al. Enhancer transcription profiling reveals an enhancer RNA-driven ferroptosis and new therapeutic opportunities in prostate cancer. *Signal Transduct. Target. Ther.* **2025**, *10*, 87. [CrossRef]
39. Rodriguez, A.; Griffiths-Jones, S.; Ashurst, J.L.; Bradley, A. Identification of mammalian microRNA host genes and transcription units. *Genome Res.* **2004**, *14*, 1902–19010. [CrossRef]
40. Yin, Z.; Shen, H.; Gu, C.M.; Zhang, M.Q.; Liu, Z.; Huang, J.; Zhu, Y.; Zhong, Q.; Huang, Y.; Wu, F.; et al. MiRNA-142-3P and FUS can be Sponged by Long Noncoding RNA DUBR to Promote Cell Proliferation in Acute Myeloid Leukemia. *Front. Mol. Biosci.* **2021**, *8*, 754936. [CrossRef]
41. Orang, A.V.; Safaralizadeh, R.; Kazemzadeh-Bavili, M. Mechanisms of miRNA-Mediated Gene Regulation from Common Downregulation to mRNA-Specific Upregulation. *Int. J. Genom.* **2014**, *2014*, 970607. [CrossRef]
42. Machowska, M.; Galka-Marciniak, P.; Kozlowski, P. Consequences of genetic variants in miRNA genes. *Comput. Struct. Biotechnol. J.* **2022**, *20*, 6443–6457. [CrossRef]
43. Lee, R.C.; Feinbaum, R.L.; Ambros, V. The *C. elegans* heterochronic gene lin-4 encodes small RNAs with antisense complementarity to lin-14. *Cell* **1993**, *75*, 843–854. [CrossRef] [PubMed]

44. Liu, B.; Shyr, Y.; Cai, J.; Liu, Q. Interplay between miRNAs and host genes and their role in cancer. *Brief. Funct. Genom.* **2018**, *18*, 255–266. [CrossRef]
45. Vilimova, M.; Pfeffer, S. Post-transcriptional regulation of polycistronic microRNAs. *Wiley Interdiscip. Rev.* **2023**, *14*, e1749. [CrossRef]
46. Le, T.A.H.; Lao, T.D. Circulating microRNAs as the Potential Diagnostic and Prognostic Biomarkers for Nasopharyngeal Carcinoma. *Genes* **2022**, *13*, 1160. [CrossRef]
47. Santovito, D.; Weber, C. Non-canonical features of microRNAs: Paradigms emerging from cardiovascular disease. *Nat. Rev. Cardiol.* **2022**, *19*, 620–638. [CrossRef]
48. Macfarlane, L.A.; Murphy, P.R. MicroRNA: Biogenesis, Function and Role in Cancer. *Curr. Genom.* **2010**, *11*, 537–561. [CrossRef]
49. Shademan, B.; Karamad, V.; Nourazarian, A.; Masjedi, S.; Isazadeh, A.; Sogutlu, F.; Avci, C.B. MicroRNAs as Targets for Cancer Diagnosis: Interests and Limitations. *Adv. Pharm. Bull.* **2023**, *13*, 435–445. [CrossRef]
50. Cai, X.; Hagedorn, C.H.; Cullen, B.R. Human microRNAs are processed from capped, polyadenylated transcripts that can also function as mRNAs. *RNA* **2004**, *10*, 1957–1966. [CrossRef]
51. Lee, D.; Shin, C. Emerging roles of DROSHA beyond primary microRNA processing. *RNA Biol.* **2018**, *15*, 186–193. [CrossRef]
52. Han, J.; Pedersen, J.S.; Kwon, S.C.; Belair, C.D.; Kim, Y.K.; Yeom, K.H.; Yang, W.Y.; Haussler, D.; Blelloch, R.; Kim, V.N. Posttranscriptional crossregulation between Drosha and DGCR8. *Cell* **2009**, *136*, 75–84. [CrossRef]
53. Kwon, S.C.; Nguyen, T.A.; Choi, Y.G.; Jo, M.H.; Hohng, S.; Kim, V.N.; Woo, J.S. Structure of Human DROSHA. *Cell* **2016**, *164*, 81–90. [CrossRef]
54. Bofill-De Ros, X.; Vang Ørom, U.A. Recent progress in miRNA biogenesis and decay. *RNA Biol.* **2024**, *21*, 1–8. [CrossRef] [PubMed]
55. Stavast, C.J.; Erkeland, S.J. The Non-Canonical Aspects of MicroRNAs: Many Roads to Gene Regulation. *Cells* **2019**, *8*, 1465. [CrossRef] [PubMed]
56. Weng, Y.T.; Chang, Y.M.; Chern, Y. The Impact of Dysregulated microRNA Biogenesis Machinery and microRNA Sorting on Neurodegenerative Diseases. *Int. J. Mol. Sci.* **2023**, *24*, 3443. [CrossRef]
57. Michlewski, G.; Cáceres, J.F. Post-transcriptional control of miRNA biogenesis. *RNA* **2019**, *25*, 1–16. [CrossRef] [PubMed]
58. Nakanishi, K. When Argonaute takes out the ribonuclease sword. *J. Biol. Chem.* **2024**, *300*, 105499. [CrossRef]
59. Grenda, A.; Budzyński, M.; Filip, A.A. Biogenesis of microRNAs and their role in the development and course of selected hematologic disorders. *Postępy Hig. Med. Doświadczalnej Online* **2013**, *8*, 174–185. [CrossRef] [PubMed]
60. Hammond, S.M. An overview of microRNAs. *Adv. Drug Deliv. Rev.* **2015**, *87*, 3–14. [CrossRef]
61. Jungers, C.F.; Djuranovic, S. Modulation of miRISC-Mediated Gene Silencing in Eukaryotes. *Front. Mol. Biosci.* **2022**, *9*, 832916. [CrossRef] [PubMed]
62. Kehl, T.; Backes, C.; Kern, F.; Fehlmann, T.; Ludwig, N.; Meese, E.; Lenhof, H.P.; Keller, A. About miRNAs, miRNA seeds, target genes and target pathways. *Oncotarget* **2017**, *8*, 107167–107175. [CrossRef]
63. Li, L.; Sheng, P.; Li, T.; Fields, C.J.; Hiers, N.M.; Wang, Y.; Li, J.; Guardia, C.M.; Licht, J.D.; Xie, M. Widespread microRNA degradation elements in target mRNAs can assist the encoded proteins. *Genes Dev.* **2021**, *35*, 1595–1609. [CrossRef]
64. Hackl, L.M.; Fenn, A.; Louadi, Z.; Baumbach, J.; Kacprowski, T.; List, M.; Tsoy, O. Alternative splicing impacts microRNA regulation within coding regions. *NAR Genom. Bioinform.* **2023**, *5*, lqad081. [CrossRef]
65. Naeli, P.; Winter, T.; Hackett, A.P.; Alboushi, L.; Jafarnejad, S.M. The intricate balance between microRNA-induced mRNA decay and translational repression. *FEBS J.* **2023**, *290*, 2508–2524. [CrossRef]
66. Kuzuoğlu-Öztürk, D.; Bhandari, D.; Huntzinger, E.; Fauser, M.; Helms, S.; Izaurralde, E. miRISC and the CCR4-NOT complex silence mRNA targets independently of 43S ribosomal scanning. *EMBO J.* **2016**, *35*, 1186–1203. [CrossRef]
67. Amorim, I.S.; Lach, G.; Gkogkas, C.G. The Role of the Eukaryotic Translation Initiation Factor 4E (eIF4E) in Neuropsychiatric Disorders. *Front. Genet.* **2018**, *9*, 561. [CrossRef] [PubMed]
68. Eulalio, A.; Huntzinger, E.; Nishihara, T.; Rehwinkel, J.; Fauser, M.; Izaurralde, E. Deadenylation is a widespread effect of miRNA regulation. *RNA* **2009**, *15*, 21–32. [CrossRef] [PubMed]
69. Arber, D.A.; Orazi, A.; Hasserjian, R.P.; Borowitz, M.J.; Calvo, K.R.; Kvasnicka, H.-M.; Wang, S.A.; Bagg, A.; Barbui, T.; Branford, S.; et al. International Consensus Classification of Myeloid Neoplasms and Acute Leukemias: Integrating morphologic, clinical, and genomic data. *Blood* **2022**, *140*, 1200–1228. [CrossRef]
70. Grove, C.S.; Vassiliou, G.S. Acute myeloid leukemia: A paradigm for the clonal evolution of cancer? *Dis. Models Mech.* **2014**, *7*, 941–951. [CrossRef]
71. American Cancer Society. Available online: https://www.cancer.org/cancer/types/acute-myeloid-leukemia/causes-risks-prevention/what-causes.html (accessed on 11 September 2024).
72. American Cancer Society. Available online: https://www.cancer.org/cancer/types/acute-myeloid-leukemia/treating.html (accessed on 11 September 2024).

73. Jimenez-Chillon, C.; Dillon, R.; Russell, N. Optimal Post-Remission Consolidation Therapy in Patients with AML. *Acta Haematol. Pol.* **2024**, *147*, 147–158. [CrossRef] [PubMed]
74. Abuelgasim, K.A.; Albuhayri, B.; Munshi, R.; Mugairi, A.A.; Alahmari, B.; Gmati, G.; Salama, H.; Alzahrani, M.; Alhejazi, A.; Alaskar, A.; et al. Impact of age and induction therapy on outcome of 180 adult patients with acute myeloid leukemia; retrospective analysis and literature review. *Leuk. Res. Rep.* **2020**, *9*, 100206. [CrossRef] [PubMed]
75. National Cancer Institute. Available online: https://seer.cancer.gov/statfacts/html/amyl.html (accessed on 11 September 2024)
76. Peng, Y.; Croce, C.M. The role of MicroRNAs in human cancer. *Signal Transduct. Target. Ther.* **2016**, *28*, 15004. [CrossRef]
77. Wallace, J.A.; O'Connell, R.M. MicroRNAs and acute myeloid leukemia: Therapeutic implications and emerging concepts. *Blood* **2017**, *14*, 1290–1301. [CrossRef]
78. Fletcher, D.; Brown, E.; Javadala, J.; Uysal-Onganer, P.; Guinn, B.A. microRNA expression in acute myeloid leukemia: New targets for therapy? *eJHaem* **2022**, *3*, 596–608. [CrossRef]
79. Kim, T.; Croce, C.M. MicroRNA: Trends in clinical trials of cancer diagnosis and therapy strategies. *Exp. Mol. Med.* **2023**, *55*, 1314–1321. [CrossRef] [PubMed]
80. Target Scan Human. Available online: https://www.targetscan.org/vert_80/ (accessed on 28 March 2025).
81. miRDB. Available online: https://mirdb.org/ (accessed on 28 March 2025).
82. ENCORI. Available online: https://rnasysu.com/encori/ (accessed on 28 March 2025).
83. Ramsingh, G.; Jacoby, M.A.; Shao, J.; De Jesus Pizzaro, R.E.; Shen, D.; Trissal, M.; Getz, A.H.; Ley, T.J.; Walter, M.J.; Link, D.C. Acquired copy number alterations of miRNA genes in acute myeloid leukemia are uncommon. *Blood* **2013**, *122*, e44–e51. [CrossRef] [PubMed]
84. Starczynowski, D.T.; Morin, R.; McPherson, A.; Lam, J.; Chari, R.; Wegrzyn, J.; Kuchenbauer, F.; Hirst, M.; Tohyama, K.; Humphries, R.K.; et al. Genome-wide identification of human microRNAs located in leukemia-associated genomic alterations. *Blood* **2011**, *117*, 595–607. [CrossRef]
85. Eyholzer, M.; Schmid, S.; Wilkens, L.; Mueller, B.U.; Pabst, T. The tumor-suppressive miR-29a/b1 cluster is regulated by CEBPA and blocked in human AML. *Br. J. Cancer* **2010**, *103*, 275–284. [CrossRef] [PubMed]
86. Li, W.; Wang, Y.; Liu, R.; Kasinski, A.L.; Shen, H.; Slack, F.J.; Tang, D.G. MicroRNA-34a: Potent Tumor Suppressor, Cancer Stem Cell Inhibitor, and Potential Anticancer Therapeutic. *Front. Cell Dev. Biol.* **2021**, *9*, 640587. [CrossRef]
87. Bousquet, M.; Quelen, C.; Rosati, R.; Mansat-De Mas, V.; La Starza, R.; Bastard, C.; Lippert, E.; Talmant, P.; Lafage-Pochitaloff, M.; Leroux, D.; et al. Myeloid cell differentiation arrest by miR-125b-1 in myelodysplastic syndrome and acute myeloid leukemia with the t(2;11)(p21;q23) translocation. *J. Exp. Med.* **2008**, *205*, 2499–2506. [CrossRef]
88. Mi, S.; Li, Z.; Chen, P.; He, C.; Cao, D.; Elkahloun, A.; Lu, J.; Pelloso, L.A.; Wunderlich, M.; Huang, H.; et al. Aberrant overexpression and function of the miR-17-92 cluster in MLL-rearranged acute leukemia. *Proc. Natl. Acad. Sci. USA* **2010**, *107*, 3710–3715. [CrossRef]
89. Chen, P.; Price, C.; Li, Z.; Li, Y.; Cao, D.; Wiley, A.; He, C.; Gurbuxani, S.; Kunjamma, R.B.; Huang, H.; et al. miR-9 is an essential oncogenic microRNA specifically overexpressed in mixed lineage leukemia-rearranged leukemia. *Proc. Natl. Acad. Sci. USA* **2013**, *110*, 11511–11516. [CrossRef]
90. Senyuk, V.; Zhang, Y.; Liu, Y.; Ming, M.; Premanand, K.; Zhou, L.; Chen, P.; Chen, J.; Rowley, J.D.; Nucifora, G.; et al. Critical role of miR-9 in myelopoiesis and EVI1-induced leukemogenesis. *Proc. Natl. Acad. Sci. USA* **2013**, *110*, 5594–5599. [CrossRef]
91. Li, Z.; Lu, J.; Sun, M.; Mi, S.; Zhang, H.; Luo, R.T.; Chen, P.; Wang, Y.; Yan, M.; Qian, Z.; et al. Distinct microRNA expression profiles in acute myeloid leukemia with common translocations. *Proc. Natl. Acad. Sci. USA* **2008**, *105*, 15535–15540. [CrossRef]
92. Gao, X.N.; Lin, J.; Li, Y.H.; Gao, L.; Wang, X.R.; Wang, W.; Kang, H.Y.; Yan, G.T.; Wang, L.L.; Yu, L. MicroRNA-193a represses c-kit expression and functions as a methylation-silenced tumor suppressor in acute myeloid leukemia. *Oncogene* **2011**, *30*, 3416–3428. [CrossRef]
93. Pulikkan, J.A.; Dengler, V.; Peramangalam, P.S.; Peer Zada, A.A.; Müller-Tidow, C.; Bohlander, S.K.; Tenen, D.G.; Behre, G. Cell-cycle regulator E2F1 and microRNA-223 comprise an autoregulatory negative feedback loop in acute myeloid leukemia. *Blood* **2010**, *115*, 1768–1778. [CrossRef] [PubMed]
94. Vineetha, R.C.; Raj, J.A.G.; Devipriya, P.; Mahitha, M.S.; Hariharan, S. MicroRNA-based therapies: Revolutionizing the treatment of acute myeloid leukemia. *Int. J. Lab. Hematol.* **2024**, *46*, 33–41. [CrossRef]
95. Ghazaryan, A.; Wallace, J.A.; Tang, W.W.; Barba, C.; Lee, S.-H.; Bauer, K.M.; Nelson, M.C.; Kim, C.N.; Stubben, C.; Voth, W.P.; et al. miRNA-1 promotes acute myeloid leukemia cell pathogenesis through metabolic regulation. *Front. Genet.* **2023**, *14*, 1192799. [CrossRef] [PubMed]
96. Salehi, A. A novel therapeutic strategy: The significance of exosomal miRNAs in acute myeloid leukemia. *Med. Oncol.* **2024**, *41*, 62. [CrossRef] [PubMed]
97. Iacomino, G. miRNAs: The Road from Bench to Bedside. *Genes* **2023**, *14*, 314. [CrossRef]
98. Wang, C.; Li, L.; Li, M.; Wang, W.; Liu, Y.; Wang, S. Silencing long non-coding RNA XIST suppresses drug resistance in acute myeloid leukemia through down-regulation of MYC by elevating microRNA-29a expression. *Mol. Med.* **2020**, *26*, 114. [CrossRef]

99. Tang, Y.-J.; Wu, W.; Chen, Q.-Q.; Liu, S.-H.; Zheng, Z.-Y.; Cui, Z.-L.; Xu, J.-P.; Xue, Y.; Lin, D.-H. miR-29b-3p suppresses the malignant biological behaviors of AML cells via inhibiting NF-κB and JAK/STAT signaling pathways by targeting HuR. *BMC Cancer* **2022**, *22*, 909. [CrossRef]
100. Randazzo, V.; Salemi, D.; Agueli, C.; Cannella, S.; Marfia, A.; Bica, M.G.; Randazzo, G.; Russo Lacerna, C.; Di Raimondo, F.; Fabbiano, F.; et al. Upregulation of Mir-29 in Normal Karyotype Aml Showing Dnmt3a Mutation. *J. Hematol. Transfus.* **2016**, *4*, 1048.
101. Ngankeu, A.; Ranganathan, P.; Havelange, V.; Nicolet, D.; Volinia, S.; Powell, B.L.; Kolitz, J.E.; Uy, G.L.; Stone, R.M.; Kornblau, S.M.; et al. Discovery and functional implications of a miR-29b-1/miR-29a cluster polymorphism in acute myeloid leukemia. *Oncotarget* **2017**, *9*, 4354–4365. [CrossRef]
102. Abdellateif, M.S.; Hassan, N.M.; Kamel, M.M.; El-Meligui, Y.M. Bone marrow microRNA-34a is a good indicator for response to treatment in acute myeloid leukemia. *Oncol. Res.* **2024**, *32*, 577–584. [CrossRef] [PubMed]
103. Wang, X.; Li, J.; Dong, K.; Lin, F.; Long, M.; Ouyang, Y.; Wei, J.; Chen, X.; Weng, Y.; He, T.; et al. Tumor suppressor miR-34a targets PD-L1 and functions as a potential immunotherapeutic target in acute myeloid leukemia. *Cell. Signal.* **2015**, *27*, 443–452. [CrossRef]
104. Hu, Y.; Ma, X.; Wu, Z.; Nong, Q.; Liu, F.; Wang, Y.; Dong, M. MicroRNA-34a-mediated death of acute myeloid leukemia stem cells through apoptosis induction and exosome shedding inhibition via histone deacetylase 2 targeting. *IUBMB Life* **2020**, *72*, 1481–1490. [CrossRef] [PubMed]
105. Huang, Y.; Zou, Y.; Lin, L.; Ma, X.; Chen, H. Identification of serum miR-34a as a potential biomarker in acute myeloid leukemia. *Cancer Biomark.* **2018**, *22*, 799–805. [CrossRef] [PubMed]
106. Wen, J.; Chen, Y.; Liao, C.; Ma, X.; Wang, M.; Li, Q.; Wang, D.; Li, Y.; Zhang, X.; Li, L.; et al. Engineered mesenchymal stem cell exosomes loaded with miR-34c-5p selectively promote eradication of acute myeloid leukemia stem cells. *Cancer Lett.* **2023**, *575*, 216407. [CrossRef]
107. Peng, D.; Wang, H.; Li, L.; Ma, X.; Chen, Y.; Zhou, H.; Luo, Y.; Xiao, Y.; Liu, L. miR-34c-5p promotes eradication of acute myeloid leukemia stem cells by inducing senescence through selective RAB27B targeting to inhibit exosome shedding. *Leukemia* **2018**, *32*, 1180–1188. [CrossRef]
108. Yang, D.-Q.; Zhou, J.-D.; Wang, Y.-X.; Deng, Z.-Q.; Yang, J.; Yao, D.-M.; Qian, Z.; Yang, L.; Lin, J.; Qian, J. Low miR-34c expression is associated with poor outcome in de novo acute myeloid leukemia. *Int. J. Lab. Hematol.* **2017**, *39*, 42–50. [CrossRef]
109. Hong, D.S.; Kang, Y.-K.; Borad, M.; Sachdev, J.; Ejadi, S.; Lim, H.Y.; Brenner, A.J.; Park, K.; Lee, J.-L.; Kim, T.-Y.; et al. Phase 1 study of MRX34, a liposomal miR-34a mimic, in patients with advanced solid tumors. *Br. J. Cancer* **2020**, *122*, 1630–1637. [CrossRef]
110. Zhang, Y.; Liu, Y.; Xu, X. Upregulation of miR-142-3p Improves Drug Sensitivity of Acute Myelogenous Leukemia through Reducing P-Glycoprotein and Repressing Autophagy by Targeting HMGB1. *Transl. Oncol.* **2017**, *10*, 410–418. [CrossRef]
111. Jiang, Z.; Liu, T.; Wang, Y.; Li, J.; Guo, L. Effect of lncRNA XIST on acute myeloid leukemia cells via miR-142-5p-PFKP axis. *Hematology* **2024**, *29*, 2306444. [CrossRef]
112. Yuan, D.M.; Ma, J.; Fang, W.B. Identification of non-coding RNA regulatory networks in pediatric acute myeloid leukemia reveals circ-0004136 could promote cell proliferation by sponging miR-142. *Eur. Rev. Med. Pharmacol. Sci.* **2019**, *23*, 9251–9258. [CrossRef] [PubMed]
113. Zhang, B.; Zhao, D.; Chen, F.; Frankhouser, D.; Wang, H.; Pathak, K.V.; Dong, L.; Torres, A.; Garcia-Mansfield, K.; Zhang, Y.; et al. Acquired miR-142 deficit in leukemic stem cells suffices to drive chronic myeloid leukemia into blast crisis. *Nat. Commun.* **2023**, *14*, 5325. [CrossRef]
114. Liu, Y.; Lei, P.; Qiao, H.; Sun, K.; Lu, X.; Bao, F.; Yu, R.; Lian, C.; Li, Y.; Chen, W.; et al. miR-9 Enhances the Chemosensitivity of AML Cells to Daunorubicin by Targeting the EIF5A2/MCL-1 Axis. *Int. J. Biol. Sci.* **2019**, *15*, 579–586. [CrossRef] [PubMed]
115. Wang, G.; Yu, X.; Xia, J.; Sun, J.; Huang, H.; Liu, Y. MicroRNA-9 restrains the sharp increase and boost apoptosis of human acute myeloid leukemia cells by adjusting the Hippo/YAP signaling pathway. *Bioengineered* **2021**, *12*, 2906–2914. [CrossRef]
116. Zhu, B.; Xi, X.; Liu, Q.; Cheng, Y.; Yang, H. MiR-9 functions as a tumor suppressor in acute myeloid leukemia by targeting CX chemokine receptor 4. *Am. J. Transl. Res.* **2019**, *11*, 3384–3397. [PubMed]
117. Bao, F.; Zhang, L.; Pei, X.; Lian, C.; Liu, Y.; Tan, H.; Lei, P. MiR-20a-5p functions as a potent tumor suppressor by targeting PPP6C in acute myeloid leukemia. *PLoS ONE* **2021**, *16*, e0256995. [CrossRef]
118. Ping, L.; Jian-Jun, C.; Chu-Shu, L.; Guang-Hua, L.; Ming, Z. Silencing of circ_0009910 inhibits acute myeloid leukemia cell growth through increasing miR-20a-5p. *Blood Cells Mol. Dis.* **2019**, *75*, 41–47. [CrossRef]
119. Chen, Z.-H.; Wang, W.-T.; Huang, W.; Fang, K.; Sun, Y.-M.; Liu, S.-R.; Luo, X.-Q.; Chen, Y.-Q. The lncRNA HOTAIRM1 regulates the degradation of PML-RARA oncoprotein and myeloid cell differentiation by enhancing the autophagy pathway. *Cell Death Differ.* **2017**, *24*, 212–224. [CrossRef]
120. Cen, Q.; Chen, J.; Guo, J.; Chen, M.; Wang, H.; Wu, S.; Zhang, H.; Xie, X.; Li, Y. CLPs-miR-103a-2-5p inhibits proliferation and promotes cell apoptosis in AML cells by targeting LILRB3 and Nrf2/HO-1 axis, regulating CD8+ T cell response. *J. Transl. Med.* **2024**, *22*, 278. [CrossRef]

121. Zheng, Z.-Z.; Ma, Y.-P.; Wu, R.-H.; Rong, G.; Li, C.; Li, G.-X.; Ren, F.-G.; Xu, L.-J. Serum miR-133 as a novel biomarker for predicting treatment response and survival in acute myeloid leukemia. *Eur. Rev. Med. Pharmacol. Sci.* **2020**, *24*, 777–783. [CrossRef]
122. Yamamoto, H.; Lu, J.; Oba, S.; Kawamata, T.; Yoshimi, A.; Kurosaki, N.; Yokoyama, K.; Matsushita, H.; Kurokawa, M.; Tojo, A.; et al. miR-133 regulates Evi1 expression in AML cells as a potential therapeutic target. *Sci. Rep.* **2016**, *6*, 19204. [CrossRef] [PubMed]
123. Wang, Q.; Yue, C.; Liu, Q.; Che, X. Exploration of differentially expressed mRNAs and miRNAs for pediatric acute myeloid leukemia. *Front. Genet.* **2022**, *13*, 865111. [CrossRef]
124. Xu, H.; Wen, Q. Downregulation of miR-135a predicts poor prognosis in acute myeloid leukemia and regulates leukemia progression via modulating HOXA10 expression. *Mol. Med. Rep.* **2018**, *18*, 1134–1140. [CrossRef] [PubMed]
125. Cheng, Y.C.; Fan, Z.; Liang, C.; Peng, C.J.; Li, Y.; Wang, L.N.; Luo, J.S.; Zhang, X.L.; Liu, Y.; Zhang, L.D. miR-133a and miR-135a Regulate All-Trans Retinoic Acid-Mediated Differentiation in Pediatric Acute Myeloid Leukemia by Inhibiting CDX2 Translation and Serve as Prognostic Biomarkers. *Technol. Cancer Res. Treat.* **2024**, *23*, 15330338241248576. [CrossRef] [PubMed]
126. Liu, L.; Yu, K.; Yu, J.; Tao, W.; Wei, Y. MiR-133 promotes the multidrug resistance of acute myeloid leukemia cells (HL-60/ADR) to daunorubicin. *Cytotechnology* **2024**, *76*, 833–846. [CrossRef]
127. Hu, Y.; Dong, X.; Chu, G.; Lai, G.; Zhang, B.; Wang, L.; Zhao, Y. miR-137 downregulates c-kit expression in acute myeloid leukemia. *Leuk. Res.* **2017**, *57*, 72–77. [CrossRef]
128. Wang, S.; Zhang, B.S.; Yang, Y.; Li, Y.; Lv, J.L.; Cheng, Y. TRIM25 contributes to the malignancy of acute myeloid leukemia and is negatively regulated by microRNA-137. *Open Med.* **2020**, *16*, 95–103. [CrossRef]
129. Wang, X.-X.; Zhang, R.; Li, Y. Expression of the miR-148/152 Family in Acute Myeloid Leukemia and its Clinical Significance. *Med. Sci. Monit.* **2017**, *23*, 4768–4778. [CrossRef]
130. Wang, X.-X.; Zhang, H.; Li, Y. Preliminary study on the role of miR-148a and DNMT1 in the pathogenesis of acute myeloid leukemia. *Mol. Med. Rep.* **2019**, *19*, 2943–2952. [CrossRef]
131. Pang, B.; Mao, H.; Wang, J.; Yang, W. MiR-185-5p suppresses acute myeloid leukemia by inhibiting GPX1. *Microvasc. Res.* **2022**, *140*, 104296. [CrossRef]
132. Zhang, W.; Liu, Y.; Zhang, J.; Zheng, N. Long Non-Coding RNA Taurine Upregulated Gene 1 Targets miR-185 to Regulate Cell Proliferation and Glycolysis in Acute Myeloid Leukemia Cells in vitro. *OncoTargets Ther.* **2020**, *13*, 7887–7896. [CrossRef] [PubMed]
133. Wu, W.; Deng, J.; Chen, C.; Ma, X.; Yu, L.; Chen, L. Circ_0001602 aggravates the progression of acute myeloid leukemia by regulating the miR-192-5p/ZBTB20 axis. *Hematology* **2023**, *28*, 2240133. [CrossRef]
134. Tian, C.; Zhang, L.; Li, X.; Zhang, Y.; Li, J.; Chen, L. Low miR-192 expression predicts poor prognosis in pediatric acute myeloid leukemia. *Cancer Biomark.* **2018**, *22*, 209–215. [CrossRef] [PubMed]
135. Ke, S.; Li, R.C.; Lu, J.; Meng, F.K.; Feng, Y.K.; Fang, M.H. MicroRNA-192 regulates cell proliferation and cell cycle transition in acute myeloid leukemia via interaction with CCNT2. *Int. J. Hematol.* **2017**, *106*, 258–265. [CrossRef]
136. Chen, D.P.; Chang, S.W.; Wen, Y.H.; Wang, W.T. Association between diminished miRNA expression and the disease status of AML patients: Comparing to healthy control. *Biomed. J.* **2023**, *46*, 100518. [CrossRef] [PubMed]
137. Krakowsky, R.H.E.; Wurm, A.A.; Gerloff, D.; Katzerke, C.; Bräuer-Hartmann, D.; Hartmann, J.U.; Wilke, F.; Thiede, C.; Müller-Tidow, C.; Niederwieser, D.; et al. miR-451a abrogates treatment resistance in FLT3-ITD-positive acute myeloid leukemia. *Blood Cancer J.* **2018**, *8*, 36. [CrossRef]
138. Li, L.; Mussack, V.; Görgens, A.; Pepeldjiyska, E.; Hartz, A.S.; Aslan, H.; Rackl, E.; Rank, A.; Schmohl, J.; El Andaloussi, S.; et al. The potential role of serum extracellular vesicle derived small RNAs in AML research as non-invasive biomarker. *Nanoscale Adv.* **2023**, *5*, 1691–1705. [CrossRef]
139. Song, L.; Lin, H.-S.; Gong, J.-N.; Han, H.; Wang, X.-S.; Su, R.; Chen, M.-T.; Shen, C.; Ma, Y.-N.; Yu, J.; et al. microRNA-451-modulated hnRNP A1 takes a part in granulocytic differentiation regulation and acute myeloid leukemia. *Oncotarget* **2017**, *8*, 55453–55466. [CrossRef]
140. Su, R.; Gong, J.-N.; Chen, M.-T.; Song, L.; Shen, C.; Zhang, X.-H.; Yin, X.-L.; Ning, H.-M.; Liu, B.; Wang, F.; et al. c-Myc suppresses miR-451⊣YWTAZ/AKT axis via recruiting HDAC3 in acute myeloid leukemia. *Oncotarget* **2016**, *7*, 77430–77443. [CrossRef]
141. Wu, K.; Li, Y.; Nie, B.; Guo, C.; Ma, X.; Li, L.; Cheng, S.; Li, Y.; Luo, S.; Zeng, Y.; et al. MEF2A is a transcription factor for circPVT1 and contributes to the malignancy of acute myeloid leukemia. *Int. J. Oncol.* **2024**, *65*, 111. [CrossRef]
142. Xie, Y.; Tan, L.; Wu, K.; Li, D.; Li, C. MiR-455-3p mediates PPARα through UBN2 to promote apoptosis and autophagy in acute myeloid leukemia cells. *Exp. Hematol.* **2023**, *128*, 77–88. [CrossRef] [PubMed]
143. Zhang, F.; Li, Q.; Zhu, K.; Zhu, J.; Li, J.; Yuan, Y.; Zhang, P.; Zhou, L.; Liu, L. LncRNA LINC00265/miR-485-5p/IRF2-mediated autophagy suppresses apoptosis in acute myeloid leukemia cells. *Am. J. Transl. Res.* **2020**, *12*, 2451–2462.
144. Wang, W.-L.; Wang, H.R.; Ji, W.G.; Guo, S.L.; Li, H.X.; Xu, X.Y. MiRNA-485-5p suppresses the proliferation of acute myeloid leukemia via targeting SALL4. *Eur. Rev. Med. Pharmacol. Sci.* **2019**, *23*, 4842–4849. [CrossRef]

145. Huang, L.; Dai, J. Expression and Clinical Significance of miRNA-495 in the Peripheral Blood of Acute Myeloid Leukemia Patients. *Proc. Anticancer Res.* **2022**, *6*, 5. [CrossRef]
146. Lei, Y.; Li, X.; Zhu, L. Matrine regulates miR-495-3p/miR-543/PDK1 axis to repress the progression of acute myeloid leukemia via the Wnt/βcatenin pathway. *Chem. Biol. Drug Des.* **2024**, *103*, e14441. [CrossRef]
147. Wang, G.; Li, X.; Song, L.; Pan, H.; Jiang, J.; Sun, L. Long noncoding RNA MIAT promotes the progression of acute myeloid leukemia by negatively regulating miR-495. *Leuk. Res.* **2019**, *87*, 106265. [CrossRef]
148. Zhang, W.; Wan, B.; Liu, B.; Wu, S.; Zhao, L. Clinical significance of miR-372 and miR-495 in acute myeloid leukemia. *Oncol. Lett.* **2020**, *20*, 1938–1944. [CrossRef] [PubMed]
149. Chen, X.Y.; Qin, X.H.; Xie, X.L.; Liao, C.X.; Liu, D.T.; Li, G.W. Overexpression miR-520a-3p inhibits acute myeloid leukemia progression via targeting MUC1. *Transl. Oncol.* **2022**, *22*, 101432. [CrossRef]
150. Xiao, J.; Wan, F.; Tian, L.; Li, Y. Tumor suppressor miR-520a inhibits cell growth by negatively regulating PI3K/AKT signaling pathway in acute myeloid leukemia. *Adv. Clin. Exp. Med.* **2024**, *33*, 729–738. [CrossRef]
151. Ge, Y.; Hong, M.; Zhang, Y.; Wang, J.; Li, L.; Zhu, H.; Sheng, Y.; Wu, W.-S.; Zhang, Z. miR-30e-5p regulates leukemia stem cell self-renewal through the Cyb561/ROS signaling pathway. *Haematologica* **2024**, *109*, 411–421. [CrossRef]
152. Ye, Q.; Ren, L.; Jiang, Z.M.; Li, X.Y.; Wei, G.Y.; Ren, Y.F.; Ren, L.H. Cryptanshinone extract of Salvia miltiorrhiza stimulates pediatric acute myeloid leukemia stem cell apoptosis and the anti-inflammatory mechanism via accelerating microRNA-211-5p to supress Janus kinase 2/signal transducer and activator of transcription 3 signaling pathway activation. *J. Physiol. Pharmacol.* **2023**, *74*, 691–700. [CrossRef]
153. Xie, W.; Wang, Z.; Guo, X.; Guan, H. MiR-409-3p regulates the proliferation and apoptosis of THP-1 through targeting Rab10. *Leuk. Res.* **2023**, *132*, 107350. [CrossRef] [PubMed]
154. Wang, X.; Zhong, L.; Dan, W.; Chu, X.; Luo, X.; Liu, C.; Wan, P.; Lu, Y.; Liu, Z.; Zhang, Z.; et al. MiR-454-3p promotes apoptosis and autophagy of AML cells by targeting ZEB2 and regulating AKT/mTOR pathway. *Hematology* **2023**, *28*, 2223874. [CrossRef] [PubMed]
155. Wang, J.; Wu, C.; Zhou, W. CircPLXNB2 regulates acute myeloid leukemia progression through miR-654-3p/CCND1 axis. *Hematology* **2023**, *28*, 2220522. [CrossRef]
156. Wang, S.; Zhang, B.S.; Yang, Y.; Fu, L.L. CircFN1 promotes acute myeloid leukemia cell proliferation and invasion but refrains apoptosis via miR-1294/ARHGEF10L axis. *Kaohsiung J. Med. Sci.* **2024**, *40*, 221–230. [CrossRef]
157. Bi, L.; Sun, L.; Jin, Z.; Zhang, S.; Shen, Z. MicroRNA-10a/b are regulators of myeloid differentiation and acute myeloid leukemia. *Oncol. Lett.* **2018**, *15*, 5611–5619. [CrossRef]
158. Wang, C.J.; Zou, H.; Feng, G.F. MiR-10b regulates the proliferation and apoptosis of pediatric acute myeloid leukemia through targeting HOXD10. *Eur. Rev. Med. Pharmacol. Sci.* **2018**, *22*, 7371–7378. [CrossRef]
159. Zhi, Y.; Xie, X.; Wang, R.; Wang, B.; Gu, W.; Ling, Y.; Dong, W.; Zhi, F.; Liu, Y. Serum level of miR-10-5p as a prognostic biomarker for acute myeloid leukemia. *Int. J. Hematol.* **2015**, *102*, 296–303. [CrossRef]
160. Yuan, Z.; Wang, W. LncRNA SNHG4 regulates miR-10a/PTEN to inhibit the proliferation of acute myeloid leukemia cells. *Hematology* **2020**, *25*, 160–164. [CrossRef]
161. Li, C.; Yan, H.; Yin, J.; Ma, J.; Liao, A.; Yang, S.; Wang, L.; Huang, Y.; Lin, C.; Dong, Z.; et al. MicroRNA-21 promotes proliferation in acute myeloid leukemia by targeting Krüppel-like factor 5. *Oncol. Lett.* **2019**, *18*, 3367–3372. [CrossRef]
162. Li, X.; Zhang, X.; Ma, H.; Liu, Y.; Cheng, S.; Wang, H.; Sun, J. Upregulation of serum exosomal miR-21 was associated with poor prognosis of acute myeloid leukemia patients. *Food Sci. Technol.* **2022**, *42*, e51621. [CrossRef]
163. Riccioni, R.; Lulli, V.; Castelli, G.; Biffoni, M.; Tiberio, R.; Pelosi, E.; Lo-Coco, F.; Testa, U. miR-21 is overexpressed in NPM1-mutant acute myeloid leukemias. *Leuk. Res.* **2015**, *39*, 221–228. [CrossRef]
164. Cao, Y.; Liu, Y.; Shang, L.; Chen, H.; Yue, Y.; Dong, W.; Guo, Y.; Yang, H.; Yang, X.; Liu, Y.; et al. Overexpression of miR-17 predicts adverse prognosis and disease recurrence for acute myeloid leukemia. *Int. J. Clin. Oncol.* **2022**, *27*, 1222–1232. [CrossRef]
165. Liu, M.; Yu, B.; Tian, Y.; Li, F. Regulatory function and mechanism research for m6A modification WTAP via SUCLG2-AS1-miR-17-5p-JAK1 axis in AML. *BMC Cancer* **2024**, *24*, 98. [CrossRef]
166. Wang, W.; Liu, J.; Chen, K.; Wang, J.; Dong, Q.; Xie, J.; Yuan, Y. Vitamin D promotes autophagy in AML cells by inhibiting miR-17-5p-induced Beclin-1 overexpression. *Mol. Cell. Biochem.* **2021**, *76*, 3951–3962. [CrossRef]
167. Huang, J.; Xiao, R.; Wang, X.; Khadka, B.; Fang, Z.; Yu, M.; Zhang, L.; Wu, J.; Liu, J. MicroRNA-93 knockdown inhibits acute myeloid leukemia cell growth via inactivating the PI3K/AKT pathway by upregulating DAB2. *Int. J. Oncol.* **2021**, *59*, 81. [CrossRef] [PubMed]
168. Xu, L.; Wang, A.; Guan, H. microRNA-106b-5p and Rab10: Potential Markers of Acute Myeloid Leukemia. *Cancer Biother Radiopharm* **2024**, *39*, 492–501. [CrossRef] [PubMed]
169. Zhang, L.; Nguyen, L.X.T.; Chen, Y.C.; Wu, D.; Cook, G.J.; Hoang, D.H.; Brewer, C.J.; He, X.; Dong, H.; Li, S.; et al. Targeting miR-126 in inv(16) acute myeloid leukemia inhibits leukemia development and leukemia stem cell maintenance. *Nat. Commun.* **2021**, *12*, 6154. [CrossRef]

170. Wang, L.; Zhang, H.; Lei, D. microRNA-146a Promotes Growth of Acute Leukemia Cells by Downregulating Ciliary Neurotrophic Factor Receptor and Activating JAK2/STAT3 Signaling. *Yonsei Med. J.* **2019**, *60*, 924–934. [CrossRef]
171. Li, X.; Xu, L.; Sheng, X.; Cai, J.; Liu, J.; Yin, T.; Xiao, F.; Chen, F.; Zhong, H. Upregulated microRNA-146a expression induced by granulocyte colony-stimulating factor enhanced low-dosage chemotherapy response in aged acute myeloid leukemia patients *Exp. Hematol.* **2018**, *68*, 66–79. [CrossRef]
172. Garavand, J.; Mohammadi, M.H.; Jalali, M.T.; Saki, N. Correlation of miR-155-5p, KRAS, and CREB Expression in Patients with Acute Myeloid Leukemia. *Clin. Lab.* **2024**, *70*, 89–98. [CrossRef]
173. Xue, H.; Hua, L.M.; Guo, M.; Luo, J.M. SHIP1 is targeted by miR-155 in acute myeloid leukemia. *Oncol. Rep.* **2014**, *32*, 2253–2259 [CrossRef]
174. Palma, C.A.; Al Sheikha, D.; Lim, T.K.; Bryant, A.; Vu, T.T.; Jayaswal, V.; Ma, D.D. MicroRNA-155 as an inducer of apoptosis and cell differentiation in Acute Myeloid Leukemia. *Mol. Cancer* **2014**, *13*, 79. [CrossRef]
175. Elgohary, T.; Abu-Taleb, F.; Ghonaim, R. The Impact of miRNA-155 Expression on Treatment Outcome in Adult Acute Myeloid Leukemia Patients. *J. Cancer Ther.* **2019**, *10*, 203–214. [CrossRef]
176. Hatem, A.; Gab~Allah, A.; Ghonaim, R.; Haggag, R. Prognostic Impact of microRNAs (miR-155, miR-10a, let7a) on the Outcome of Adult Patients with Acute Myeloid Leukemia. *Zagazig Univ. Med. J.* **2021**, *27*, 810–825. [CrossRef]
177. El-Hassib, D.M.A.; Zidan, M.A.-A.; Marei, Y.M.; El Gheit, N.E.S.N.A.; Alnoury, H.A. Study of Micro RNA 181 a3p As a Biomarker for Diagnosis of Acute Myeloid Leukemia. *Egypt. J. Hosp. Med.* **2023**, *91*, 4780–4785. [CrossRef]
178. Butrym, A.; Rybka, J.; Baczyńska, D.; Poręba, R.; Mazur, G.; Kuliczkowski, K. Expression of microRNA-181 determines response to treatment with azacitidine and predicts survival in elderly patients with acute myeloid leukemia. *Oncol. Lett.* **2016**, *12*, 2296–2300. [CrossRef]
179. Su, R.; Lin, H.S.; Zhang, X.H.; Yin, X.L.; Ning, H.M.; Liu, B.; Zhai, P.F.; Gong, J.N.; Shen, C.; Song, L.; et al. MiR-181 family: Regulators of myeloid differentiation and acute myeloid leukemia as well as potential therapeutic targets. *Oncogene* **2015**, *34*, 3226–3239. [CrossRef]
180. Gao, X.; Fan, S.; Zhang, X. MiR-1306-5p promotes cell proliferation and inhibits cell apoptosis in acute myeloid leukemia by downregulating PHF6 expression. *Leuk. Res.* **2022**, *120*, 106906. [CrossRef]
181. Jiang, X.; Hu, C.; Arnovitz, S.; Bugno, J.; Yu, M.; Zuo, Z.; Chen, P.; Huang, H.; Ulrich, B.; Gurbuxani, S.; et al. miR-22 has a potent anti-tumor role with therapeutic potential in acute myeloid leukemia. *Nat. Commun.* **2016**, *7*, 11452. [CrossRef]
182. Shen, C.; Chen, M.T.; Zhang, X.H.; Yin, X.L.; Ning, H.M.; Su, R.; Lin, H.S.; Song, L.; Wang, F.; Ma, Y.N.; et al. The PU.1-Modulated MicroRNA-22 Is a Regulator of Monocyte/Macrophage Differentiation and Acute Myeloid Leukemia. *PLoS Genet.* **2016**, *12*, e1006259. [CrossRef] [PubMed]
183. Qu, H.; Zheng, G.; Cheng, S.; Xie, W.; Liu, X.; Tao, Y.; Xie, B. Serum miR-22 is a novel prognostic marker for acute myeloid leukemia. *J. Clin. Lab. Anal.* **2020**, *34*, e23370. [CrossRef] [PubMed]
184. Yao, H.; Sun, P.; Duan, M.; Lin, L.; Pan, Y.; Wu, C.; Fu, X.; Wang, H.; Guo, L.; Jin, T.; et al. microRNA-22 can regulate expression of the long non-coding RNA MEG3 in acute myeloid leukemia. *Oncotarget* **2017**, *8*, 65211–65217. [CrossRef]
185. Gu, Y.; Si, J.; Xiao, X.; Tian, Y.; Yang, S. miR-92a Inhibits Proliferation and Induces Apoptosis by Regulating Methylenetetrahydrofolate Dehydrogenase 2 (MTHFD2) Expression in Acute Myeloid Leukemia. *Oncol. Res.* **2017**, *25*, 1069–1079. [CrossRef]
186. Su, X.Y.; Zhao, Q.; Ke, J.M.; Wu, D.H.; Zhu, X.; Lin, J.; Deng, Z.Q. Circ_0002232 Acts as a Potential Biomarker for AML and Reveals a Potential ceRNA Network of Circ_0002232/miR-92a-3p/PTEN. *Cancer Manag. Res.* **2020**, *12*, 11871–11881. [CrossRef]
187. Saadi, M.I.; Arandi, N.; Yaghobi, R.; Azarpira, N.; Geramizadeh, B.; Ramzi, M. Up-Regulation of the MiR-92a and miR-181a in Patients with Acute Myeloid Leukemia and their Inhibition with Locked Nucleic acid (LNA)-antimiRNA.; Introducing c-Kit as a New Target Gene. *Int. J. Hematol. Oncol.* **2018**, *28*, 238–247. [CrossRef]
188. Rashed, R.A.; Hassan, N.M.; Hussein, M.M. MicroRNA-92a as a marker of treatment response and survival in adult acute myeloid leukemia patients. *Leuk. Lymphoma* **2020**, *61*, 2475–2481. [CrossRef] [PubMed]
189. Qi, X.; Zhang, Y. MicroRNA-199a deficiency relates to higher bone marrow blasts, poor risk stratification and worse prognostication in pediatric acute myeloid leukemia patients. *Pediatr. Hematol. Oncol.* **2022**, *39*, 500–507. [CrossRef]
190. Li, Y.; Zhang, G.; Wu, B.; Yang, W.; Liu, Z. miR-199a-5p Represses Protective Autophagy and Overcomes Chemoresistance by Directly Targeting DRAM1 in Acute Myeloid Leukemia. *J. Oncol.* **2019**, *2019*, 5613417. [CrossRef]
191. Ellson, I.; Martorell-Marugán, J.; Carmona-Sáez, P.; Ramos-Mejia, V. MiRNA expression as outcome predictor in pediatric AML: Systematic evaluation of a new model. *npj Genom. Med.* **2024**, *9*, 40. [CrossRef]
192. Favreau, A.J.; McGlauflin, R.E.; Duarte, C.W.; Sathyanarayana, P. miR-199b, a novel tumor suppressor miRNA in acute myeloid leukemia with prognostic implications. *Exp. Hematol. Oncol.* **2016**, *5*, 4. [CrossRef] [PubMed]
193. Alemdehy, M.F.; Haanstra, J.R.; de Looper, H.W.; van Strien, P.M.; Verhagen-Oldenampsen, J.; Caljouw, Y.; Sanders, M.A.; Hoogenboezem, R.; de Ru, A.H.; Janssen, G.M.; et al. ICL-induced miR139-3p and miR199a-3p have opposite roles in hematopoietic cell expansion and leukemic transformation. *Blood* **2015**, *125*, 3937–3948. [CrossRef]

194. Yuan, Y.; Tan, S.; Wang, H.; Zhu, J.; Li, J.; Zhang, P.; Wang, M.; Zhang, F. Mesenchymal Stem Cell-Derived Exosomal miRNA-222-3p Increases Th1/Th2 Ratio and Promotes Apoptosis of Acute Myeloid Leukemia Cells. *Anal. Cell. Pathol.* **2023**, *2023*, 4024887. [CrossRef]
195. Pei, H.Z.; Peng, Z.; Zhuang, X.; Wang, X.; Lu, B.; Guo, Y.; Zhao, Y.; Zhang, D.; Xiao, Y.; Gao, T.; et al. miR-221/222 induce instability of p53 By downregulating deubiquitinase YOD1 in acute myeloid leukemia. *Cell Death Discov.* **2023**, *9*, 249. [CrossRef]
196. Liu, Z.; Zhong, L.; Dan, W.; Chu, X.; Liu, C.; Luo, X.; Zhang, Z.; Lu, Y.; Wan, P.; Wang, X.; et al. miRNA-222-3p enhances the proliferation and suppresses the apoptosis of acute myeloid leukemia cells by targeting Axin2 and modulating the Wnt/β-catenin pathway. *Biochem. Biophys. Res. Commun.* **2022**, *620*, 83–91. [CrossRef] [PubMed]
197. Pavlovic, D.; Tosic, N.; Zukic, B.; Pravdic, Z.; Vukovic, N.S.; Pavlovic, S.; Gasic, V. Expression Profiles of Long Non-Coding RNA GAS5 and MicroRNA-222 in Younger AML Patients. *Diagnostics* **2021**, *12*, 86. [CrossRef] [PubMed]
198. Xu, D.; Jiang, J.; He, G.; Zhou, H.; Ji, C. KMT2A is targeted by miR-361-3p and modulates leukemia cell's abilities to proliferate, migrate and invade. *Hematology* **2023**, *28*, 2225341. [CrossRef]
199. Liu, S.; Xu, H.; Li, Z. Linoleic acid derivatives target miR-361-3p/BTG2 to confer anticancer effects in acute myeloid leukemia. *J. Biochem. Mol. Toxicol.* **2023**, *37*, e23481. [CrossRef] [PubMed]
200. Johnson, D.T.; Davis, A.G.; Zhou, J.H.; Ball, E.D.; Zhang, D.E. MicroRNA let-7b downregulates AML1-ETO oncogene expression in t(8;21) AML by targeting its 3′UTR. *Exp. Hematol. Oncol.* **2021**, *10*, 8. [CrossRef]
201. Fazi, F.; Racanicchi, S.; Zardo, G.; Starnes, L.M.; Mancini, M.; Travaglini, L.; Diverio, D.; Ammatuna, E.; Cimino, G.; Lo-Coco, F.; et al. Epigenetic silencing of the myelopoiesis regulator microRNA-223 by the AML1/ETO oncoprotein. *Cancer Cell* **2007**, *12*, 457–466. [CrossRef]
202. Fu, L.; Shi, J.; Liu, A.; Zhou, L.; Jiang, M.; Fu, H.; Xu, K.; Li, D.; Deng, A.; Zhang, Q.; et al. A minicircuitry of microRNA-9-1 and RUNX1-RUNX1T1 contributes to leukemogenesis in t(8;21) acute myeloid leukemia. *Int. J. Cancer* **2017**, *140*, 653–661. [CrossRef] [PubMed]
203. Li, Z.; Chen, P.; Su, R.; Li, Y.; Hu, C.; Wang, Y.; Arnovitz, S.; He, M.; Gurbuxani, S.; Zuo, Z.; et al. Overexpression and knockout of miR-126 both promote leukemogenesis. *Blood* **2015**, *126*, 2005–2015. [CrossRef]
204. Krivdova, G.; Voisin, V.; Schoof, E.M.; Marhon, S.A.; Murison, A.; McLeod, J.L.; Gabra, M.M.; Zeng, A.G.X.; Aigner, S.; Yee, B.A.; et al. Identification of the global miR-130a targetome reveals a role for TBL1XR1 in hematopoietic stem cell self-renewal and t(8;21) AML. *Cell Rep.* **2022**, *38*, 110481. [CrossRef] [PubMed]
205. Li, Y.; Ning, Q.; Shi, J.; Chen, Y.; Jiang, M.; Gao, L.; Huang, W.; Jing, Y.; Huang, S.; Liu, A.; et al. A novel epigenetic AML1-ETO/THAP10/miR-383 mini-circuitry contributes to t(8;21) leukaemogenesis. *EMBO Mol. Med.* **2017**, *9*, 933–949. [CrossRef]
206. Yuan, Z.; Zhong, L.; Liu, D.; Yao, J.; Liu, J.; Zhong, P.; Yao, S.; Zhao, Y.; Li, L.; Chen, M.; et al. MiR-15b regulates cell differentiation and survival by targeting CCNE1 in APL cell lines. *Cell. Signal.* **2019**, *60*, 57–64. [CrossRef] [PubMed]
207. Guo, B.; Qin, R.; Chen, J.J.; Pan, W.; Lu, X.C. MicroRNA-125b Accelerates and Promotes PML-RARa-driven Murine Acute Promyelocytic Leukemia. *Biomed. Environ. Sci.* **2022**, *35*, 485–493. [CrossRef]
208. Liu, D.; Zhong, L.; Yuan, Z.; Yao, J.; Zhong, P.; Liu, J.; Yao, S.; Zhao, Y.; Liu, L.; Chen, M.; et al. miR-382-5p modulates the ATRA-induced differentiation of acute promyelocytic leukemia by targeting tumor suppressor PTEN. *Cell. Signal.* **2019**, *54*, 1–9. [CrossRef]
209. Ovcharenko, D.; Stölzel, F.; Poitz, D.; Fierro, F.; Schaich, M.; Neubauer, A.; Kelnar, K.; Davison, T.; Müller-Tidow, C.; Thiede, C.; et al. miR-10a overexpression is associated with NPM1 mutations and MDM4 downregulation in intermediate-risk acute myeloid leukemia. *Exp. Hematol.* **2011**, *39*, 1030–1042. [CrossRef]
210. Gadewal, N.; Kumar, R.; Aher, S.; Gardane, A.; Gaur, T.; Varma, A.K.; Khattry, N.; Hasan, S.K. miRNA-mRNA Profiling Reveals Prognostic Impact of SMC1A Expression in Acute Myeloid Leukemia. *Oncol. Res.* **2020**, *28*, 321–330. [CrossRef]
211. Rücker, F.G.; Russ, A.C.; Cocciardi, S.; Kett, H.; Schlenk, R.F.; Botzenhardt, U.; Langer, C.; Krauter, J.; Fröhling, S.; Schlegelberger, B.; et al. Altered miRNA and gene expression in acute myeloid leukemia with complex karyotype identify networks of prognostic relevance. *Leukemia* **2013**, *27*, 353–361. [CrossRef]
212. Shahzad, M.; Amin, M.K.; Daver, N.G.; Shah, M.V.; Hiwase, D.; Arber, D.A.; Kharfan-Dabaja, M.A.; Badar, T. What have we learned about TP53-mutated acute myeloid leukemia? *Blood Cancer J.* **2024**, *14*, 202. [CrossRef] [PubMed]
213. Chen, Y.; Chen, S.; Lu, J.; Yuan, D.; He, L.; Qin, P.; Tan, H.; Xu, L. MicroRNA-363-3p promote the development of acute myeloid leukemia with RUNX1 mutation by targeting SPRYD4 and FNDC3B. *Medicine* **2021**, *100*, e25807. [CrossRef] [PubMed]
214. Barreyro, L.; Sampson, A.M.; Hueneman, K.; Choi, K.; Christie, S.; Ramesh, V.; Wyder, M.; Wang, D.; Pujato, M.; Greis, K.D.; et al. Dysregulated innate immune signaling cooperates with RUNX1 mutations to transform an MDS-like disease to AML. *iScience* **2024**, *27*, 109809. [CrossRef] [PubMed]

MDPI AG
Grosspeteranlage 5
4052 Basel
Switzerland
Tel.: +41 61 683 77 34

Genes Editorial Office
E-mail: genes@mdpi.com
www.mdpi.com/journal/genes

www.ingramcontent.com/pod-product-compliance
Lightning Source LLC
Chambersburg PA
CBHW062041220926
51850CB00010B/2627